Contents

Unit 5 Equilibria, energetics and elements

Unit 6 Practical skills in chemistry 2

Introduction

About this book

This textbook covers the OCR specification for A2 chemistry introduced in 2009. It follows the specification closely and includes all the topics in the modules of Units 4 (F324) and 5 (F325). Although it is assumed that students will be able to refer to an AS textbook to remind themselves of the chemistry they have previously studied, there is an introductory section to each unit that briefly reviews the knowledge required to tackle the new material. Unit 6 (F326) is the centre-based assessment of practical work and Chapter 16 gives important advice to help students prepare for each of the three components of the assessment.

All the units at A2 are synoptic and examination questions can call upon knowledge from the AS course. Students are therefore reminded of the links with previous work which are relevant to their current study. These links are drawn clearly throughout the text.

The introduction of an A* grade at A2 is likely to be of considerable importance to the most able students and universities are increasingly concerned to find ways of differentiating between applicants at all levels. Examination papers now include even more demanding parts to questions that are designed to stretch the best candidates.

To be able to tackle these questions confidently requires regular practice throughout the course. To meet this need this book follows the pattern established in the AS textbook by the same authors. The main body of the text covers the essential material required by all students attempting the examination. There are then two further sections:

- An additional reading section provides a broader context to some of the chemistry considered in the chapter and either elaborates on a particular issue or allows students to test their understanding in a related context. This is designed to encourage students to be flexible in their thinking and it may help students who are asked to discuss a particular topic during an interview for a higher education course.
- A second section provides some more demanding questions to stretch and challenge the most able. Where possible these questions are set to show the relevance of an idea in a more general context. It is also hoped that these questions might provide some help to those attempting the increasingly popular Chemistry Olympiad.

As was the case with AS, there is much mention in the A2 specification of 'How Science Works'. This is an important aspect of science education and there are many instances in this book where it is emphasised. However, the constant repetition of the phrase has been avoided. Nonetheless, it should be expected that the context in which questions are set in the examination could reflect aspects of the way science works.

The first eight chapters describe and explain the content of **Unit 4: Rings, polymers and analysis**. There are three modules within the unit. The first of these introduces aromatic chemistry and extends the organic chemistry to include the $-NH_2$ functional group. The second module covers polyesters, polyamides and proteins and also includes a section on synthesis. The specification excludes most of the reactions of organic molecules that cannot be performed readily in the laboratory and also those that are largely of historical interest only. However, it is likely that the more demanding questions in the examination will provide extra information, allowing a wider range of conversions to be tested. Some examples of this are included in the relevant chapters. The third module is devoted wholly to the study of analytical techniques used by chemists. Practice at combining the information provided by the various spectra is included.

Unit 5: Equilibria, energetics and elements is addressed in detail in Chapters 9–15. This unit also consists of three modules. The first is an extensive section covering rates of reaction, a quantitative approach to equilibrium and the behaviour of acids and bases. Examiners frequently encounter confusion over the experimental methods used to determine orders of reaction. Particular care is taken in this book to emphasise the difference between using the 'initial rate' approach and the 'continuous run' method. The mathematical requirements of the various sections of this module are introduced and a range of examples is included. The second module focuses on energy cycles and introduces the important ideas behind entropy and free energy. These are new to the OCR specification and can be approached in a number of ways. The emphasis here is placed on the central concept of energy rather than on 'disorder', which can lead to confusion if the topic is pursued at a deeper level. The final module covers transition metals.

Unit 6: Practical skills is considered in Chapter 16 in which aspects of the three tasks that must be completed during the course are reviewed. There are essential reminders on the calculation of percentage errors and the importance of significant figures. Attention is drawn to mistakes that are commonplace reasons for the loss of marks during the assessments.

Margin comments are provided throughout the book. These comprise valuable reminders and snippets of information and include examiners tips (indicated by an e symbol).

Online resources

To access your free online resources go to **www.hodderplus.co.uk/philipallan**. These comprise:

- end-of-chapter worksheets for students to print off and complete in order to create sets of summary notes
- exam style questions with graded student answers supported by examiner's comments

Scheme of assessment

Candidates at A2 take two written papers and a practical component that is internally assessed.

Unit 4: Rings, polymers and analysis is examined by a written paper, lasting 1 hour. The paper is worth 60 marks and carries 30% of the total marks for A2. All the questions are compulsory.

Unit 5: Equilibria, energetics and elements is examined by a written paper, lasting 1 hour 45 minutes. The paper is worth 100 marks and carries 50% of the total marks at A2. All the questions are compulsory.

The papers cover the range of topics within those units and are constructed to test a candidate's knowledge and understanding as well as the ability to apply knowledge in an unfamiliar situation. The latter particularly applies to those questions or parts of questions that are designed to 'stretch and challenge' students. These are a particular feature of the A2 examination and will contribute towards the award of an A* grade.

The internally assessed practical component is identical in structure to that encountered at AS and requires candidates to show evidence of:

- accuracy in observation and recording
- ability to interpret results, both qualitatively and quantitatively
- judgement in evaluating an experiment and providing possible improvements to the method employed

The practical component carries 20% of the marks at A2.

A2 does not exist as a qualification in its own right and the marks on these units are combined with those achieved at AS to award an A-level grade based on the total mark.

Advice to candidates

Each examination question on the written papers at A2 must be answered in the booklet provided. Therefore, a particular number of lines are given for your anticipated response. The first thing to do is to check the number of marks available for an answer. This is given in brackets at the end of each question or part-question. This is more important than the number of lines, because it indicates how many separate points you should make in your response. Some candidates express themselves concisely in their answers (or just have small writing), so the size of space provided is only a rough guide as to how much you should write. Try not to exceed the space provided as the examiners have thought carefully about the response required and will have provided more than enough space for a full response. Remember that there is always the risk that if you write too much, you could contradict yourself and lose a mark that would have been awarded. Using only the space provided is now particularly important because the examination papers are scanned and sent for marking electronically. If you write outside the permitted area, it might not be scanned fully and part of your response could be missed.

You may want to cross out an answer that you believe is incorrect, but it is unwise to do so unless you are sure you are replacing it with something more accurate. It is surprising how often examiners see something worthy of a mark within a crossed-out passage.

Finally, heed the obvious advice to read the question carefully. It is distressing for an examiner to read an answer that would have scored many marks if it had been a response to a question other than the one asked. Remember that if a question has a long introduction, it usually contains information that is either essential in order to answer the question or is designed to make it easier. It is not uncommon to read incorrect formulae in an answer even though the formulae concerned were given in the question.

Terms used in examination questions

You will be asked precise questions in the examinations, so you can save a lot of valuable time, as well as ensuring you score as many marks as possible, by knowing what is expected. Terms used most commonly are explained below.

- **Define** — a formal statement or equivalent paraphrase is required. Precise definitions are given in this book. If you learn these, you should score all the marks available for definitions.
- **Explain what is meant by** — this normally implies that a definition should be given, together with some relevant comment on the significance or context of the term(s) concerned, particularly if two or more terms are included in the question. The amount of supplementary comment intended should be interpreted according to the number of marks available.
- **State** — give a concise answer with little or no supporting argument or explanation.
- **Describe** — state in words (using diagrams and/or equations as appropriate) the main points of the topic. It is often used with reference either to particular phenomena or to particular experiments. It may be used when referring to a mechanism, in which case it is important to show the appropriate curly arrow(s), lone pairs of electrons and dipoles. The amount of description required depends on the number of marks available.
- **Name** — give the full systematic name, not the formula.
- **Identify** — give the name, formula or structure.
- **Deduce/predict** — you are not expected to recall the answer, rather to make a logical connection between other pieces of information/data. Such information may be given wholly in the question or may depend on answers extracted in an earlier part of the question. 'Predict' also implies a concise answer with no supporting statement required.
- **Outline** — restrict the answer to essential detail only.
- **Suggest** — this may either imply that there is no unique answer or that you are expected to apply your knowledge to a 'novel' situation that is an extension of the specification content.
- **Calculate** — a numerical answer is required. Working should be shown. Your answer should be set out in such a way that if a mistake is made, this can be tracked by the examiner and marks can be awarded for any consequential steps.
- **Sketch** — applied to diagrams, this means that a simple, freehand drawing is acceptable. Nevertheless, care should be taken over proportions and the important details should be labelled clearly.

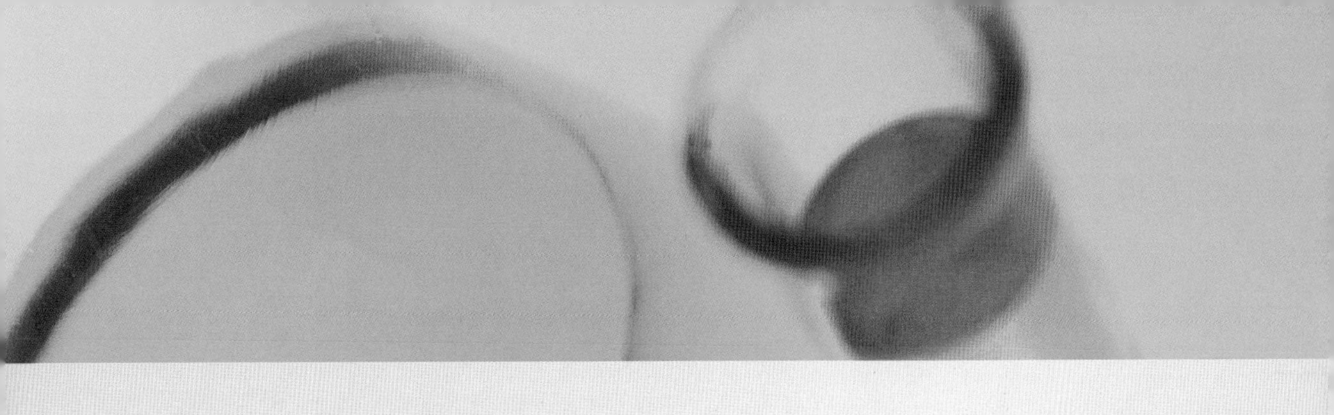

Unit 4

Rings, polymers and analysis

Introduction to Unit 4

Unit 4 (F324) of the OCR specification provides a deeper knowledge of organic chemistry by extending the study of the functional groups covered at AS in **Unit 2: Chains, energy and resources**. It also includes the application of modern instrumental techniques to the analysis and detection of organic molecules. Also covered are many important materials that are derived from organic molecules, and how organic chemistry shapes the world.

The main areas of the unit are:

- Module 1: Rings, acids and amines
  - arenes
  - carbonyl compounds
  - carboxylic acids and esters
  - amines
- Module 2: Polymers and synthesis
  - amino acids and proteins
  - polyesters and polyamides
  - synthesis
- Module 3: Analysis
  - chromatography
  - spectroscopy

This book is specific and follows the OCR specification carefully. It attempts to transform the precise 'teacher speak' used in the specification into a more 'user-friendly' language. It provides extension material designed to stretch and challenge the most able students, as well as providing support and material for students of all abilities.

Summary of essential AS chemistry required for Unit 4

This unit aims to build upon the concepts developed during AS chemistry, particularly those met in Unit 2 (F322): Chains, energy and resources. You are expected to use the concepts developed in AS chemistry and questions may be set that require both a knowledge and understanding of Unit 2.

Summary of Unit 2 chemistry

You are expected to be able to recall that organic chemicals can be grouped together in homologous series in which each member of the series differs from the next by CH_2. Each homologous series has a group of atoms that determine the chemistry of that series. This group of atoms is referred to as the 'functional group'.

A general formula can be written for each homologous series. You need to be able to use displayed formulae, molecular formulae, empirical formulae, structural formulae and skeletal formulae. You should be able name individual chemicals, including structural isomers and *E/Z (cis–trans)* isomers.

Name	Displayed formula	Molecular formula	Empirical formula	Structural formula	Skeletal formula
Butane	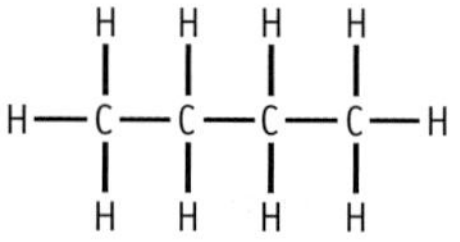 H—C—C—C—C—H (H above and below each C)	C_4H_{10}	C_2H_5	$CH_3CH_2CH_2CH_2$	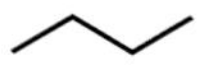
Z-but-2-ene	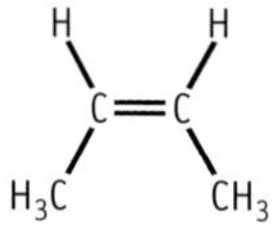 H, H / C=C / H_3C, CH_3	C_4H_8	CH_2	$CH_3CHCHCH_3$	
E-but-2-ene	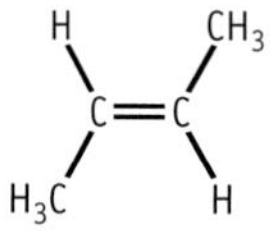 H, CH_3 / C=C / H_3C, H	C_4H_8	CH_2	$CH_3CHCHCH_3$	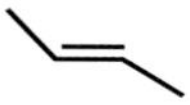
2-methyl propan-2-ol	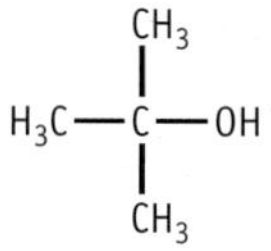 CH_3 / H_3C—C—OH / CH_3	C_4H_9OH	C_4H_9OH	$(CH_3)_3COH$	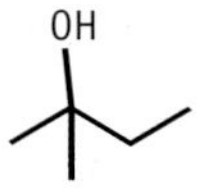 OH
1-chloro-2-methylpropane	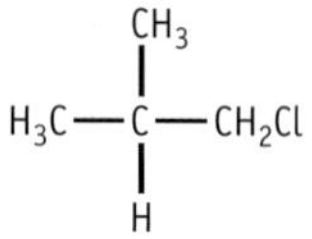 CH_3 / H_3C—C—CH_2Cl / H	C_4H_9Cl	C_4H_9Cl	$(CH_3)_2CHCH_2Cl$	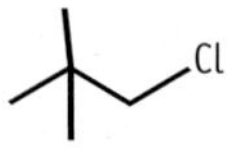 Cl

You will be expected to perform calculations to determine empirical formulae and to be able to calculate the percentage yield from experimental data and also the atom economy of a reaction. Terms first introduced in AS chemistry will be used and it is essential that these are fully understood. They include:

- a **radical** has a single unpaired electron — for example Cl•
- a **nucleophile** is an electron pair donor — for example OH^-
- an **electrophile** is an electron pair acceptor — for example H^+

It is unlikely that you will be tested on the particular mechanisms covered in Unit 2: Chains, energy and resources. However, the concept of showing the movement of electrons by the use of 'curly arrows' is developed further in this unit. Therefore, you are advised to revise the three mechanisms encountered in AS organic chemistry:

- **radical substitution** in reactions between alkanes and halogens
- **electrophilic addition** in reactions between alkenes and halogens
- **nucleophilic substitution** in the hydrolysis of halogenoalkanes

Polymerisation is studied at greater depth in this A2 unit. You should understand fully the basic ideas and be able to construct the formula of a polymer from a given monomer. Some common monomers and their corresponding polymers are shown below.

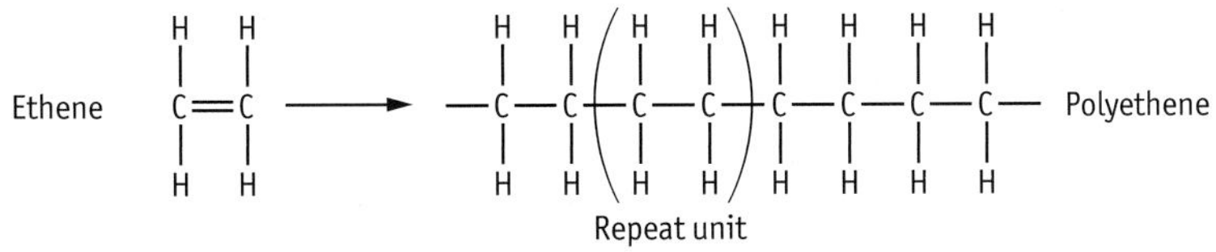

Ethene → Polyethene

Repeat unit

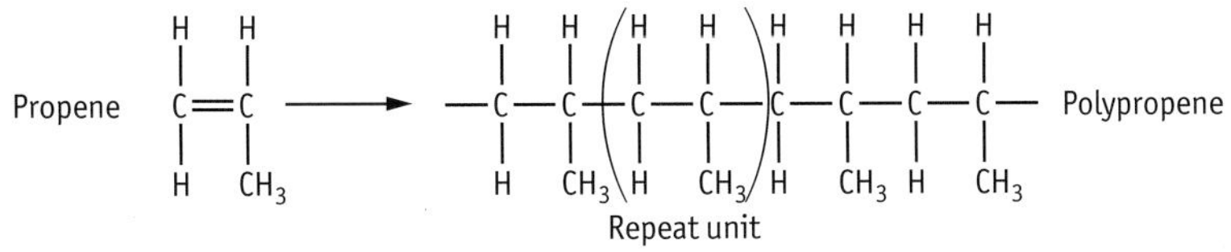

Propene → Polypropene

Repeat unit

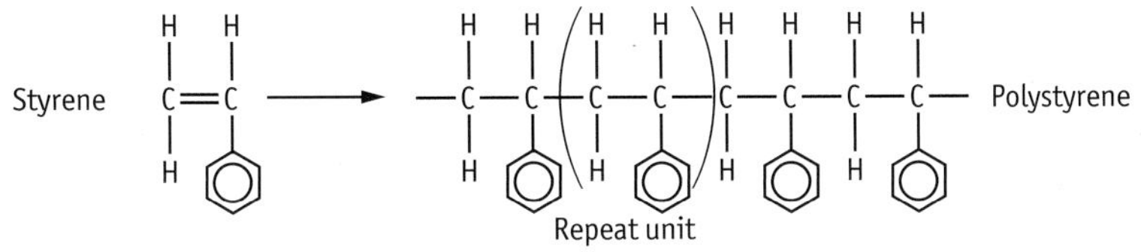

Styrene → Polystyrene

Repeat unit

The section devoted to 'alcohols' is particularly relevant. The oxidation of alcohols can produce aldehydes, ketones and carboxylic acids. All three functional groups are studied in A2 and it is, therefore, essential that you can recall suitable equations, reagents and conditions as listed below.

- Oxidation of a primary alcohol to an **aldehyde**:
  Methanol is oxidised to methanal:

  $CH_3OH + [O] \rightarrow HCHO + H_2O$
  Methanol → Methanal

  Ethanol is oxidised to ethanal:

  $CH_3CH_2OH + [O] \rightarrow CH_3CHO + H_2O$
  Ethanol → Ethanal

- Oxidation of a secondary alcohol to a **ketone**:
  Propan-2-ol is oxidised to propanone (propan-2-one):

  $CH_3CHOHCH_3 + [O] \rightarrow CH_3COCH_3 + H_2O$
  Propan-2-ol → Propanone

  Butan-2-ol is oxidised to butanone (butan-2-one):

  $CH_3CH_2CHOHCH_3 + [O] \rightarrow CH_3CH_2COCH_3 + H_2O$
  Butan-2-ol → Butanone

- Oxidation of a primary alcohol to a **carboxylic acid**:
  Methanol is oxidised to methanoic acid:

  $CH_3OH + 2[O] \rightarrow HCOOH + H_2O$
  Methanol → Methanoic acid

Ethanol is oxidised to ethanoic acid:

$$CH_3CH_2OH + 2[O] \rightarrow CH_3COOH + H_2O$$

Ethanol　　　　　　　　Ethanoic acid

- Tertiary alcohols are resistant to oxidation.

The oxidising agent, [O], is an acidified solution of potassium dichromate, $K_2Cr_2O_7$. It is usually acidified using sulphuric acid, H_2SO_4, and the oxidising mixture is often written simply as $Cr_2O_7^{2-}/H^+$.

The oxidising mixture is bright orange and changes to dark green during the redox process.

Oxidation of a primary alcohol can lead to the formation of either an aldehyde or a carboxylic acid. The choice of apparatus determines which is formed. Refluxing produces a carboxylic acid and distillation produces an aldehyde as the major product. The use of reflux or distillation therefore influences which product is obtained.

Alcohols and carboxylic acids both contain the O–H bond and it follows therefore that both have intermolecular hydrogen bonds (see pages 92–94 in the AS book). Aldehydes (and ketones) do not form hydrogen bonds. Aldehydes are, therefore, more volatile than either alcohols or carboxylic acids and have lower boiling points. Reflux allows continuous evaporation and condensation without the loss of any volatile components. Distillation also involves evaporation and condensation but the volatile components are separated out.

Alcohols react with carboxylic acids to produce esters and water. These reaction were studied at AS but are also very much part of A2 chemistry.

Reactions of alcohols that may be tested in Unit 4

$$CH_3CH_2OH + HOOCCH_3 \longrightarrow CH_3CH_2OCOCH_3 + H_2O$$

Loss of H_2O

Oxidation The oxidising agent is $H^+/Cr_2O_7^{2-}$ and we see a colour change of orange to green. If the ethanol is set up under reflux, the ethanol is completely oxidised to ethanoic acid, but if it is set up under distillation, ethanal can be separated out.

$$CH_3CH_2OH + [O] \longrightarrow CH_3CHO + H_2O$$

When refluxed

$$2\,CH_3CH_2OH + 2[O] \longrightarrow 2\,CH_3COOH$$

When distilled

Arenes and phenols

Chapter 1

This chapter covers the bonding and structure of benzene and related compounds. Arenes undergo electrophilic substitution reactions, the ease of which depends on the structure of the arene.

Arenes

The simplest arene is benzene. Benzene has a composition by mass of 92.3% carbon and 7.7% hydrogen. Its relative molecular mass is 78.0. This information shows that the empirical formula of benzene is CH and its molecular formula is C_6H_6.

Benzene is an important compound that is essential to many industrial processes. Most benzene is used to make alkylbenzenes, for example:

- ethylbenzene — used to make phenylethene, which is the monomer for poly(phenylethene) or polystyrene
- (1-methylethyl)benzene/cumene — used to make phenol and propanone
- dodecylbenzene — used to make detergents

Benzene is also an important feedstock for the production of cyclohexane in the manufacture of nylon. It is also used in the manufacture of a variety of dyes, medicines and explosives.

Some uses of benzene are shown in Figure 1.1.

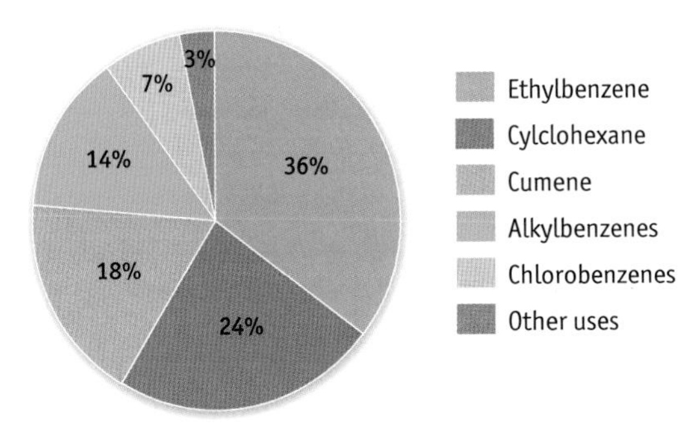

Figure 1.1 *Uses of benzene*

Benzene was first isolated in 1825 by Michael Faraday. By the late nineteenth century it was realised that benzene and substances related to it form a natural chemical family, called **aromatics** after a characteristic property of many of the compounds. The cyclic structure of benzene was not established fully until the late nineteenth century and the exact structure of benzene is still open to interpretation today.

Structure of benzene

The French chemist August Kekulé suggested that benzene was a cyclic molecule with alternating C=C double bonds and C–C single bonds. The C=C double bonds are formed by the sideways overlap of adjacent *p*-orbitals (see page 158 in the AS book).

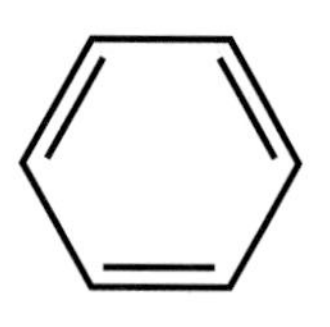

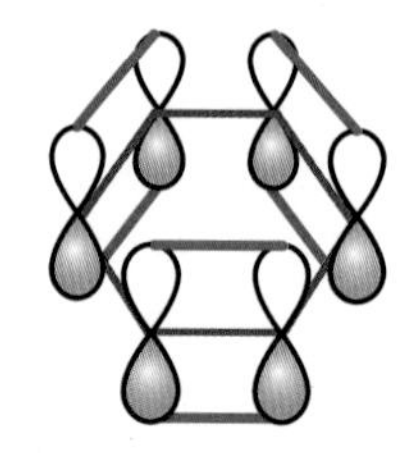

August Kekulé (1829–1896)

His full name was Friedrich August Kekulé von Stradonitz.

Kekulé suggested that benzene would react in the same way as cyclohexene, C_6H_{10} and undergo electrophilic addition reactions. However, there are three major pieces of evidence against the Kekulé structure.

Kekulé claims to have discovered the ring shape of the benzene molecule when dreaming about a snake seizing its own tail. This is a common symbol in some ancient cultures and is known as 'ouroboros'.

First, compounds that contain a C=C double bond readily decolorise bromine. Benzene only reacts with bromine when boiled and exposed to ultraviolet light or in the presence of a suitable catalyst. This casts doubt on the existence of C=C double bonds in benzene.

Second, experimental data from measuring the enthalpy change when unsaturated cyclohexene reacts with hydrogen to produce cyclohexane shows that ΔH is $-120\,kJ\,mol^{-1}$.

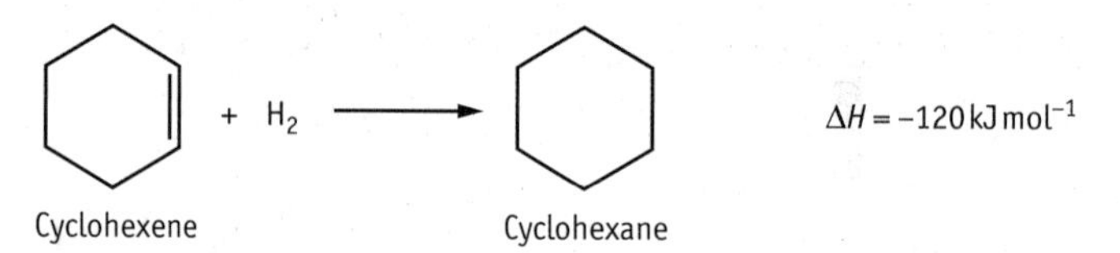

It follows that the enthalpy of hydrogenation of benzene ought to be three times ($-360\,kJ\,mol^{-1}$) that of cyclohexene.

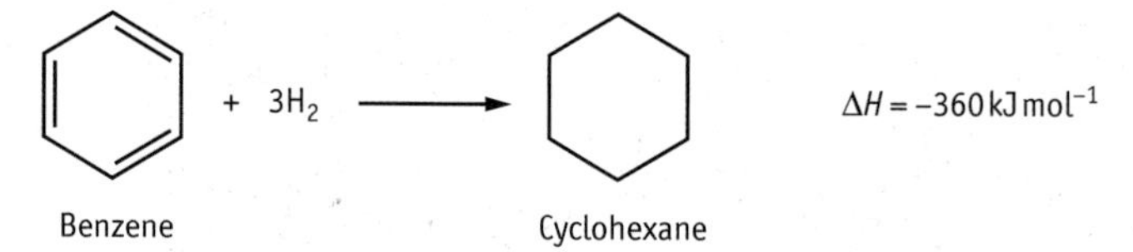

However, when measured experimentally, the enthalpy of hydrogenation of benzene was found to be $-208\,kJ\,mol^{-1}$, which is about $150\,kJ\,mol^{-1}$ less than that predicted from the theoretical alternating double bond–single bond model.

These enthalpy data (Figure 1.2) cast further doubts on the validity of the Kekulé structure.

The third piece of evidence comes from X-ray diffraction data. Using X-ray diffraction techniques it is possible to measure the bond lengths within a molecule. It was found that the average C–C single bond has a length of 154 pm while the average C=C double bond length is 134 pm (pm = picometre = 1×10^{-12} m). If Kekulé's structure is correct, the bond lengths in benzene should alternate between long (154 pm) and short (134 pm). However, all the bonds in benzene are the same length — 139 pm. This suggests an intermediate bond somewhere between a double bond and a single bond.

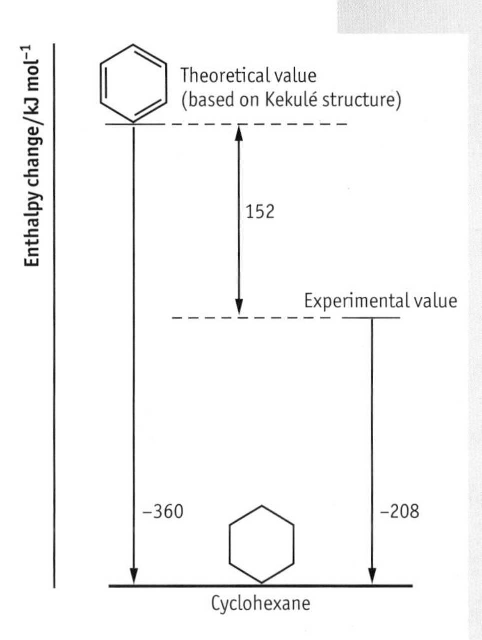

***Figure 1.2** Enthalpy of formation data relating to benzene*

There is sufficient doubt about the Kekulé model that an alternative model has been suggested. The current model of the structure of benzene suggests that each carbon atom contributes one electron to a π-delocalised ring of electrons above and below the plane of atoms. Each carbon has one *p*-orbital at right angles to the plane of atoms. Each *p*-orbital overlaps with adjacent *p*-orbitals in such a way that the delocalisation is extended over all six carbon atoms.

The π-delocalised ring accounts for the increased stability of benzene and explains its low reactivity with bromine. In addition, it explains why all six

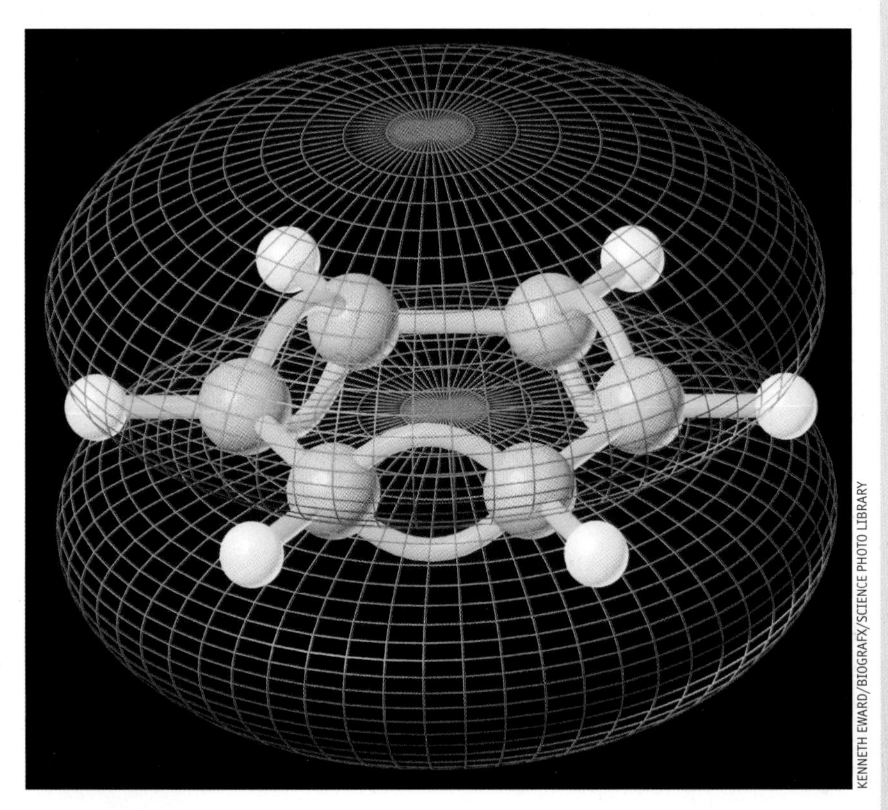

Model illustrating the structure of benzene

carbon–carbon bond lengths are identical. The structure of benzene can be represented in a number of ways:

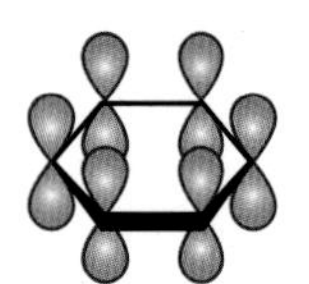

Each C atom has a *p*-orbital at right angles to the plane of atoms

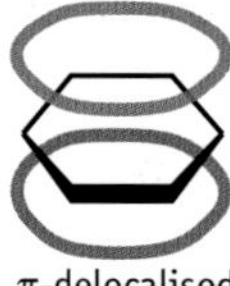

π-delocalised ring above and below the plane

Skeletal formula of benzene

Students may represent the structure of benzene as:

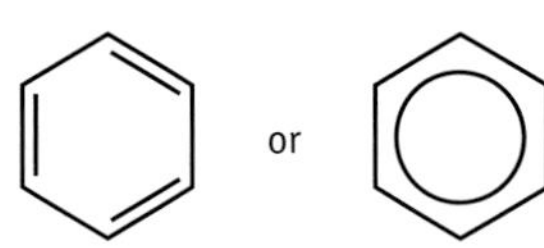

The right-hand structure is used in equations; both are used when describing mechanisms.

Electrophilic substitution

As discussed on page 8, benzene does not readily react with and decolorise bromine. However, it does react with bromine in the presence of a suitable catalyst.

Analysis of the product shows that benzene undergoes substitution reactions rather than addition reactions.

Benzene is stable and retains the π-delocalised ring. The product is formed by replacing one (or more) of the hydrogen atoms in C_6H_6, so that the product is of the form C_6H_5X, where X is an electrophile.

> An electrophile is a particle that can accept an electron pair, resulting in the formation of a (dative) covalent bond.

The most common electrophile is H^+ but it is clearly pointless reacting benzene, C_6H_6, with H^+. Catalysts are used to generate different electrophiles such as NO_2^+, Cl^+, Br^+ and $^+CH_3$.

> Metal ions such as Na^+ cannot behave as electrophiles because they do not form covalent bonds.

The general equation is:

$$C_6H_6 + X^+ \rightarrow C_6H_5X + H^+$$

where X^+ is the electrophile.

Formation of nitroarenes

Nitroarenes are versatile compounds that are used in the preparation of drugs and dyes. The explosive trinitrotoluene, TNT, is made by the nitration of methylbenzene.

Nitration of benzene

The nitration of benzene requires the use of a nitrating mixture consisting of concentrated nitric acid and concentrated sulphuric acid.

$$C_6H_6 + HNO_3 \xrightarrow[H_2SO_4]{conc.} C_6H_5NO_2 + H_2O$$

The reaction is heated gently at a temperature of between 50°C and 55°C. It is essential to control the temperature to minimise the possibility of forming dinitro- and trinitro-benzenes. (The temperature required varies from one arene to another.)

The mechanism for the reaction has three distinct steps.

Step 1 Generation of the electrophile

The electrophile in the reaction is the nitronium ion, NO_2^+, which is generated by the reaction between sulphuric acid and nitric acid. This complex reaction can be summarised as:

$$H_2SO_4 + HNO_3 \rightarrow HSO_4^- + H_2NO_3^+$$
$$H_2NO_3^+ \rightarrow H_2O + NO_2^+$$
$$H_2SO_4 + H_2O \rightarrow H_3O^+ + HSO_4^-$$

The net reaction is:

$$2H_2SO_4 + HNO_3 \rightarrow 2HSO_4^- H_3O^+ + NO_2^+$$

Step 2 Electrophilic attack at the benzene ring

The mechanism can be written using either the delocalised model of benzene:

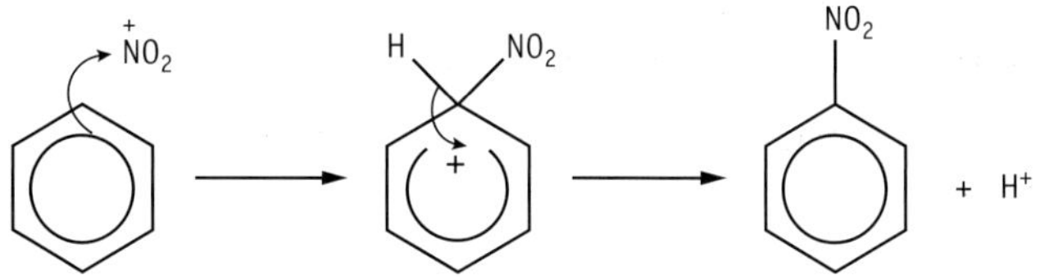

or the Kekulé model of benzene:

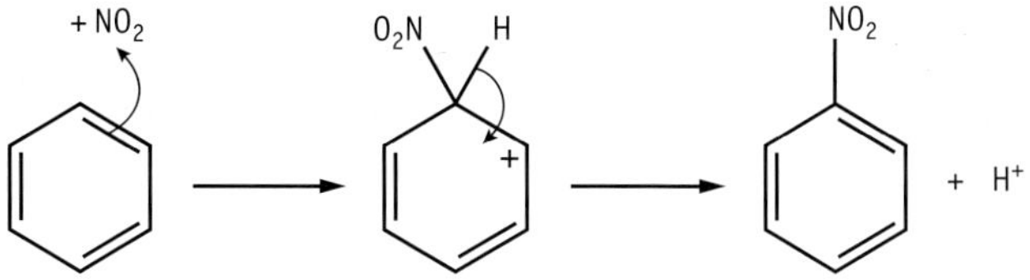

Step 3 Regeneration of the catalyst

The H^+ formed in the second stage of the mechanism then reacts with the hydrogen sulphate ion, HSO_4^-, to reform sulphuric acid, H_2SO_4. Therefore, the H_2SO_4 can be regarded as a catalyst in the reaction.

$$H^+ + HSO_4^- \rightarrow H_2SO_4$$

Formation of haloarenes

Halogenation of benzene

Benzene undergoes an electrophilic substitution reaction with both chlorine and bromine. If chlorine is bubbled through benzene at room temperature in the presence of a halogen carrier, chlorobenzene is formed.

The mechanism for the reaction has three distinct steps.

Step 1 Generation of the electrophile

The halogen carrier is either iron, iron(III) chloride, $FeCl_3$, or aluminium chloride, $AlCl_3$. If iron is used it reacts initially with the chlorine to form $FeCl_3$.

Halogen carriers work by polarising the Cl–Cl bond and promoting heterolytic fission of the halogen bond:

$$Cl{-}Cl \quad AlCl_3 \rightarrow \overset{\delta+}{Cl}{-}\overset{\delta-}{Cl}AlCl_3$$

The Cl–Cl bond is now polarised and the $Cl^{\delta+}$ can behave as an electrophile.

Halogen carriers such as $AlCl_3$ and $FeCl_3$ have vacant orbitals that can accept an electron pair and form a dative covalent bond.

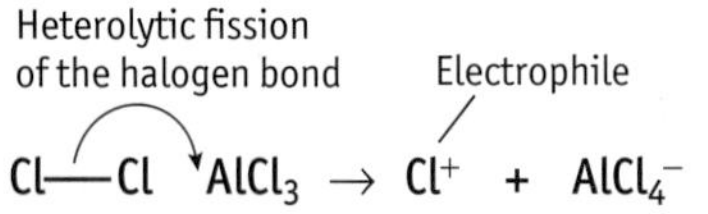

Heterolytic fission of the halogen bond

Electrophile

$$Cl{-}Cl \quad AlCl_3 \rightarrow Cl^+ + AlCl_4^-$$

Step 2 Electrophilic attack at the benzene ring

As with the nitration of benzene, the mechanism can be written using either the delocalised model of benzene:

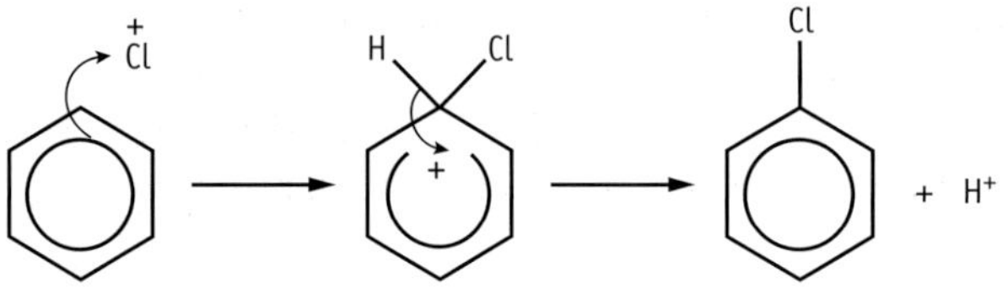

+ Cl

H Cl

+

Cl

+ H^+

or the Kekulé model of benzene

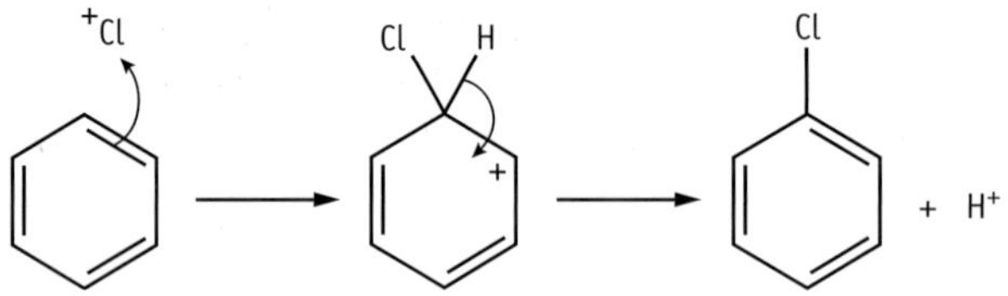

+Cl

Cl H

+

Cl

+ H^+

Step 3 Regeneration of the catalyst

The H^+ formed in the second stage of the mechanism then reacts with the $AlCl_4^-$ ion to reform aluminium chloride, $AlCl_3$. Therefore, $AlCl_3$ can be regarded as a catalyst in the reaction:

$$H^+ + AlCl_4^- \rightarrow AlCl_3 + HCl$$

The chlorination of benzene is a Friedel–Crafts reaction in which $AlCl_3$ behaves as a halogen carrier. Halogen carriers are able to accept a halide ion and to 'carry it' through the reaction. At the end of the reaction the halide ion is released and the hydrogen halide is formed.

$AlCl_3$, $AlBr_3$, $FeCl_3$, $FeBr_3$ and Fe can all behave as halogen carriers.

Friedel–Crafts reactions were developed by Charles Friedel and James Crafts in the late nineteenth century. There are three main types: halogenation, alkylation and acylation. They are all electrophilic substitution reactions.

Bromination of alkenes and arenes

You will recall from AS *Chains, energy and resources* that alkenes, such as cyclohexene, react readily with bromine, in the absence of sunlight, and undergo electrophilic addition reactions. The reaction is rapid and is initiated by the induced dipole in bromine. Alkenes like cyclohexene are able to induce a dipole in bromine because of the high electron density in the carbon-to-carbon double bond, C=C, which is sufficient to induce a dipole in the Br–Br bond:

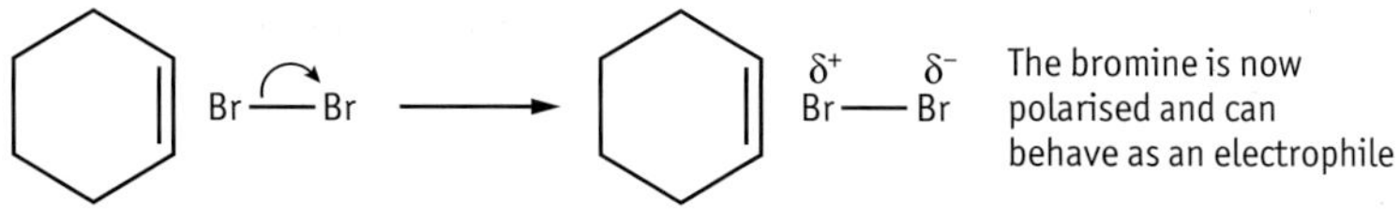

The bromine is now polarised and can behave as an electrophile

The C=C double bond has high electron density because the overlapping *p*-electrons are localised between the two carbons

The mechanism for the reaction between cyclohexene and bromine is shown below:

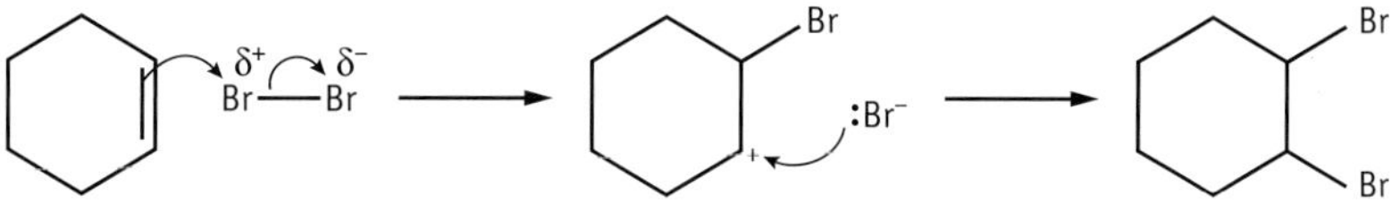

Benzene also reacts with bromine but it is more resistant and reacts much less readily. The electron density in benzene is insufficient to induce a dipole in the Br–Br bond because the *p*-electrons are delocalised over the six carbon atoms in the ring. It follows that polarisation of the Br–Br bond requires the presence of a halogen carrier to generate the electrophile. The resultant reaction is electrophilic substitution, *not* electrophilic addition. This is explained by the stability of the π-delocalised ring of electrons that is retained in most reactions of arenes.

Uses of arenes

Arenes such as benzene, methylbenzene and 1,4-dimethylbenzene are used as additives to improve the performance of petrol. They are manufactured by reforming straight-chain alkanes (see *AS Chemistry*, page 149).

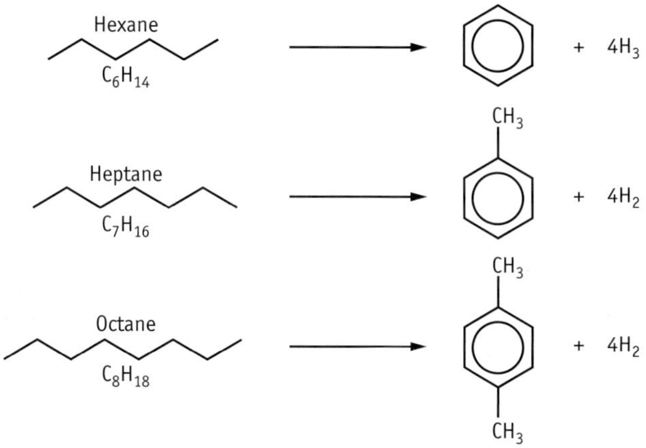

Hexane C_6H_{14} → benzene + $4H_3$

Heptane C_7H_{16} → methylbenzene (CH_3) + $4H_2$

Octane C_8H_{18} → 1,4-dimethylbenzene (CH_3, CH_3) + $4H_2$

Benzene is the feedstock for a great variety of products ranging from medicines such as aspirin and benzocaine to explosives such as 2,4,6-trinitromethylbenzene (TNT) and including a wide range of azo dyes. Phenylethene (styrene) is also manufactured from benzene and is the monomer used to make the polymer poly(phenylethene) or polystyrene (see the additional reading at the end of this chapter).

The products made from benzene are of great value. However, benzene itself is carcinogenic and it is believed that it can cause leukaemia. For this reason you will not use it in the laboratory. Some chlorinated benzene compounds are also extremely toxic.

Wine is made by the conversion of sugar in grapes to ethanol by yeast. Wine contains a complex mixture of amino acids, phenols, carbohydrates and other compounds.

Phenols

Phenols occur widely in nature and, like alcohols, contain the hydroxyl group, –OH. However, in a phenol the –OH group is attached directly to a benzene ring.

OH, H_3C — Compound **A**

OH, CH_2 — Compound **B**

The formula of both compounds A and B is C_7H_8O. They are isomers. Compound A is a phenol; compound B is an alcohol.

The acidity of phenols

Phenols are weak acids, but alcohols are not acidic. In fact, ethanol ionises less than water. This difference in acidity is best explained by the relative inductive effect of aryl and alkyl groups and the relative stability of the phenoxide ion and the ethoxide ion.

The inductive effect can be regarded as the movement of electrons along a σ-bond. It is caused by differences in electronegativity and electron density. Alkyl groups, for example methyl, $-CH_3$, and ethyl, $-C_2H_5$, release electrons along the σ-bond and have what is known as a positive inductive effect:

$$CH_3CH_2 \rightarrow OH$$

In ethanol, this has the effect of increasing the electron density in the O–H bond and this means that ethanol is unlikely to donate H^+. In addition, the ethoxide ion, $C_2H_5O^-$, which would have been formed if ethanol were able to donate a proton, is made unstable because the positive inductive effect pushes electrons towards the oxygen making it more likely to accept H^+ and form ethanol once more.

Equilibrium lies to the left

$$CH_3CH_2OH \rightleftharpoons CH_3CH_2O^- + H^+$$

In phenol, the $-C_6H_5$ ring is electron deficient and one of the lone pairs of electrons on the oxygen in the –OH group is incorporated and delocalised into the ring. This weakens the O–H bond and stabilises the phenoxide ion, $C_6H_5O^-$, that is formed.

The overlap of the *p*-orbitals leads to a delocalisation that extends from the ring out over the oxygen atom. As a result, the negative charge is no longer entirely localised on the oxygen, but is spread out around the whole ion.

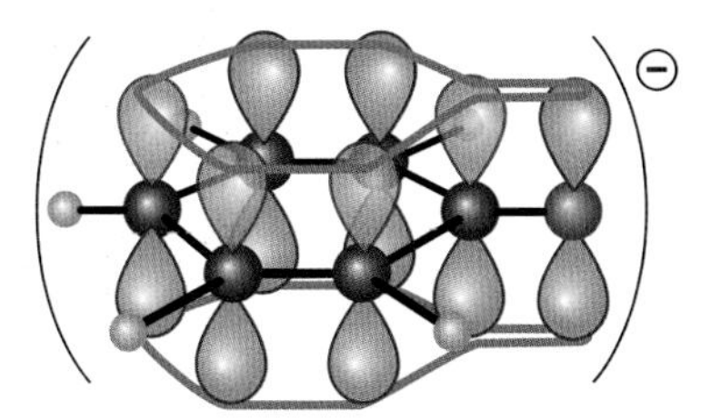

Consequently phenol behaves as a weak acid. The increased electron density in the π-delocalised system means that phenol undergoes electrophilic substitution reactions much more readily than benzene does.

Reactions of phenols

Reaction with sodium hydroxide and with sodium

Phenol is weakly acidic and forms salts when it reacts with both sodium hydroxide and with sodium.

$$C_6H_5OH + NaOH \longrightarrow C_6H_5O^-Na^+ + H_2O$$

$$C_6H_5OH + Na \longrightarrow C_6H_5O^-Na^+ + \tfrac{1}{2}H_2$$

Reaction with bromine

Phenol also reacts readily with bromine. The bromine is decolorised and white crystals of 2,4,6-tribromophenol are formed.

$$C_6H_5OH + 3Br_2 \longrightarrow C_6H_2Br_3OH + 3HBr$$

White solid

Unlike benzene, phenol does not require a halogen carrier and reacts instantly with bromine. This occurs because the ring is activated by one of the lone pairs of electrons on the oxygen atom in the –OH group. A lone pair of electrons from the oxygen is delocalised into the ring. This increases the electron density in the ring so that it induces a dipole in the Br–Br bond, thereby generating an electrophile, $Br^{\delta+}$. The increased electron density of the ring attracts the electrophile, leading to a rapid reaction at room temperature.

Substitution occurs at the 2, 4 and 6 positions on the ring. This can be best illustrated using the Kekulé model.

Each of the intermediate ions A, B and C has a negative charge on the ring and hence these positions (2, 4 and 6) are most likely to be attacked by an electrophile.

The bromination of benzene takes place only in the presence of a suitable halogen carrier.

e You will not be asked in the exam why bromine is always directed to the 2, 4 and 6 positions in phenol. A common question is to ask you to compare the relative ease of bromination of phenol and benzene.

Uses of phenols

- In the mid-nineteenth century Lister used a dilute aqueous solution of phenol as an antiseptic.
- Phenols have a characteristic pine-tar odour and turn milky in water.
- Phenols are effective antibacterial agents. They are also effective against fungi and many viruses. They retain more activity in the presence of organic material than iodine or chlorine-containing disinfectants. Examples of the phenol class include: lysol, Pine-Sol, Cresi-400, Environ, and chlorophenols such as 2,4,6-trichlorophenol, TCP.

Questions

1 a Name each of the compounds **A–F**:

b What is the molecular formula of **F**?

c What is the empirical formula of **B**?

d Draw and name two isomers of **A** and **B**.

e Write a balanced equation for the complete combustion of **C**.

f Starting with benzene, outline how you could prepare **D**.

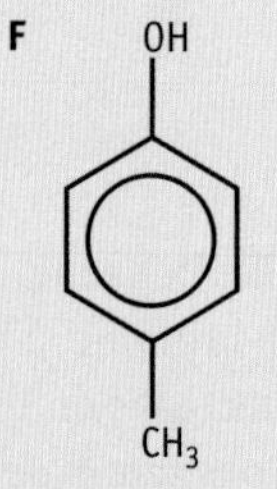

2 Explain why phenol reacts with NaOH(aq) but benzene and ethanol do not.

3 Benzene, phenol and cyclohexene all react with bromine.

a For each reaction:
- name the type of reaction
- state the reagents and conditions (if any)
- identify the organic product and state any observations made

b Explain the different rates of reaction between bromine and each of the following pairs of compounds:
- benzene and phenol
- benzene and cyclohexene

4 2,4,6-trinitromethylbenzene (TNT) can be used as an explosive. When detonated, it forms $CO_2(g)$, $H_2O(l)$ and $N_2(g)$. Calculate the volume of gas produced when 9.08 g of TNT is exploded. Assume all measurements were taken at s.t.p. (1 mol of gas occupies 24.0 dm^3)

5 TCP, 2,4,6-trichlorophenol, can be used as an antiseptic. A 100 cm^3 solution of TCP was titrated with a 0.400 mol dm^{-3} NaOH(aq) solution. Exactly 25.0 cm^3 NaOH(aq) were required.

a Calculate the mass of TCP dissolved in the 100 cm^3 solution of TCP.

b Calculate the concentration (mol dm^{-3}) of the TCP.

Summary

You should now be able to explain:

- the bonding in benzene and all arenes
- that arene chemistry is dominated by substitution reactions
- that hydrogen atoms on the ring can be replaced by electrophiles
- that phenols contain the –OH group which is attached directly to the benzene ring
- that phenols can behave as weak acids
- that the increased electron density in the ring enhances the reactivity of the ring in that it polarises Br_2 and enables reaction without the need for a halogen-carrier catalyst

Additional reading

Benzene is an important industrial chemical. It is particularly versatile and is used in the manufacture of a wide range of important industrial chemicals, some of which are shown below.

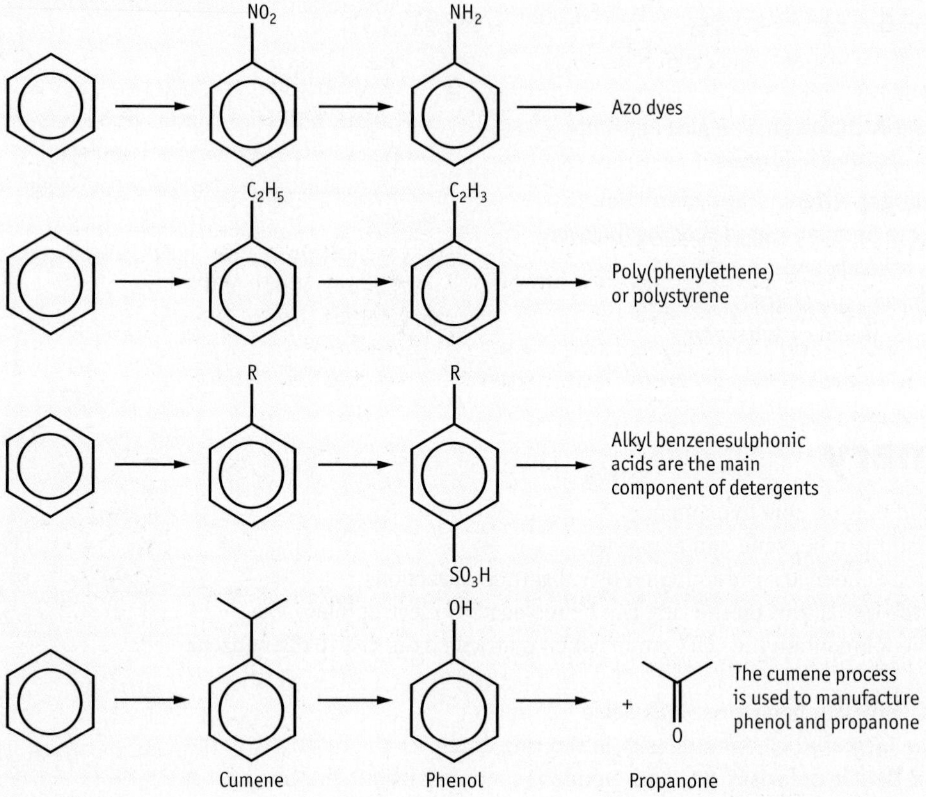

NO_2 NH_2 → Azo dyes

C_2H_5 C_2H_3 → Poly(phenylethene) or polystyrene

R R SO_3H → Alkyl benzenesulphonic acids are the main component of detergents

OH + O → The cumene process is used to manufacture phenol and propanone

Cumene Phenol Propanone

Production of dyes

The preparation of azo dyes is part of the OCR specification. It is covered fully in Chapter 4, pages 53–54.

Production of polymers

Phenylethene is known commonly as styrene. It can be made from benzene by reaction with ethene in the presence of an acid. The initial step is the reaction between ethene and H^+:

$$H_2C{=}CH_2 + H^+ \longrightarrow H_3C{-}\overset{+}{C}H_2$$

The $CH_3CH_2^+$ then behaves as an electrophile and reacts with benzene to produce phenylethane, $C_6H_5CH_2CH_3$. This provides the correct carbon skeleton for the formation of phenylethene, which is formed by dehydrogenation in the presence of a suitable catalyst:

$$C_6H_5CH_2CH_3 \rightarrow C_6H_5CHCH_2 + H_2$$

Phenylethene then undergoes addition polymerisation to form poly(phenylethene) or polystyrene:

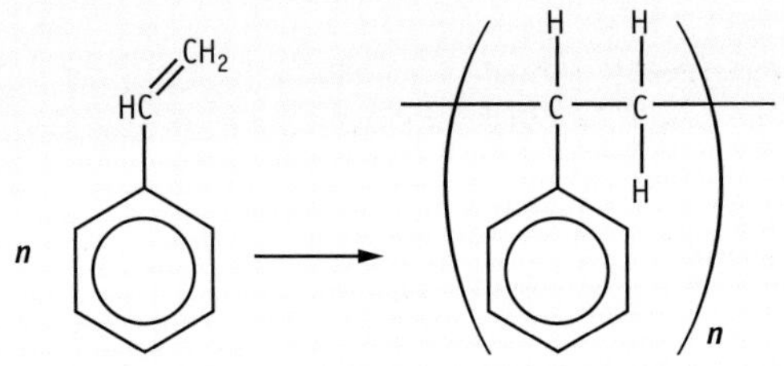

Polystyrene is used widely in packaging, insulation and everyday objects such as drinking cups, ceiling tiles and telephones.

Preparation of detergents

Alkylbenzenesulphonic acids and alkybenzenesulphonates are the essential ingredients in many detergents. In order to manufacture these, benzene has to be both alkylated (an alkyl group is substituted for one of the hydrogen atoms in the ring) and sulphonated (a sulphonic acid group, $-SO_3H$, is substituted for one of the hydrogens in the ring).

It is usual to alkylate the ring first by reacting benzene with a long-chain alkene in the presence of a suitable catalyst:

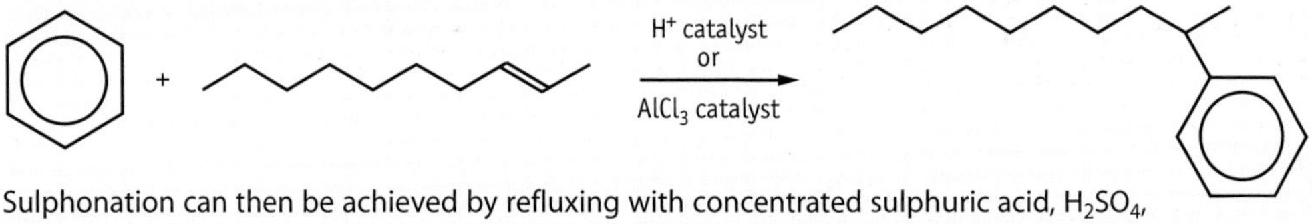

H^+ catalyst or $AlCl_3$ catalyst

Sulphonation can then be achieved by refluxing with concentrated sulphuric acid, H_2SO_4, or with oleum, $H_2S_2O_7$. The electrophile is thought to be SO_3. The mechanism for the sulphonation of benzene is shown below:

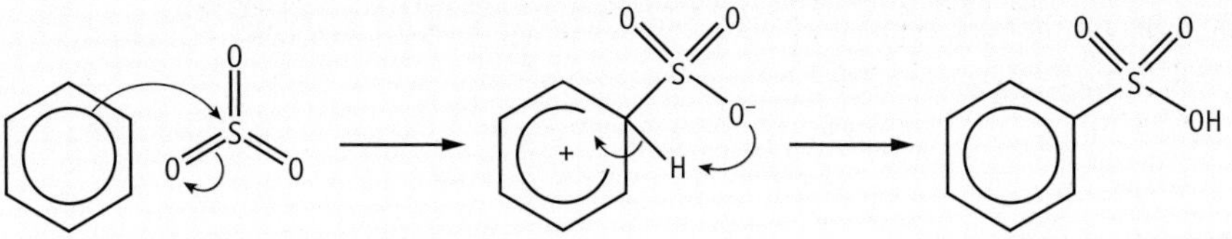

The alkyalted benzene is sulphonated in the same way. The product is then treated with NaOH to produce a sulphonate salt.

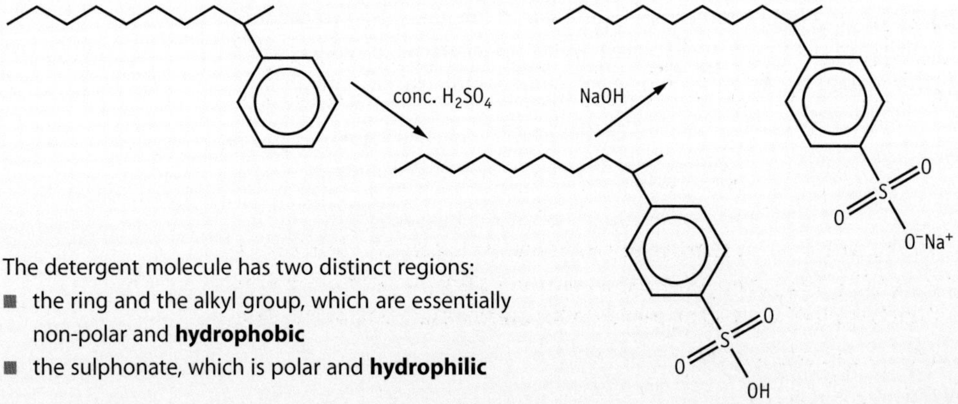

conc. H_2SO_4 NaOH O^-Na^+

The detergent molecule has two distinct regions:

- the ring and the alkyl group, which are essentially non-polar and **hydrophobic**
- the sulphonate, which is polar and **hydrophilic**

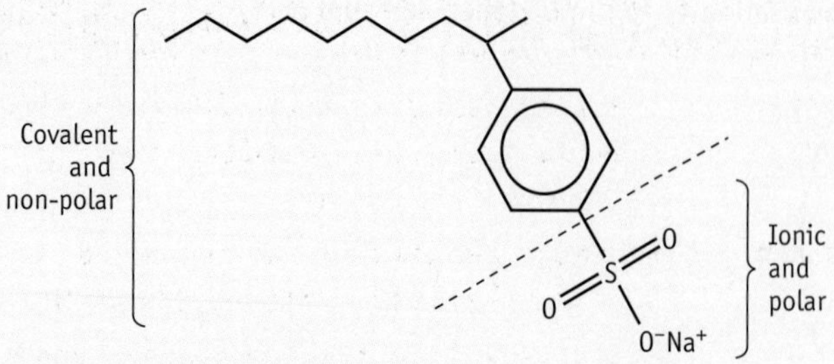

Covalent and non-polar

Ionic and polar

O^-Na^+

Detergents are extremely versatile and can remove dirt and grease from a wide variety of substances without forming a scum.

Preparation of phenol by the cumene process

Benzene and propene, two relatively cheap feedstocks, are used in the industrial manufacture of phenol. Propanone is produced as a co-product. The overall chemical process is summarised as:

$$C_6H_6 + CH_3CH{=}CH_2 + O_2 \rightarrow C_6H_5OH + CH_3COCH_3$$

Benzene and propene react in the presence of an acid catalyst. The product is (1-methylethyl)benzene (cumene).

H^+ catalyst

Cumene

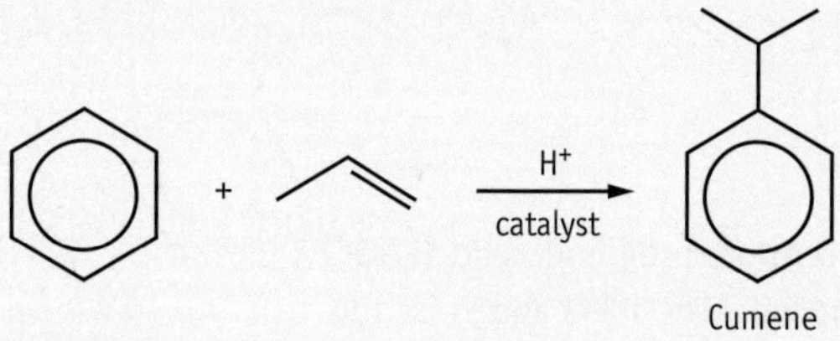

Air is passed through the (1-methylethyl)benzene (cumene) to form an intermediate that is then decomposed in warm dilute sulphuric acid.

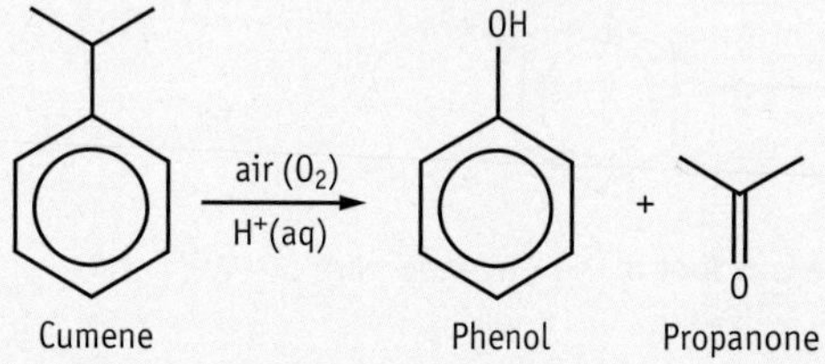

OH

air (O_2)

$H^+(aq)$

O

Cumene Phenol Propanone

Phenol and propanone are both extremely useful chemicals. Phenol is used to produce disinfectants, antiseptics and dyes whilst propanone (or acetone) is widely used as a solvent. Both products are valuable and so the cumene process has a high atom economy.

Stretch and challenge

Benzene reacts with alkenes in the presence of an acid catalyst to form:

$$C_6H_6 + C_nH_{2n} \longrightarrow C_6H_5C_nH_{2n+1}$$

The initial step in the reaction is the formation of a carbonium ion when the H^+ catalyst reacts with the alkene.

Benzene reacts with propene in the presence of an acid catalyst to form (1-methylethyl)benzene, $C_6H_5CH(CH_3)_2$:

$$C_6H_6 + C_3H_6 \xrightarrow{H^+} C_6H_5CH(CH_3)_2$$

a Draw the displayed formula of (1-methylethyl) benzene, $C_6H_5CH(CH_3)_2$, and predict all the bond angles.

b Outline, with the aid of curly arrows, the reaction between benzene, propene and H^+.

c In the initial reaction between propene and H^+ it is possible to form two different carbonium ions, one primary and the other secondary. Use your understanding of inductive effects to explain which of these is more stable and which is, therefore, more likely to react.

Carbonyl compounds Chapter 2

In AS Unit 2: Chains, energy and resources you encountered carbonyl compounds in the form of aldehydes and ketones and it is essential that you revisit some of the chemistry of alcohols. You will be expected to recall the preparation of carbonyl compounds from alcohols.

Aldehyde and ketone groups occur widely in sugars such as glucose and fructose, and they also contribute to the distinctive odours of many plants and foods.

The carbonyl group is C=O and, like the C=C double bond, is formed by the overlap of adjacent *p*-orbitals. Like the alkenes, carbonyl compounds are unsaturated molecules. However, unlike alkenes, carbonyls are polar because of the difference in electronegativity between the carbon and oxygen atoms.

$$>C-O \longrightarrow >\overset{\delta+}{C}=\overset{\delta-}{O}$$

The *p*-orbitals overlap to form a π-bond

The position of the C=O on the carbon chain determines whether or not the compound is classified as an aldehyde or a ketone. An aldehyde has the carbonyl on the end of the carbon chain; in a ketone, the carbonyl group can be anywhere, other than on the terminal carbon atom.

An aldehyde contains the functional group:

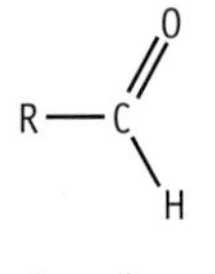

$$R-C(=O)-H$$

Ketones look similar, but the functional group is:

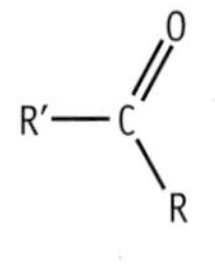

$$R'-C(=O)-R$$

(R and R′ = alkyl group, for example $-CH_3$ or $-CH_3CH_2$)

Formation of carbonyl compounds and carboxylic acids

These are formed by the oxidation of alcohols. The ease and extent of oxidation depends on the type of alcohol:

- Tertiary alcohols are resistant to oxidation.
- Primary and secondary alcohols undergo oxidation when warmed with an oxidising mixture such as acidified dichromate, $Cr_2O_7^{2-}/H^+$ (e.g. $K_2Cr_2O_7/H_2SO_4$).
  - A secondary alcohol is oxidised to a ketone.
  - A primary alcohol can be oxidised partially to an aldehyde or oxidised completely to a carboxylic acid.

The balanced equations for the oxidation of a primary alcohol to an aldehyde and of a secondary alcohol to a ketone are similar. Water is always formed and the equation can be represented as:

primary/secondary alcohol + [O] $\rightarrow$ carbonyl compound + water

For exam purposes, the oxidising agent can be represented as [O] and the equation balanced as shown below. Full equations can be deduced using oxidation numbers and half-equations. This is covered in the additional reading and stretch-and-challenge questions at the end of this chapter.

Oxidation of a primary alcohol to an aldehyde

Methanol is oxidised to methanal:

$CH_3OH + [O] \rightarrow HCHO + H_2O$
Methanol — Methanal

Ethanol is oxidised to ethanal:

$CH_3CH_2OH + [O] \rightarrow CH_3CHO + H_2O$
Ethanol — Ethanal

Oxidation of a secondary alcohol to a ketone

Propan-2-ol is oxidised to propanone (propan-2-one):

$CH_3CHOHCH_3 + [O] \rightarrow CH_3COCH_3 + H_2O$
Propan-2-ol — Propanone

Butan-2-ol is oxidised to butanone (butan-2-one):

$CH_3CH_2CHOHCH_3 + [O] \rightarrow CH_3CH_2COCH_3 + H_2O$
Butan-2-ol — Butanone

Oxidation of a primary alcohol to a carboxylic acid

The primary alcohol is oxidised in two steps:

- The first step creates an aldehyde.
- In the second step, the second C–H bond is converted into a C–O–H.

Overall, the carboxylic acid functional group is generated:

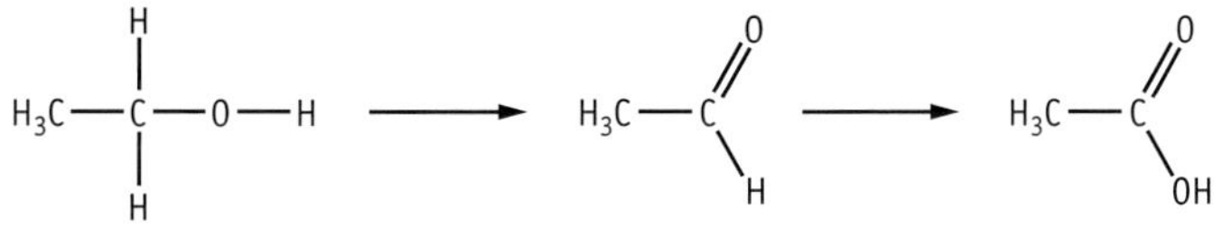

$H_3C-CH_2-O-H \longrightarrow H_3C-C(=O)H \longrightarrow H_3C-C(=O)OH$

Primary alcohol — Aldehyde — Carboxylic acid

All balanced equations for the oxidation of a primary alcohol to a carboxylic acid are similar. Water is always formed and the reaction can be represented as:

primary alcohol + 2[O] $\rightarrow$ carboxylic acid + water

Methanol is oxidised to methanoic acid:

$CH_3OH + 2[O] \rightarrow HCOOH + H_2O$
Methanol — Methanoic acid

Ethanol is oxidised to ethanoic acid:

$$\underset{\text{Ethanol}}{CH_3CH_2OH} + 2[O] \rightarrow \underset{\text{Ethanoic acid}}{CH_3COOH} + H_2O$$

When oxidising a primary alcohol, such as propan-1-ol, the choice of apparatus is important. The alcohol and the oxidising mixture have to be heated. This can be achieved by either distillation (Figure 2.1) or reflux (Figure 2.2).

If propan-1-ol is oxidised it is possible to produce either propanal or propanoic acid. The boiling points are shown in Table 2.1.

***Table 2.1** Boiling points of propan-1-ol, propanal and propanoic acid*

Name	Formula	Boiling point/°C
Propan-1-ol	$CH_3CH_2CH_2OH$	97
Propanal	CH_3CH_2CHO	49
Propanoic acid	CH_3CH_2COOH	141

In general, alcohols and carboxylic acids have considerably higher boiling points than aldehydes. Alcohols and carboxylic acids both contain –O–H groups and therefore both form intermolecular hydrogen bonds.

e You will not be expected to recall any reactions of alcohols other than oxidation.

In order to isolate the aldehyde a distillation is usually carried out. Distillation involves evaporation followed by condensation and allows the most volatile component to be separated.

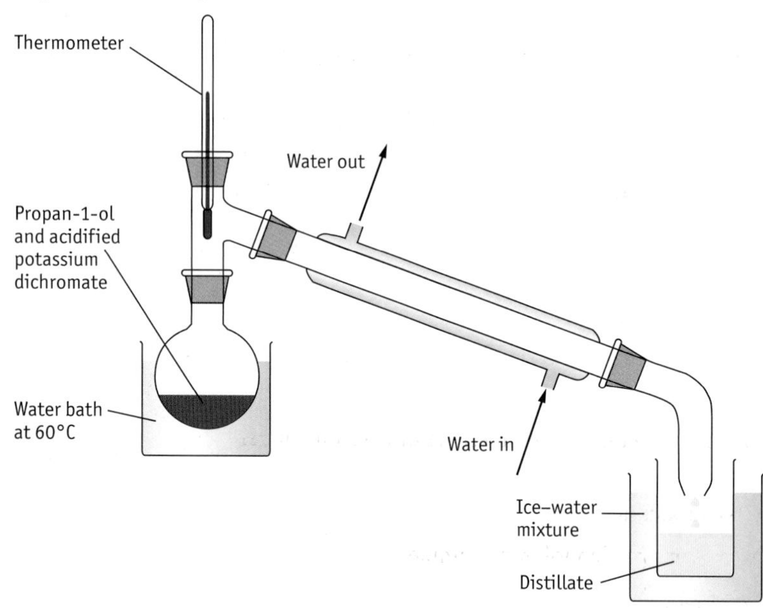

***Figure 2.1** Apparatus for distillation*

When the reaction mixture is heated, the volatile components evaporate first. Propanal has the lowest boiling point and vaporises most readily. The outer sleeve of the condenser contains circulating cold water, so when the propanal reaches the condenser, it condenses and is separated from the reaction mixture.

Reflux involves a process of continuous evaporation and condensation that prevents any volatile components from escaping.

If a reflux apparatus is used, the components undergo continuous evaporation and condensation and volatile components will not escape or be separated out. As with distillation, when the reaction mixture is heated the volatile components evaporate first. Propanal has the lowest boiling point and is vaporised most readily. The outer sleeve of the condenser contains circulating cold water, so when the propanal reaches the condenser, it condenses (as with distillation). However, it falls back into the oxidising mixture and is *not* separated from the reaction mixture. The propanal is then oxidised to form propanoic acid.

Any oxidation reaction that utilises the mixture of dichromate and sulphuric acid is accompanied by a distinctive colour change from orange to green. This occurs because orange dichromate ions, $Cr_2O_7^{2-}$, are reduced to green Cr^{3+} ions.

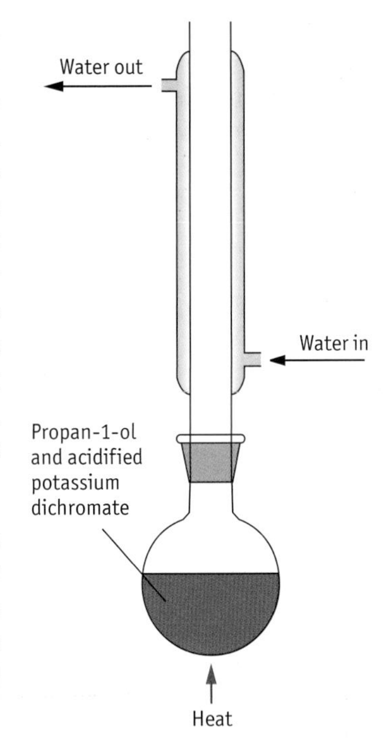

Figure 2.2 *Apparatus for reflux*

Reactions of carbonyl compounds

Aldehydes and ketones both contain the C=O group and therefore have some reactions in common. However, aldehydes can be oxidised easily whereas ketones cannot, so there are reactions that can be used to distinguish between them.

Reactions common to both aldehydes and ketones

Reduction

Aldehydes or ketones can be reduced to their respective alcohols using an aqueous solution of sodium tetrahydridoborate(III), $NaBH_4$, as the reducing agent.

[H] is used to represent the reducing agent in equations.

All balanced equations for the reduction of an aldehyde or a ketone are similar. The reaction can be represented as:

aldehyde/ketone +2[H] → alcohol

An aldehyde is reduced to a primary alcohol, for example:

$$CH_3CH_2CHO + 2[H] \rightarrow CH_3CH_2CH_2OH$$

Propanal — Propan-1-ol

A ketone is reduced to a secondary alcohol:

$$CH_3COCH_3 + 2[H] \rightarrow CH_3CH(OH)CH_3$$

Propanone — Propan-2-ol

The reduction can be regarded as a nucleophilic addition reaction with the reducing agent, $NaBH_4$, providing the hydride ion, $:H^-$, which behaves as a nucleophile.

A nucleophile can be defined as an electron-pair donor.

Nucleophilic addition reactions

$^{\delta+}C{=}O^{\delta-}$ is a polar bond because the carbon and oxygen atoms have different electronegativities. The carbon in the carbonyl group is δ+ and can therefore be attacked by a nucleophile. The carbonyl group, C=O, is unsaturated and hence undergoes addition reactions. The mechanism is, therefore, nucleophilic addition.

This is best illustrated by addition reactions with the hydride ion, H^-.

An example is the reduction of ethanal by aqueous $NaBH_4$:

H_3C—C(=O$^{\delta-}$)(H) + :H^- ⟶ H_3C—C(O^-)(H)—OH ... $H^{\delta+}$—$O^{\delta-}$—H ⟶ H_3C—C(OH)(H)—H + $^-$:O—H

Ethanal

Ethanol

e When describing this mechanism you should show the movement of electrons by using curly arrows. You should also include relevant dipoles and lone pairs of electrons.

Ketones behave similarly — for example, the reduction of propanone by aqueous $NaBH_4$:

H_3C—C(=O$^{\delta-}$)(CH_3) + :H^- ⟶ H_3C—C(O^-)(CH_3)—OH ... $H^{\delta+}$—$O^{\delta-}$—H ⟶ H_3C—C(OH)(CH_3)—H + $^-$:O—H

Propanone

Propan-2-ol

Reaction with 2,4-dinitrophenylhydrazine

Aldehydes and ketones both react with 2,4-dinitrophenylhydrazine.

Carbonyl compounds react with an excess of 2,4-dinitrophenylhydrazine to produce a bright red, orange or yellow precipitate. The precipitates formed are derivatives of 2,4-dinitrophenylhydrazine and are known as 2,4-dinitrophenylhydrazones.

The reactions of carbonyl compounds with 2,4-dinitrophenylhydrazine (2,4-DNPH) are important for several reasons:

NIGEL EVANS

A precipitate of 2,4-dinitrophenylhydrazine

e You are not expected to recall the formula of 2,4-dinitrophenylhydrazine. The abbreviation 2,4-DNPH is acceptable.

The formula of 2,4-dinitrophenylhydrazine is used in the additional reading at the end of this chapter, where the reactions of aldehydes and ketones are explained in more detail

- 2,4-DNPH reacts with a carbonyl compound to produce a distinctive precipitate. The organic product is a bright red, orange or yellow precipitate. Therefore, this reaction can be used to identify the presence of a carbonyl group (aldehyde and ketone).
- The organic product (the 2,4-DNPH derivative) is relatively easy to purify by recrystallisation. The melting point of the brightly coloured precipitate can then be determined.
- Each derivative has a different melting point, which may be used to identify a specific carbonyl compound. Table 2.2 shows the melting points of the derivatives of a few common carbonyl compounds.

Table 2.2
Melting points of some carbonyl compounds

Carbonyl compound	Boiling point/°C	Melting point of the 2,4-DNPH derivative/°C
Ethanal	20	148
Propanal	49	156
Butanal	75	123
Methylpropanal	64	182
Propanone	56	128
Butanone	80	115
Pentan-2-one	102	142
Pentan-3-one	102	153
3-Methylbutan-2-one	94	124

The preparation and purification of a 2,4-dinitrophenylhydrazine derivative can be used to identify a specific aldehyde or ketone. The process involves the initial preparation of the derivative, which is filtered and then recrystallised and dried. The melting point of the derivative is then determined and checked against data values to identify the original carbonyl compound. This is particularly useful when trying to distinguish between isomers such as pentan-2-one and pentan-3-one, which have the same boiling point.

Reactions for aldehydes only

Aldehydes and ketones can be distinguished by a series of redox reactions. Aldehydes are oxidised readily to carboxylic acids; ketones are not oxidised easily.

There are three common oxidising mixtures that can be used.

(1) **Oxidising mixture**: acidified dichromate ($H^+/Cr_2O_7^{2-}$)

Conditions: warm

Observation: when reacted with an aldehyde there is a colour change from orange to green

(2) **Oxidising mixture**: alkaline solution of Cu^{2+} ions with an alkaline solution of potassium tartrate (Fehling's solution)

Conditions: warm gently in a water bath at about 60°C

Observation: Fehling's solution is a dark blue solution which when reacted with an aldehyde produces a red precipitate of copper(I) oxide, $Cu_2O(s)$

(3) **Oxidising mixture**: an aqueous solution of Ag^+ ions in excess ammonia, $Ag(NH_3)_2^+$ (Tollens' reagent)

Conditions: warm gently in a water bath at about 60°C

Observation: silver metal is precipitated forming a silver mirror

The OCR specification recommends that you use Tollens' reagent to distinguish between an aldehyde and a ketone.

Tollens' reagent is prepared by adding a few drops of dilute sodium hydroxide solution to a solution of silver nitrate. A brown precipitate of silver oxide, $Ag_2O(s)$, is produced immediately. Dilute ammonia is then added until the $Ag_2O(s)$ just redissolves. This creates an ammoniacal solution of silver nitrate, $Ag(NH_3)_2^+$, which acts as the oxidising agent.

In this reaction, the Ag^+ ions are reduced to silver metal (electron gain) and the aldehyde is oxidised to a carboxylic acid. The reaction with ethanal is given below:

$Ag^+ + e^-$ —reduction→ Ag (silver mirror)

$H_3C{-}CHO + [O]$ —oxidation→ $H_3C{-}COOH$

Ethanal — An aldehyde is oxidised to a carboxylic acid — Ethanoic acid

The oxidation of an aldehyde to a carboxylic acid can be identified using infrared spectroscopy.

Infrared spectroscopy was first introduced in Unit 2: Chains, energy and resources (see the AS textbook, page 187).

The relevant absorptions to pick out are those of the C=O group which is present in both aldehydes and ketones and which has a strong absorption in the range 1640–1750 cm^{-1}. Carboxylic acids also have a C=O absorption but in addition have a characteristic broad absorption within the region 2500–3300 cm^{-1}, which is due to the O–H group in the carboxylic acid.

The absorption due to the carbonyl group can be seen clearly in the IR spectrum of ethanal:

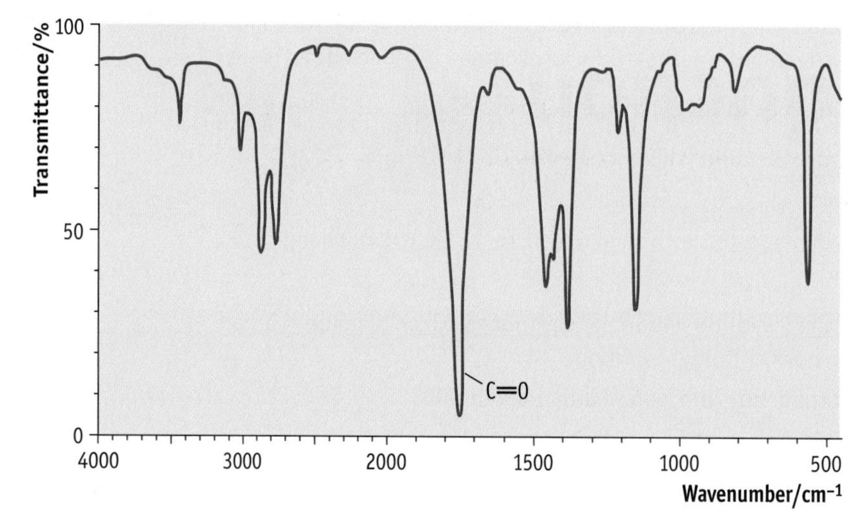

The spectrum below shows the absorption for both the C=O and the O–H groups, confirming that the ethanal has been oxidised to ethanoic acid:

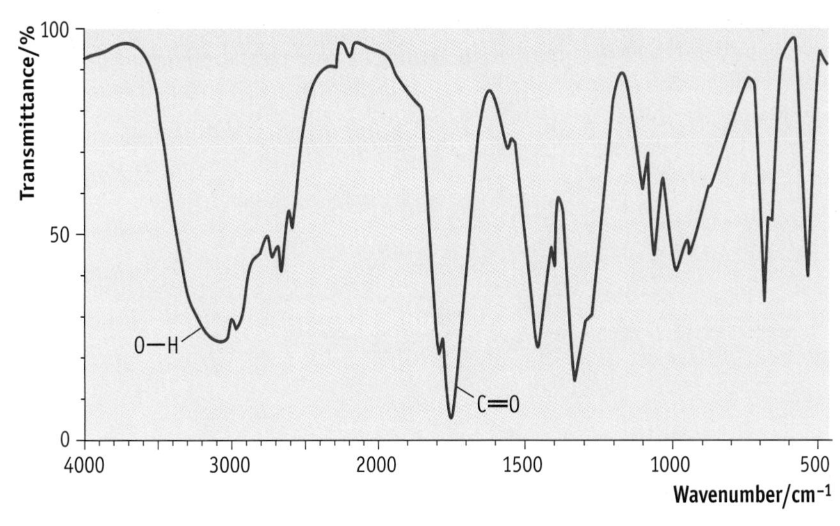

Questions

1 a Name each of the compounds **A–F**.

b Classify each of **A** to **F** as either an aldehyde or a ketone.

c What is the molecular formula of compound **F**?

d Identify the alcohols from which **C** and **D** could be prepared.

e Compounds **B** and **E** are isomers of each other. Draw the structures of three other isomers of **B** and **E**.

A $H_3C—CH(CH_3)—C(=O)H$

B $H_3C—CH(CH_3)—C(=O)CH_3$

C $C_6H_5—CH_2—C(=O)H$

D $C_6H_5—C(=O)—CH_3$

E $H_3C—C(CH_3)_2—C(=O)H$

F O=(cyclohexane ring)=O

2 Butan-2-ol can be prepared by the reduction of a carbonyl compound. Identify the compound, state the reagents and conditions required, write a balanced equation and describe, with the aid of curly arrows, the mechanism.

3 Cinnamaldehyde, C_9H_8O, is a yellow liquid with a powerful odour.

a Complete the reaction grid below for the reactions of cinnamaldehyde.

b Write a balanced equation for the oxidation of cinnamaldehyde by Tollens' reagent.

c There are two possible isomers of cinnamaldehyde. Draw their structures and state the type of isomerism.

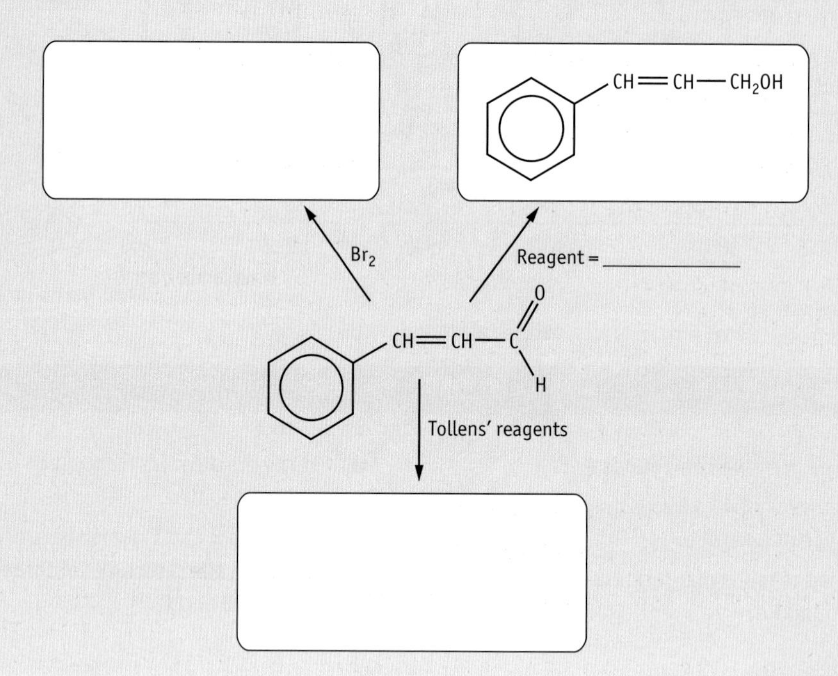

Summary

You should now be able to:

- distinguish between aldehydes and ketones
- write equations for the preparation of aldehydes and ketones by the oxidation of primary and secondary alcohols respectively
- describe the nucleophilic addition mechanism when aldehydes and ketones are reduced by $NaBH_4(aq)$
- describe the reactions of aldehydes and ketones with 2,4-dinitrophenylhydrazine and explain how the derivatives formed can be used to identify the carbonyl compounds
- describe how aldehydes react with Tollens' reagent

Additional reading

You are expected to recall the preparation of aldehydes and ketones by the oxidation of alcohols and to be able to write balanced equations using [O] to represent the oxidising agent. It is perfectly acceptable to simply learn these equations. However, with a little extension to your understanding of oxidation number, you should be able to work out the full equation from first principles using ionic half-equations.

You will recall that oxidation number can be assigned using a simple set of rules.

Rule number	Rule	Example
1	All elements in their natural state have the oxidation number zero	Hydrogen, H_2; oxidation number = 0
2	Oxidation numbers of the atoms of any molecule add up to zero	Water, H_2O; sum of oxidation numbers = 0
3	Oxidation numbers of the components of any ion add up to the charge on that ion	Sulphate, SO_4^{2-}, sum of oxidation numbers = −2

When calculating the oxidation numbers of elements in either a molecule or an ion you should apply the following order of priority:

1	The oxidation numbers of elements in groups 1, 2 and 3 are always +1, +2 and +3 respectively.
2	The oxidation number of fluorine is always −1.
3	The oxidation number of hydrogen is usually +1.
4	The oxidation number of oxygen is usually −2.
5	The oxidation number of chlorine is usually −1.

By applying these rules and priorities in sequence it is possible to deduce any oxidation number.

Consider the oxidation of ethanol to ethanal using acidified dichromate, $H^+/Cr_2O_7^{2-}$, in terms of oxidation number. Whenever we use an oxidising agent in the presence of an acid, water is formed. During the oxidation, there is a colour change from orange to green. The orange colour is due to the $Cr_2O_7^{2-}$ ion and the green colour is due to the Cr^{3+} ion. The oxidation number of Cr in $Cr_2O_7^{2-}$ is +6; the oxidation number of Cr in Cr^{3+} is +3. Using this, we can construct an ionic half-equation for the changes that take place for Cr.

When writing ionic half-equations, we must balance both symbols and charge.

It is sensible to first balance the charge. We know that the oxidation number of Cr changes from +6 to +3 and that for each Cr atom this involves gaining three electrons. Since there are two Cr atoms in $Cr_2O_7^{2-}$, a total of six electrons is required. We can now write:

$$Cr_2O_7^{2-} + 6e^- \rightarrow 2Cr^{3+}$$

We now have to balance the symbols. You will recall that whenever an oxidising agent is used in the presence of an acid, water is formed. Since $Cr_2O_7^{2-}$ contains seven oxygens, $7H_2O$ are formed. This requires $14H^+$. Therefore, the ionic half-equation for the oxidation is:

$$14H^+ + Cr_2O_7^{2-} + 6e^- \rightarrow 2Cr^{3+} + 7H_2O$$

For $Cr_2O_7^{2-}$, **rule 3** tells us that the oxidation numbers have to add up to −2 and **priority 4** tells us that O has priority over Cr.

Oxidation number = −2

↓

$Cr_2O_7^{2-}$

There are seven oxygen atoms giving a total of −14. Hence, the contribution of the two Cr atoms must add up +12. Therefore, the oxidation number of each Cr atom is +6.

We can use this concept of oxidation number and do the same for the oxidation of ethanol to ethanal. First, we have to work out the oxidation numbers. This is easier if we write the molecular formula:

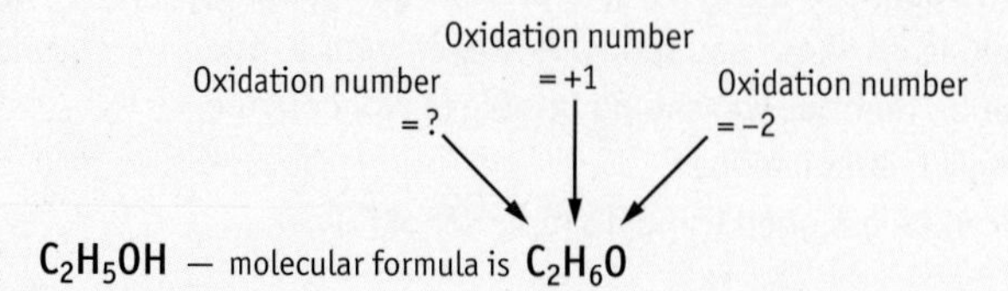

e The oxidation numbers assigned to carbon are theoretical. It is worth remembering that carbon almost always forms four bonds.

There are six hydrogens, giving a total of +6 and one oxygen (−2). Hence, the contribution from the two carbon atoms must add up to −4, which means that the oxidation number of each carbon is −2.

For ethanal:

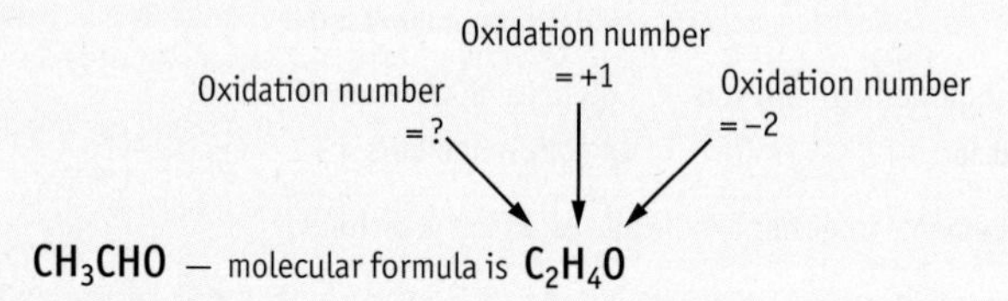

There are four hydrogens, giving a total of +4 and one oxygen (−2). Hence, the contribution from the two carbon atoms must add up to −2, which means that the oxidation number of each carbon is −1.

Each carbon loses one electron:

$$C_2H_5OH \rightarrow CH_3CHO + 2e^-$$

In order to balance for symbol and charge we must add two H^+ to the right-hand side. So, the half-equation for the reduction is:

$$C_2H_5OH \rightarrow CH_3CHO + 2e^- + 2H^+$$

We can now combine the two ionic half-equations:

The half-equation for reduction is:

$$14H^+ + Cr_2O_7^{2-} + 6e^- \rightarrow 2Cr^{3+} + 7H_2O$$

The half-equation for oxidation is:

$$C_2H_5OH \rightarrow CH_3CHO + 2e^- + 2H^+$$

It is essential that there is the same number of electrons in each equation. It follows that the second half-equation has to be multiplied by 3 to give:

$$3C_2H_5OH \rightarrow 3CH_3CHO + 6e^- + 6H^+$$

Adding the two equations gives:

$$14H^+ + Cr_2O_7^{2-} + 6e^- + 3C_2H_5OH \rightarrow 2Cr^{3+} + 7H_2O + 3CH_3CHO + 6e^- + 6H^+$$

This can be simplified to:

$$8H^+ + Cr_2O_7^{2-} + 3C_2H_5OH \rightarrow 2Cr^{3+} + 7H_2O + 3CH_3CHO$$

The $Cr_2O_7^{2-}$ came from $K_2Cr_2O_7$ and the H^+ came from H_2SO_4 so the full equation is:

$$4H_2SO_4 + K_2Cr_2O_7 + 3C_2H_5OH \rightarrow Cr_2(SO_4)_3 + 7H_2O + 3CH_3CHO + K_2SO_4$$

e This is a difficult concept but it enables you to deduce complex equations. If you understand the process it will help greatly when you come to study electrode potentials in Unit 5.

2,4-dinitrophenylhydrazine

The formula of 2,4-dinitrophenylhydrazine is:

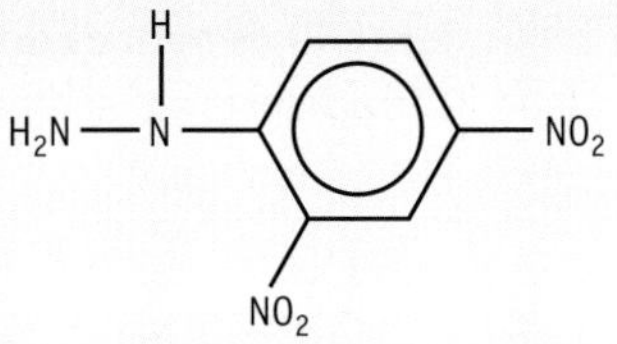

The specification does not require you to recall the formula or structure of 2,4-dinitrophenylhydrazine. However, knowledge of both may help with stretch-and-challenge questions.

It reacts with carbonyl compounds to produce brightly coloured precipitates.

The reaction between ethanal and propanone with 2,4-dinitrophenylhydrazine is shown below:

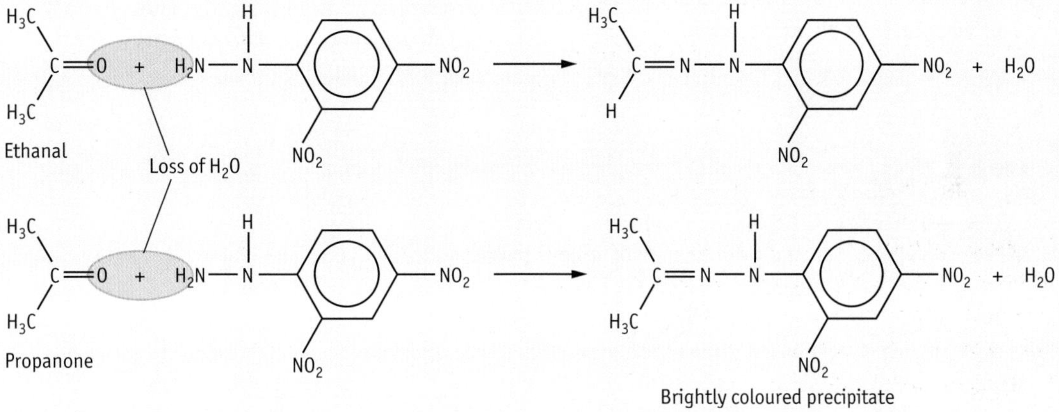

Ethanal

Loss of H_2O

Propanone

Brightly coloured precipitate

In this reaction, water is lost and it can be considered to be a condensation reaction.

Stretch and challenge

1 Write equations for the reaction between excess ethanal and

a phenylamine, $C_6H_5NH_2$

b hydroxylamine, NH_2OH

c hydrazine, NH_2NH_2

2 Use the information on page 31 and your understanding of oxidation numbers to deduce the ionic and fully balanced equation for the oxidation of ethanal to ethanoic acid using acidified potassium dichromate(VI).

3 In the presence of NaOH, benzaldehyde, C_6H_5CHO, is both oxidised and reduced. The products are sodium benzoate and phenylmethanol:

$$2C_6H_5CHO + NaOH \rightarrow C_6H_5COO^-Na^+ + C_6H_5CH_2OH$$

Sodium benzoate Phenylmethanol

The mechanism for this reaction is shown below with the curly arrows missing. Copy the mechanism and add the curly arrows to track the movement of the electron pairs in the reaction.

Step 1

Step 2

Step 3

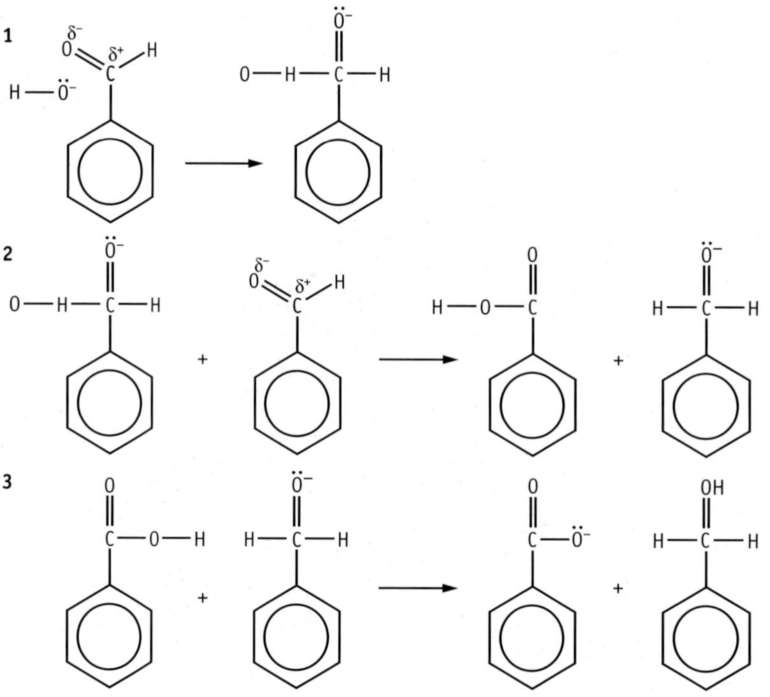

Carboxylic acids and esters

Chapter 3

Carboxylic acids

You should recall from AS that carboxylic acids can be prepared by the oxidation of a primary alcohol.

All carboxylic acids contain the functional group –COOH. Carboxylic acids are present in many foods. Ethanoic acid and citric acid are responsible for the sharp tastes of vinegar and lemons, respectively. Benzoic acid is used as a preservative and as a flavouring in fizzy drinks. Ethanedioic acid (oxalic acid) in found in the leaves of many plants. Food such as rhubarb and spinach that are rich in oxalic acid can cause the formation of kidney stones (calcium oxalate). Of course, it is the leaves of rhubarb that are particularly rich in oxalic acid and these should not be eaten.

> In examinations, you may be asked to write a balanced equation for the formation of a carboxylic acid from an alcohol. Unfortunately, some students fail to do this correctly. Therefore, you are urged strongly to read through your AS notes before proceeding.

The leaves of rhubarb are rich in oxalic acid

The formulae of some carboxylic acids are given below.

$H_3C-COOH$ — Ethanoic acid

$HOOC-CH_2-C(OH)(COOH)-CH_2-COOH$ — Citric acid

C_6H_5-COOH — Benzoic acid

$HOOC-COOH$ — Oxalic acid (ethanediodic acid)

The carboxylic acid group, –COOH is polar

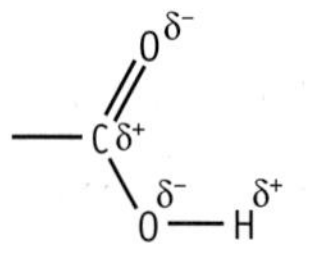

and gives rise to hydrogen bonding.

Oxygen is more electronegative than either carbon or hydrogen. This results in both the C=O and the O–H, in the –COOH group, being polar. Carboxylic acids form hydrogen bonds with other carboxylic acids and with water.

In the absence of water, carboxylic acids form dimers. This reduces their volatility and hence increases their boiling points.

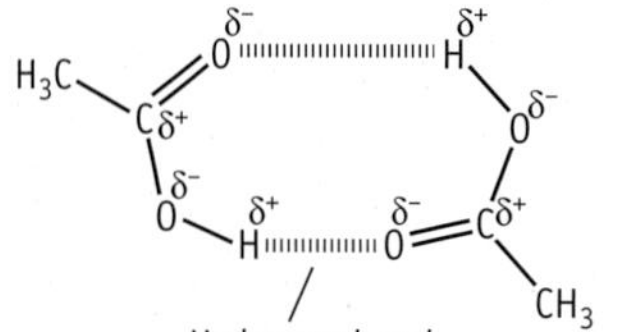

Hydrogen bond

The ability to form hydrogen bonds and to undergo dipole–dipole interaction also explains why methanoic acid and ethanoic acid are soluble in water. Solubility decreases with increasing molar mass — benzoic acid, C_6H_5COOH, is soluble in hot water but insoluble in cold water.

Carboxylic acids are acidic and can, therefore, donate protons. However, they are weak acids and dissociate only partially into their ions.

$$CH_3COOH(aq) \rightleftharpoons CH_3COO^-(aq) + H^+(aq)$$

In an examination, you might be expected to recall the definitions of acids, bases and salts.

The carboxylic acid group can be attached to either a chain (aliphatic) or to a ring (aromatic), for example:

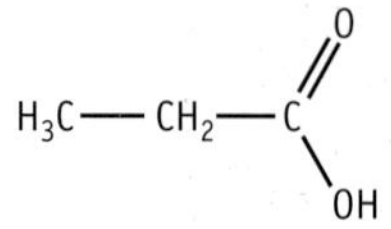

Propanoic acid — aliphatic

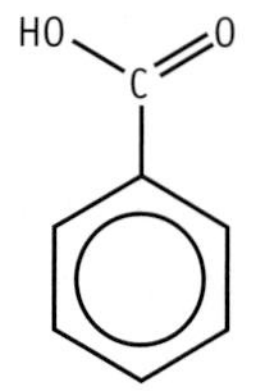

Benzoic acid — aromatic

Reaction with acids

Carboxylic acids display typical reactions of an acid and can form salts (carboxylates). Salt formation can occur by any of the following reactions, illustrated by the formation of sodium ethanoate ($CH_3COO^-Na^+$) from ethanoic acid (CH_3COOH):

- acid + base → salt + water

$$CH_3COOH(aq) + NaOH(aq) \rightarrow CH_3COO^-Na^+(aq) + H_2O(l)$$

- acid + (reactive) metal → salt + hydrogen

$$CH_3COOH(aq) + Na(s) \rightarrow CH_3COO^-Na^+ + \tfrac{1}{2}H_2$$

- acid + carbonate → salt + water + carbon dioxide

$$2CH_3COOH(aq) + Na_2CO_3(aq) \rightarrow 2CH_3COO^-Na^+(aq) + H_2O(l) + CO_2(g)$$

MARTYN F. CHILLMAID/SCIENCE PHOTO LIBRARY

Bubbles of carbon dioxide are produced when ethanoic acid is added to sodium hydrogen carbonate

The reaction with a carbonate can be used as a test for a carboxylic acid. When an acid is added to a solution of a carbonate, bubbles (effervescence) of carbon dioxide are seen.

Reaction with an alcohol

A carboxylic acid can react with an alcohol to form an ester. This type of reaction is known as **esterification**. It is a reversible reaction that is carried out usually in the presence of a concentrated acid catalyst, such as sulphuric acid. The general reaction can be summarised as follows:

$$\underset{\text{Carboxylic acid}}{R{-}C({=}O){-}O{-}H} + \underset{\text{Alcohol}}{R'{-}OH} \xrightleftharpoons[\text{as catalyst}]{H^+} \underset{\text{Ester}}{R{-}C({=}O){-}O{-}R'} + \underset{\text{Water}}{H_2O}$$

Think back to the AS section on equilibrium and to Le Chatelier's principle. Why is a concentrated acid, rather than a dilute acid, used as a catalyst?

Esters

An ester can be prepared by the reaction of a carboxylic acid and an alcohol. The name of an ester is derived from the names of the carboxylic acid and the alcohol from which it is formed. The first part of the name relates to the alcohol and the second part of the name relates to the acid — for example:

- methylethanoate

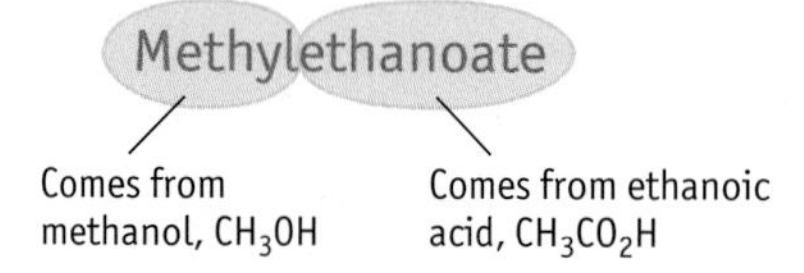

- ethylmethanoate

Ethylmethanoate

Comes from ethanol, CH_3H_2OH

Comes from methanoic acid, HCO_2H

In an organic reaction it is the functional groups that react. In the reaction between a carboxylic acid and an alcohol it is helpful to write the formulae so that the functional groups 'face each other'. The alcohol group reacts with the carboxylic acid group to produce water and the ester can be deduced by simply joining the two organic parts together:

$$CH_3COOH + HOCH_3 \rightleftharpoons CH_3COOCH_3 + H_2O$$

$$H_3C{-}C({=}O){-}O{-}H + H{-}O{-}CH_3 \rightleftharpoons H_3C{-}C({=}O){-}O{-}CH_3 + H_2O$$

There are two ways in which the ester could be formed because there are two different bonds in the alcohol that might break. The bond between the hydrogen and the oxygen in the alcohol might break:

$$H_3C{-}C({=}O){-}O{-}H \quad H{-}O{-}CH_3$$

Loss of water

The bond between the carbon and the oxygen in the alcohol could also break:

$H_3C-C(=O)-O-H \quad H-O-CH_3$

Loss of water

To decide which is correct, the reaction has been studied using alcohols containing the ^{18}O isotope. The two possible routes are:

A

$H_3C-C(={}^{16}O)-{}^{16}O-H \quad H-{}^{18}O-CH_3 \rightleftharpoons H_3C-C(={}^{16}O)-{}^{18}O-CH_3 + H_2{}^{16}O$

B

$H_3C-C(={}^{16}O)-{}^{16}O-H \quad H-{}^{18}O-CH_3 \rightleftharpoons H_3C-C(={}^{16}O)-{}^{16}O-CH_3 + H_2{}^{18}O$

When the ester is analysed using mass spectrometry the ^{18}O isotope is always found in the ester and not in the water.

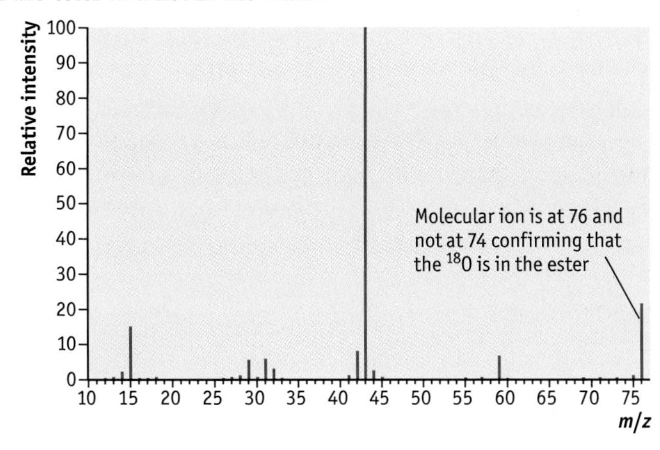

This indicates that the reaction proceeds as shown in route **A**.

Preparation of esters

A range of esters can be formed by changing either the alcohol or the carboxylic acid, or both — for example:

$H_3C-C(=O)-OH \quad HO-C_2H_5 \rightleftharpoons H_3C-C(=O)-O-C_2H_5 + H_2O$

Ethanoic acid; Loss of water; Ethanol; Ethyl ethanoate

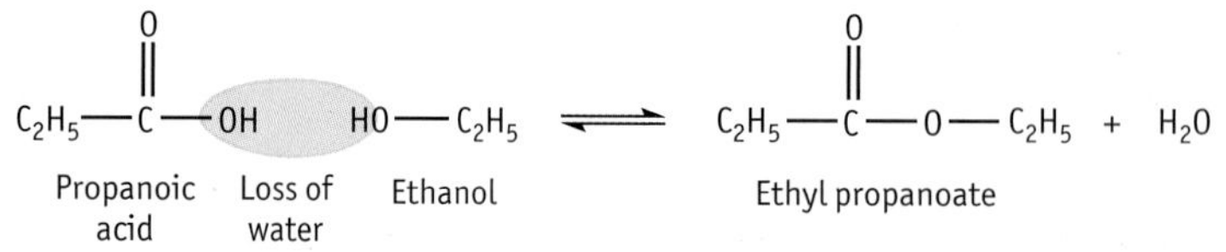

$$C_2H_5{-}C({=}O){-}OH + HO{-}C_2H_5 \rightleftharpoons C_2H_5{-}C({=}O){-}O{-}C_2H_5 + H_2O$$

Propanoic acid | Loss of water | Ethanol | Ethyl propanoate

An ester can also be prepared by the reaction of an acid anhydride with an alcohol. Acid anhydrides contain the –OCOCO– functional group. Ethanoic anhydride is the most common. Equations for the preparation of two esters from ethanoic anhydride are shown below.

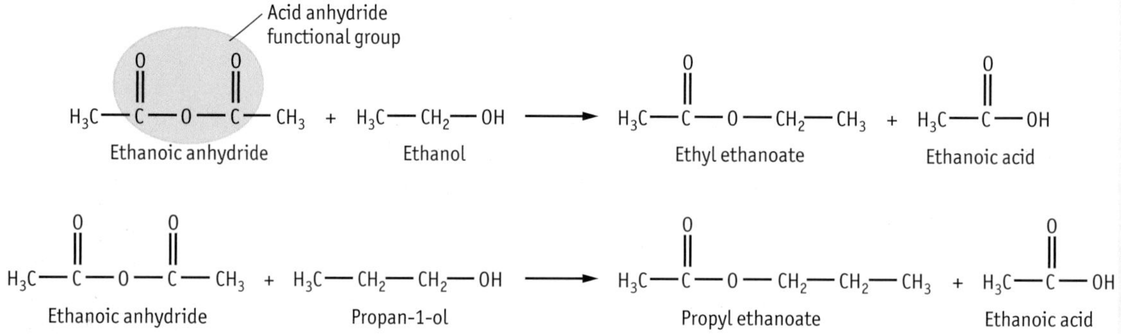

Acid anhydride functional group

$$H_3C{-}C({=}O){-}O{-}C({=}O){-}CH_3 + H_3C{-}CH_2{-}OH \longrightarrow H_3C{-}C({=}O){-}O{-}CH_2{-}CH_3 + H_3C{-}C({=}O){-}OH$$

Ethanoic anhydride | Ethanol | Ethyl ethanoate | Ethanoic acid

$$H_3C{-}C({=}O){-}O{-}C({=}O){-}CH_3 + H_3C{-}CH_2{-}CH_2{-}OH \longrightarrow H_3C{-}C({=}O){-}O{-}CH_2{-}CH_2{-}CH_3 + H_3C{-}C({=}O){-}OH$$

Ethanoic anhydride | Propan-1-ol | Propyl ethanoate | Ethanoic acid

The advantages of preparing an ester from an acid anhydride are that the reaction is not reversible and it occurs readily.

Physical properties

Esters have a characteristic 'fruity' smell. They are used in both perfumes and artificial flavourings. Naturally occurring fruits contain a complex mix of chemicals, many of which contribute to the overall smell or scent. By mixing together different chemicals, including esters, it is possible to manufacture artificial flavours. The esters shown below are used to generate the flavours pineapple, pear and apple.

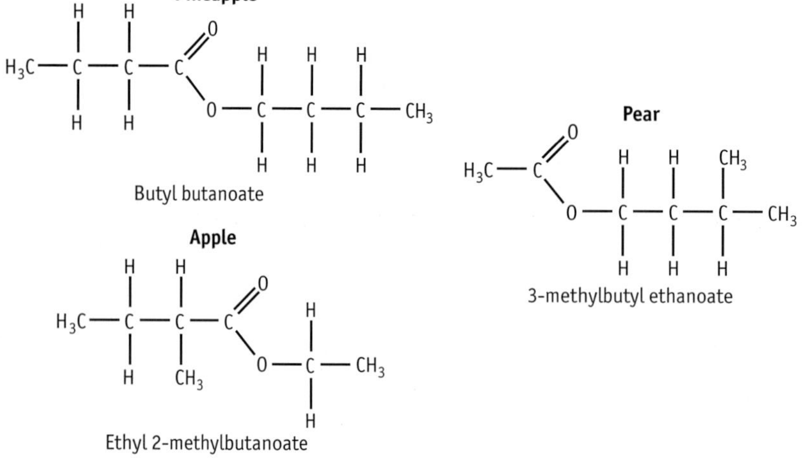

Pineapple

$$H_3C{-}CH_2{-}CH_2{-}C({=}O){-}O{-}CH_2{-}CH_2{-}CH_2{-}CH_3$$

Butyl butanoate

Pear

$$H_3C{-}C({=}O){-}O{-}CH_2{-}CH_2{-}CH(CH_3){-}CH_3$$

3-methylbutyl ethanoate

Apple

$$H_3C{-}CH_2{-}CH(CH_3){-}C({=}O){-}O{-}CH_2{-}CH_3$$

Ethyl 2-methylbutanoate

Hydrolysis of esters

Esters react with water (hydrolysis), breaking down to form two new products.

ⓔ Do not confuse *hydrolysis* with *hydration*. When a chemical is hydrated it reacts with water to form a single new product. You met this in Unit 2 — the formation of ethanol by the hydration of ethene (pages 160 and 171 of the AS book).

$$H_2C{=}CH_2 + H_2O \longrightarrow H{-}CH_2{-}CH_2{-}OH$$

The hydrolysis of an ester is a slow reaction that is carried out in the presence of either an acid, $H^+(aq)$, or a base, $OH^-(aq)$. Acid-catalysed hydrolysis leads to the formation of the carboxylic acid and the alcohol — for example, using dilute sulphuric acid as the catalyst:

$$H_3C{-}C({=}O){-}O{-}CH_3 + H_2O \underset{\text{as catalyst}}{\overset{H^+(aq)}{\rightleftharpoons}} H_3C{-}C({=}O){-}O{-}H + H{-}O{-}CH_3$$

Methyl ethanoate | Water | Ethanoic acid | Methanol

ⓔ Think back to the AS section on equilibrium and to Le Chatelier's principle. Why is a dilute acid, rather than a concentrated acid, used as the catalyst?

Base-catalysed hydrolysis leads to the formation of the salt of the carboxylic acid (the carboxylate) and the alcohol — for example, using an aqueous solution of NaOH as the catalyst:

$$H_3C{-}C({=}O){-}O{-}CH_3 + H_2O \underset{\text{as catalyst}}{\overset{NaOH(aq)}{\rightleftharpoons}} H_3C{-}C({=}O){-}O{-}H + H{-}O{-}CH_3$$

Methyl ethanoate | Water | Ethanoic acid | Methanol

The carboxylic acid then reacts with the NaOH catalyst

$$\downarrow$$

$$H_3C{-}C({=}O){-}O^-Na^+ + H_2O$$

Sodium ethanoate

The net reaction of the base-catalysed hydrolysis of methyl ethanoate can be written as:

$$H_3C{-}C({=}O){-}O{-}CH_3 + NaOH \underset{\text{as catalyst}}{\overset{\text{Aqueous base}}{\rightleftharpoons}} H_3C{-}C({=}O){-}O^-Na^+ + CH_3OH$$

Methyl ethanoate | Sodium ethanoate | Methanol

Triglycerides

Many esters occur naturally in plants and animals. The esters found in vegetable oils (derived from plants) and in animal fats are esters of fatty acids.

Fatty acids

Fatty acids have a range of lengths and shapes, two of which are shown below.

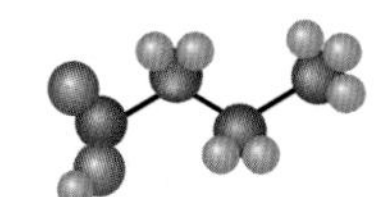

Butanoic acid $CH_3(CH_2)_2COOH$

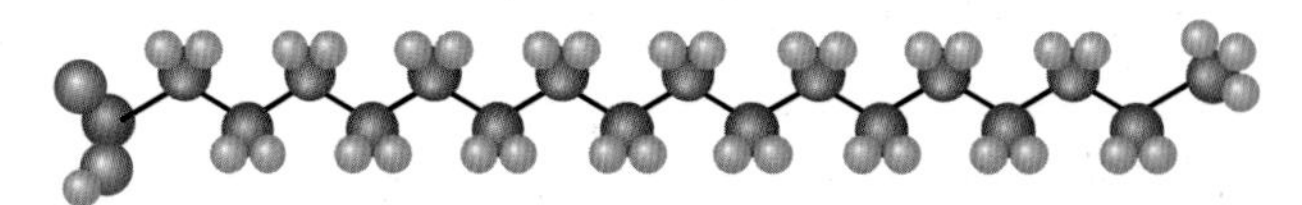

Octadecanoic acid (stearic acid) $CH_3(CH_2)_{16}COOH$

The simplest fatty acid is butanoic acid, $CH_3CH_2CH_2COOH$, which gives the characteristic unpleasant odour to rancid butter.

Fatty acids have trivial names – for example oleic acid (found in olive oil) — that are in common use, but they also have systematic names. The carbon chain is numbered from the carbon atom in the carboxylic acid group (carbon 1). This is known as Δ-notation. The systematic name of oleic acid is *Z*-octadec-9-enoic acid, 18, 1(9), which sounds much more difficult than oleic acid. However, the numbers at the end of the name indicate that a molecule of oleic acid contains 18 carbon atoms and that there is one *Z* (or *cis*) double bond on the ninth carbon atom, starting from the carboxylic acid end. Therefore, the formula is $CH_3(CH_2)_7CH{=}CH(CH_2)_7COOH$.

Z is the *cis*-isomer.

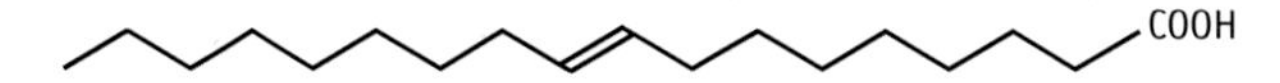

The systematic name of linoleic acid is all-*Z*-octadec-9,12-enoic acid, 18,2(9,12). The numbers at the end indicate that a molecule of linoleic acid contains 18 carbon atoms and that there are two double bonds (both *Z* (or *cis*)), one on the ninth carbon and one on the twelfth carbon starting from the carboxylic acid end. Therefore, the formula is:

$CH_3(CH_2)_4CH{=}CHCH_2CH{=}CH(CH_2)_7COOH$

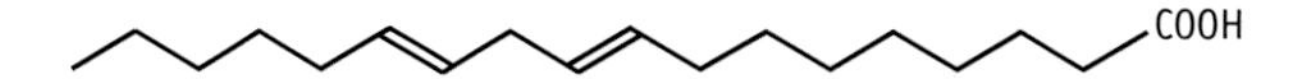

Fatty acids are an essential part of a healthy diet and are crucial to good health. For example, arachidonic acid is used by cells to produce eicosanoids such as prostaglandins and leukotrienes that are important in temperature regulation.

Fatty acids can be saturated, mono-unsaturated or polyunsaturated. A saturated fatty acid molecule contains no C=C double bonds and is essentially a straight chain.

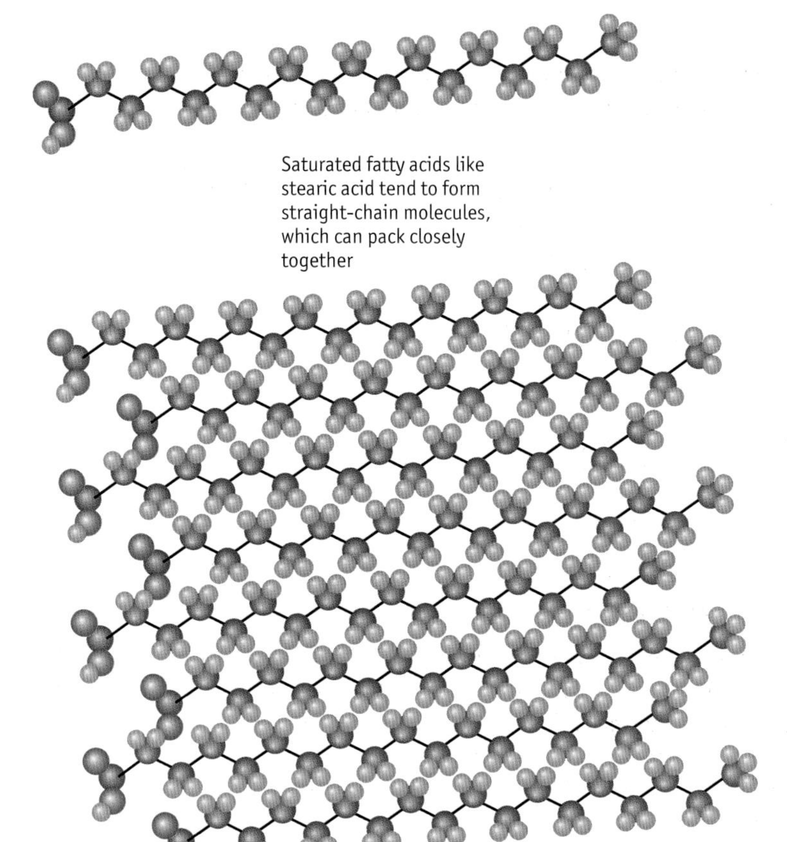

Saturated fatty acids like stearic acid tend to form straight-chain molecules, which can pack closely together

Saturated fatty acid molecules are straight chains that can pack together closely. This means that there are many points of contact, resulting in significant intermolecular forces. This increases viscosity and reduces volatility. Therefore saturated fatty acids are relatively dense and are solid at room temperature.

A mono-unsaturated fatty acid molecule, for example oleic acid $CH_3(CH_2)_7CH=CH(CH_2)_7COOH$, contains one C=C double bond. Oleic acid is found in most animal and vegetable fats that occur naturally as the *Z* (*cis*) isomer. The double bond creates a 'kink', so unsaturated fatty acids do not stack together as well as saturated fatty acids do. Hence, there are fewer points of contact, fewer intermolecular forces and the compounds tend to be liquid at room temperature.

Z-oleic acid (*cis*)

Oleic acid is a liquid at room temperature, but at lower temperature tends to form globules

NIGEL EVANS

Oleic acid, in olive oil, solidifies when cooled in a refrigerator.

A molecule of a polyunsaturated fatty acid contains more than one C=C double bond and hence has more than one kink (or bend). The consequence is that polyunsaturated fatty acids stay fluid even at low temperature. They are synthesised by plankton and seaweed, which are eaten by fish and become incorporated into their fatty tissues. This makes fish a ready source of polyunsaturated fatty acids.

Naturally occurring fatty acids, found in both animals and plants, are mostly the *Z* (*cis*) form and, therefore, have the distinctive kinked shape. Food manufacturers partially hydrogenate these polyunsaturated fatty acids (add hydrogen to some of the C=C double bonds) because this prolongs the shelf life of the product — the partially hydrogenated fatty acids are less likely to become rancid. The hydrogenation process converts the *Z* (*cis*) form into the *E* (*trans*) form, which straightens out the kink.

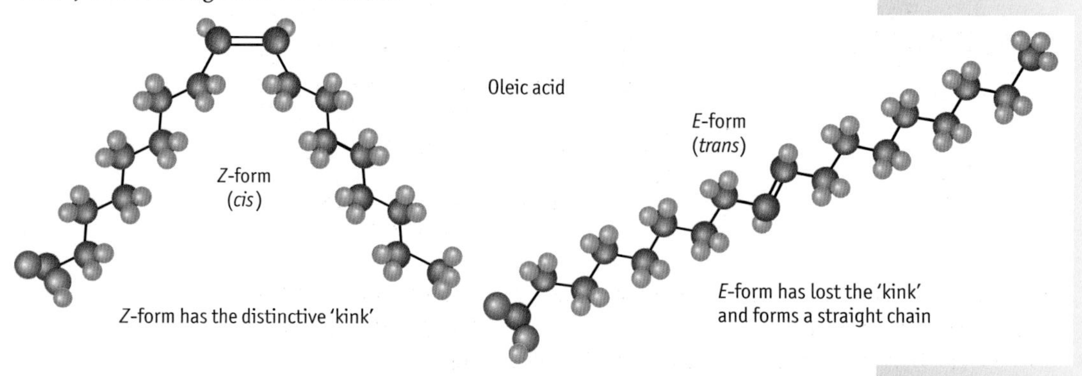

Exposure to prolonged heat, such as using oil repeatedly for deep-frying, also creates *E* (*trans*) fatty acids. The chemical structure of the *Z* (*cis*) and *E* (*trans*) fats are identical, but their geometric differences affect important biological processes.

Straight-chain fatty acids (saturated fatty acids and *E* (*trans*) fatty acids) can stack together, which can lead to a build up of plaques on the insides of the walls of arteries causing them to thicken. A diet high in saturated fats, and possibly in *E* (*trans*) fatty acids, can also lead to increased blood levels of cholesterol and low-density lipoprotein, LDL. It has been known for many years that such a diet can lead to atherosclerosis and heart disease and we have been encouraged to lower the amount of saturated fat in our diets.

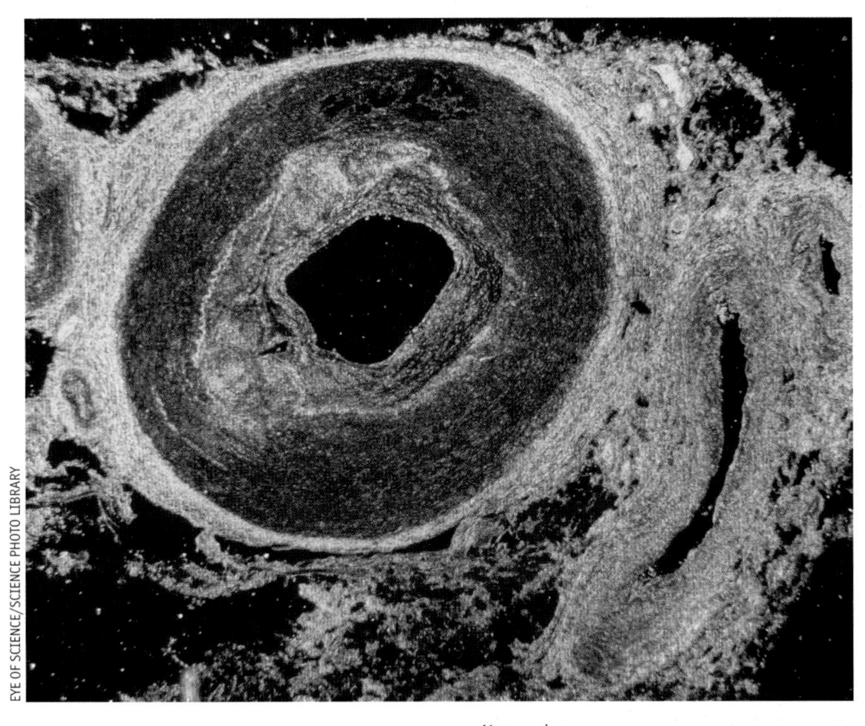

Light micrograph of a cross-section through an artery obstructed by an atheroma plaque (the dark grainy deposit on the inner wall, reducing the size of the lumen)

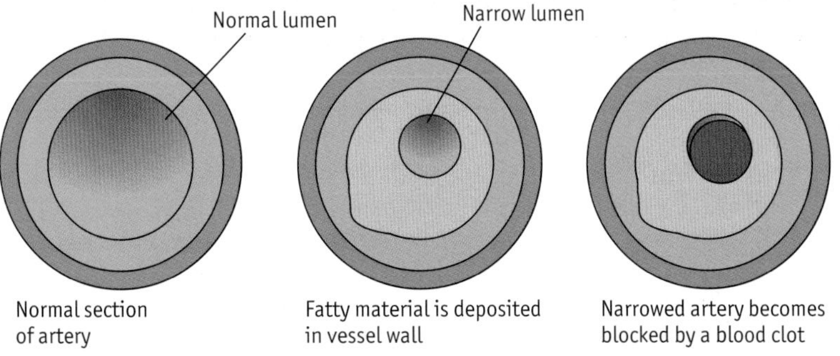

Cholesterol is essential for the formation of cell membranes and several body hormones. It is insoluble in water (and therefore in blood) and is transported around the body by lipoproteins. These fall into two categories:

- high density lipoprotein, HDL, carries about one-third of the cholesterol and is known commonly as 'good' cholesterol because it carries cholesterol away from the arteries and back to the liver
- low density lipoprotein, LDL, which is known commonly as 'bad' cholesterol and which can lead to a build up of plaques and to atherosclerosis

Recent studies show a correlation between the amount of *E* (*trans*) fatty acids and LDL. People worried about their blood cholesterol levels should avoid eating *E* (*trans*) fatty acids by using liquid vegetable oils.

Glycerides

Glycerides are formed when one or more fatty acid molecules react with a glycerol molecule. Glycerol contains three alcohol groups and is known as a triol; its systematic name is propane-1,2,3-triol and its formula is $CH_2(OH)CH(OH)CH_2OH$. Glycerol can react with fatty acids to produce esters. Reaction of a molecule of glycerol with one fatty acid molecule produces a monoglyceride, with two fatty acid molecules produces a diglyceride and with three fatty acid molecules produces a triglyceride:

For exam purposes, a triglyceride is described as 'a triester of glycerol and fatty acids'.

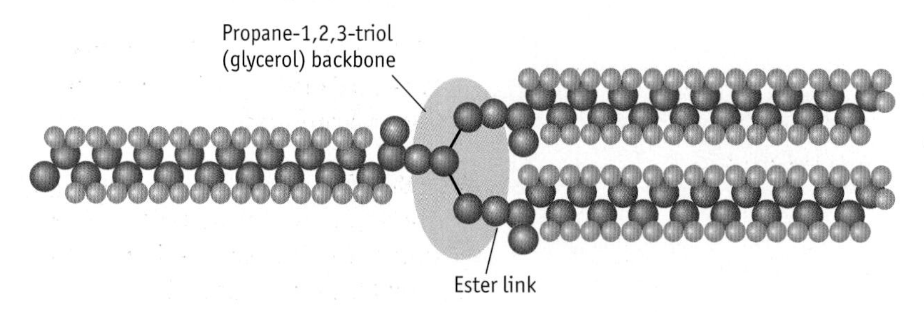

Hydrolysis of triglycerides

Naturally occurring fats and oils (triglycerides) can be hydrolysed by refluxing with a base. This produces propane-1,2,3-triol and the salts of the fatty acids. This is known as **saponification**, which means 'the forming of soap'. Modern soaps are made from blends of oils. The base hydrolysis of a triglyceride is shown below:

$$\begin{array}{l} H-\underset{|}{\overset{H}{\overset{|}{C}}}-O-\overset{O}{\overset{\|}{C}}-(CH_2)_{16}CH_3 \\ CH_3(CH_2)_{16}-\overset{O}{\overset{\|}{C}}-O-\underset{|}{C}-H \\ H-\underset{\underset{H}{|}}{C}-O-\underset{\underset{O}{\|}}{C}-(CH_2)_{16}CH_3 \end{array} \xrightarrow[\text{hydrolysis}]{\text{Base}} \begin{array}{l} H-\underset{|}{\overset{H}{\overset{|}{C}}}-OH \\ HO-\underset{|}{C}-H \\ H-\underset{\underset{H}{|}}{C}-OH \end{array} + 3\,CH_3(CH_2)_{16}-\overset{O}{\overset{\|}{C}}-O^-Na^+$$

Biodiesel

Vegetable oils and animal fats and their derivatives, particularly methyl esters, are being used increasingly as alternative diesel fuels, known as **biodiesel**. Biodiesel is defined as 'a monoalkyl ester of long-chain fatty acids from renewable feedstocks such as vegetable oil or animal fats'. When a fuel is injected into the combustion chamber of a diesel engine there is a time delay before ignition. This time delay is the basis of the cetane number. The shorter the time delay, the higher is the cetane number. Hexadecane, $C_{16}H_{34}$, is a high quality standard with a short ignition time and it has been assigned a cetane number of 100. A diesel engine run on a fuel with a lower cetane number than it was designed for will be harder to start, noisier, operate roughly and have higher emissions.

Biodiesels have higher cetane numbers than their hydrocarbon counterparts.

Questions

1 a Name each of the following compounds:

A H—C(=O)—OH

B H_3C—CH_2—C(=O)—O—CH_3

C H—C(=O)—O—CH_2—CH_2—CH_3

D C_6H_5—C(=O)—O—CH_2—CH_3

E C_6H_5—C(=O)—OH

F CH_3—C(=O)—O—C_6H_5

b What is the molecular formula of **E**?

c What is the empirical formula of **F**?

d Calculate the percentage of carbon in **D**.

e Explain, with the aid of a diagram, why compound **A** is soluble in water.

f An ester can be prepared by reacting an alcohol with a carboxylic acid. Write an equation for the formation of **C**.

g Write a balanced equation for the reaction of compound **A** with:

(i) Na(s)

(ii) $NaHCO_3(aq)$

2 Propyl ethanoate can be hydrolysed by refluxing with NaOH(aq).

a Explain what is meant by 'refluxing'.

b Write a balanced equation for the reaction. Name the products.

3 Draw the triglyceride that could be formed from propane-1,2,3-triol and octadecanoic acid (stearic acid). Use a skeletal formula to represent the carboxylic acid.

4 Linoleic acid is named systematically as (all *Z* (*cis*))octadec-9,12-enoic acid, 18,2(9,12). Draw the structure of linoleic acid.

5 Explain the difference between *cis* and *trans* fatty acids. State one disadvantage of a diet high in *trans* fatty acids.

6 12.0 g of ethanoic acid was reacted with methanol in the presence of concentrated sulphuric acid. 3.70 g of the ester, methyl ethanoate, was isolated.

a Write a balanced equation for this reaction and explain the role of the concentrated sulphuric acid.

b Calculate the percentage yield.

c Calculate the atom economy.

d Suggest why the percentage yield is so low.

7 a The triglyceride shown below is hydrolysed by heating with aqueous sodium hydroxide, NaOH. Identify the products.

b By referring to the products from part **a**, explain the terms:

(i) saturated

(ii) unsaturated and

(iii) polyunsaturated

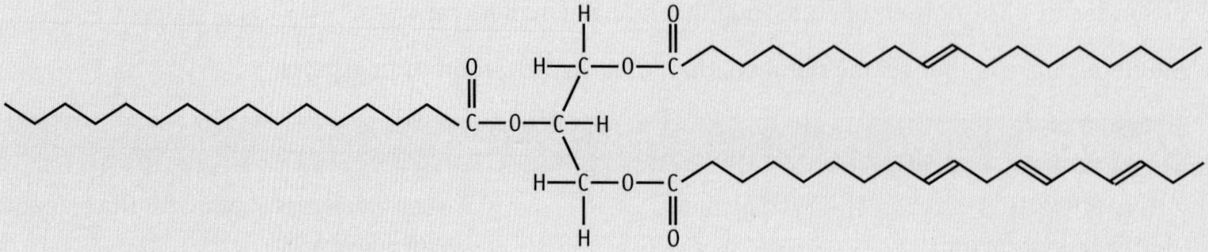

Summary

You should now be able to:

- explain the water solubility of some carboxylic acids
- describe the reactions of carboxylic acids with metals, carbonates and bases
- describe the esterification of alcohols with carboxylic acids and with acid anhydrides
- describe the acid and the base hydrolysis of esters
- describe a triglyceride and compare the structures of fatty acids
- compare the link between *trans* fatty acids and 'bad' cholesterol
- state the uses of esters in perfumes and flavourings and the increased use of esters as biodiesel

Additional reading

Bonding in the carboxylic acid group

The carboxylic acid group, –COOH, is usually drawn as a carbon atom with a double bond to an oxygen atom and a single bond to the hydroxyl group, OH.

Closer inspection of bond lengths reveals that this might not be the case:

- The bond length of C=O in an aldehyde or a ketone is of the order 121 pm; in a carboxylic acid it is longer, of the order 125 pm.
- The bond length of C–O in an alcohol is around 143 pm; in a carboxylic acid it is around 131 pm.

The lengthening of the C=O bond and the shortening of the C–O bond can be explained by looking at the interaction of the *p*-orbitals in the –COOH group.

You will recall from Chapter 1, pages 8–9, when investigating the structure of benzene, that C–C bonds (154 pm) are longer than C=C bonds (134 pm) and that the actual carbon-to-carbon bond length in benzene is somewhere in between at 139 pm. Something similar seems to be happening with the C=O in carboxylic acids. The increase in C=O bond length in a carboxylic acid indicates that the electron density between the carbon and the oxygen atoms is less. The reverse appears to happen with the C–O bond — an increase in electron density results in a decrease in bond length.

The variation in bond length in benzene can be explained by delocalisation and the same concept can be applied to carboxylic acids. The *p*-orbitals on the O–C–O– in the carboxylic acid overlap, creating a delocalised electron cloud above and below the plane as shown:

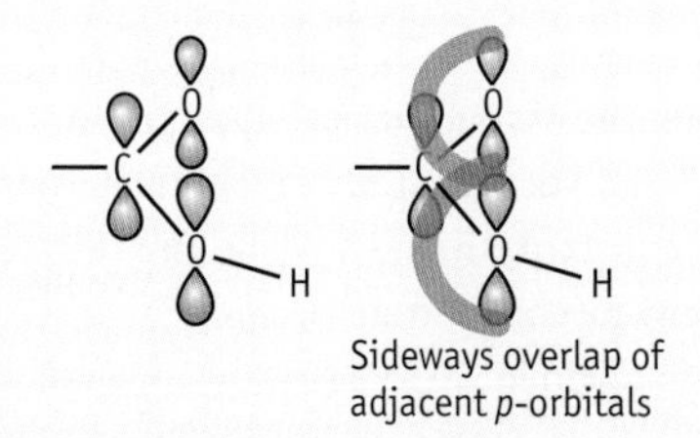

Sideways overlap of adjacent *p*-orbitals

The carboxylate ion, COO^-, is probably best represented as:

rather than

Addition–elimination mechanism

Carboxylic acids, esters and acid anhydrides can all react by the addition–elimination mechanism. Indeed, all acid derivatives, including acid chlorides (also known as acyl chlorides), RCOCl, and amides, $RCONH_2$, also react by the same mechanism.

You are not expected to know this mechanism and will not be tested on it, but it may improve your understanding

The general mechanism is:

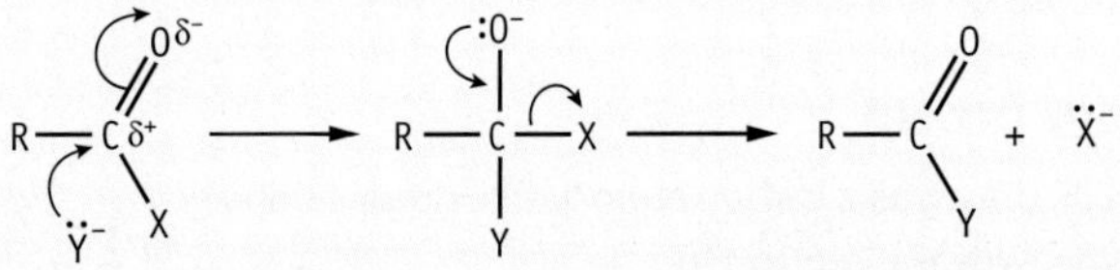

where R = an alkyl group (e.g. CH_3–, C_2H_5–) or an arene group (e.g. C_6H_5–)

X and Y = –OH, –OR, –OCOR, – Cl, $-NH_2$ or –NHR

The mechanism can, therefore, be applied to carboxylic acids, esters, acid anhydrides and to other acid derivatives.

Acid anhydrides can be used to make both carboxylic acids and esters. The mechanism below shows the formation of a methyl ester:

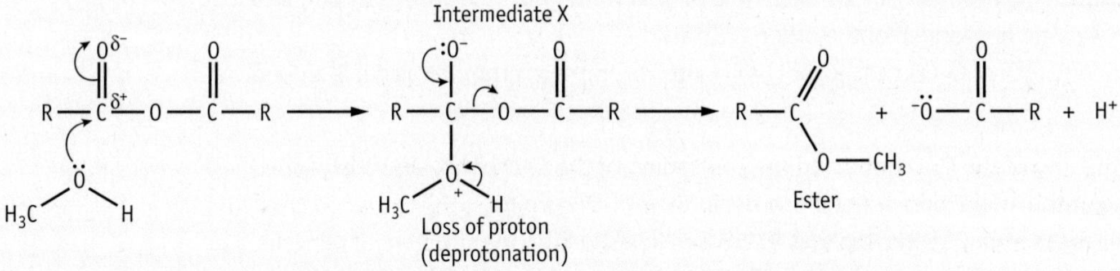

Intermediate X

Ester

Loss of proton (deprotonation)

Methanol reacts with an acid anhydride to form the intermediate **X**. This is followed by the re-forming of the C=O bond with the loss of either the CH_3O^- or $ROCO^-$ group. The group that breaks away is known as the **leaving group**. The leaving group depends on a number of factors, including the stability of the resultant ion.

$ROCO^-$ is more stable than CH_3O^- and so is a better leaving group. $ROCO^-$ is a better leaving group than both HO^- and CH_3O^-.

Variation in the acidity of carboxylic acids

Carboxylic acids are weak acids and dissociate partially into their ions.

$$RCOOH(aq) \rightleftharpoons RCOO^-(aq) + H^+(aq)$$

You will recall from Unit 2, that Le Chatelier's principle states that at equilibrium, the amount of each component remains the same. In Unit 5 you will study equilibria in more detail and you will learn about the acid dissociation constant, K_a. The K_a value can be used to calculate the pH of an acid provided that you know its concentration, *c*.

The K_a values for some simple carboxylic acids are given in Table 3.1.

Carboxylic acid		K_a/mol dm^{-3}
Methanoic acid	$HCOOH$	1.6×10^{-4}
Ethanoic acid	CH_3COOH	1.7×10^{-5}
Propanoic acid	CH_3CH_2COOH	1.3×10^{-5}
Methylpropanoic acid	$(CH_3)_2CHCOOH$	6.9×10^{-6}
Benzoic acid	C_6H_5COOH	6.3×10^{-5}
Phenylethanoic acid	$C_6H_5CH_2COOH$	4.9×10^{-5}

Table 3.1
K_a values for some simple carboxylic acids

The pH can be calculated using the formula:

$$pH = -\log \sqrt{cK_a}$$

$$x = \sqrt{cK_a}$$

Therefore, the pH can be calculated for any acid at any concentration, *c*.

At a concentration of 1.00 mol dm^{-3}, the pH of the carboxylic acids in Table 3.1 is given in Table 3.2.

Table 3.2

Carboxylic acid		K_a/mol dm^{-3}	pH
Methanoic acid	$HCOOH$	1.6×10^{-4}	1.90
Ethanoic acid	CH_3COOH	1.7×10^{-5}	2.38
Propanoic acid	CH_3CH_2COOH	1.3×10^{-5}	2.44
Methylpropanoic acid	$(CH_3)_2CHCOOH$	6.9×10^{-6}	2.58
Benzoic acid	C_6H_5COOH	6.3×10^{-5}	2.10
Phenylethanoic acid	$C_6H_5CH_2COOH$	4.9×10^{-5}	2.15

By comparing methanoic, ethanoic and propanoic acids it appears that as the alkyl group becomes larger, the pH increases (the acidity decreases). This indicates that methanoic acid releases the proton, H^+, more readily from the O–H bond.

Alkyl groups such has CH_3– and CH_3CH_2– push electrons into the –COOH group by the inductive effect, which increases the electron density in the –COOH group.

This makes it slightly more difficult for the O–H bond to break and release the proton, H^+.

The benzene ring is electron deficient and electrons are drawn into it. In benzoic acid this has the effect of decreasing the electron density in the –COOH group making it slightly easier for the O–H bond to break and release the H^+.

Stretch and challenge

1 Put the following in order of increasing acidity. Give reasons for your choice.

a
- C_6H_5COOH
- $C_6H_5CH_2COOH$
- $(C_6H_5)_2CHCOOH$

b
- CH_3COOH
- CCl_3COOH
- $CH_2ClCOOH$

2 Phenol, C_6H_5OH, is weakly acidic and is a weaker acid than benzoic acid, C_6H_5COOH. This can be explained by the dispersal of the negative charge in the ions $C_6H_5O^-$ and $C_6H_5COO^-$.

a With the aid of suitable diagrams, explain how the negative charges in $C_6H_5O^-$ and in $C_6H_5COO^-$ are dispersed.

Hint: you will find this easier if you use the Kekulé model for $C_6H_5O^-$.

b Explain why phenol is less acidic than benzoic acid.

3 Carboxylic acids contain a C=O group yet do not give a precipitate when treated with 2,4-dinitrophenylhydrazine.

Explain why not.

Amines

Chapter 4

Amines contain the functional group $-NH_2$ group. The nitrogen atom has a lone pair of electrons and the shape around it is pyramidal. You should recall from AS chemistry (see page 86 in the AS textbook) that ammonia, NH_3, is also pyramidal and that the bond angle is approximately 107°. The bond angle in amines is similar but varies depending on the group attached to the $-NH_2$.

The amine group can be attached to either a chain or a ring. If it is attached to a chain, the amine is aliphatic; if it is attached to a ring, the amine is aromatic.

Aliphatic amines include:

- CH_3NH_2 aminomethane (methylamine)
- $CH_3CH_2NH_2$ aminoethane (ethylamine)
- $CH_3CH_2CH_2NH_2$ 1-aminopropane

Aromatic amines include phenylamine, $C_6H_5NH_2$.

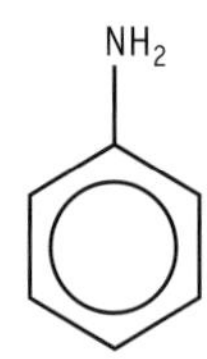

Preparation of amines

Aliphatic amines

Aliphatic amines such as aminoethane, $CH_3CH_2NH_2$, can be prepared by the reaction between a halogenoalkane and excess ethanolic ammonia.

$$CH_3CH_2Cl + NH_3(alc) \rightarrow CH_3CH_2NH_2 + HCl$$

The nitrogen in the NH_3 has a lone pair of electrons that allows the NH_3 to behave as a nucleophile — the reaction can be regarded as a nucleophilic substitution reaction. It is essential that the solvent is ethanol; if any water is present the NH_3 is protonated and the ammonium ion, NH_4^+ is formed. The nitrogen then no longer has a lone pair of electrons and cannot behave as a nucleophile.

If NH_3 is in excess, the HCl product reacts with NH_3 and NH_4Cl is produced:

$$CH_3CH_2Cl + 2NH_3(alc) \rightarrow CH_3CH_2NH_2 + NH_4Cl$$

Depending on the conditions, further reactions can result in the formation of secondary, tertiary and quaternary amines — $(CH_3CH_2)_2NH$, $(CH_3CH_2)_3N$, $(CH_3CH_2)_4N^+$ respectively (see the additional reading at the end of this chapter).

Aromatic amines

Aromatic amines such as phenylamine, $C_6H_5NH_2$, can be prepared by heating nitrobenzene, under reflux, with tin and concentrated hydrochloric acid. This is a reduction reaction in which the reducing agent is formed from the reaction between tin and concentrated hydrochloric acid.

Using [H] to represent the reducing agent, the equation for this reaction is:

$$C_6H_5NO_2 + 6[H] \xrightarrow{\text{reflux}} C_6H_5NH_2 + 2H_2O$$

Basicity of amines

Amines are weak bases since the lone pair of electrons on the nitrogen can accept a proton. Consequently, amines react with acids to form salts:

$$CH_3NH_2 + H^+ \longrightarrow CH_3NH_3^+$$

Forms a dative covalent bond with the proton

Two examples of the formation of salts from amines are:

$C_2H_5NH_2 + HCl \rightarrow C_2H_5NH_3^+Cl^-$
ethylamine

$C_6H_5NH_2 + HCl \rightarrow C_6H_5NH_3^+Cl^-$
phenylamine

Azo dyes

Aromatic amines, for example phenylamine, are used in the industrial synthesis of azo dyes. The synthesis involves two stages:

- formation of a diazonium compound
- coupling reaction with a phenol to form the azo dye

Stage 1: formation of a diazonium compound

Reagents: nitrous acid, HNO_2, made *in situ* from $NaNO_2$ and excess HCl

Conditions: temperature must be below 10°C

Nitrous acid, HNO_2, is prepared by reacting sodium nitrite, $NaNO_2$, with excess hydrochloric acid, HCl.

$NaNO_2 + HCl \rightarrow HNO_2 + NaCl$

The nitrous acid, HNO_2, and the excess HCl now react with the phenylamine:

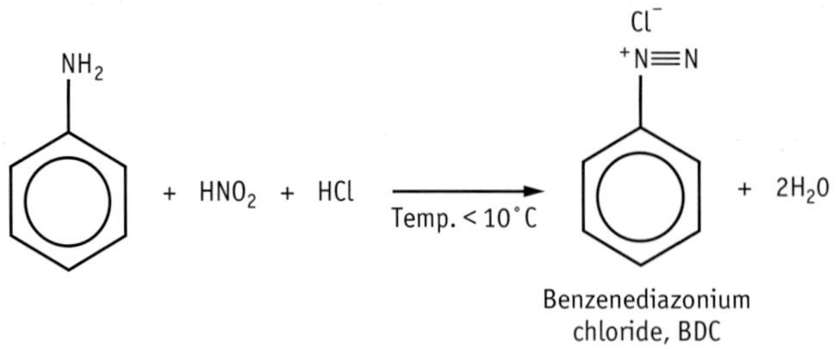

$$C_6H_5NH_2 + HNO_2 + HCl \xrightarrow[\text{Temp.} < 10^\circ C]{} C_6H_5N^+{\equiv}N\ Cl^- + 2H_2O$$

Benzenediazonium chloride, BDC

The reaction has to be kept below 10°C because benzene diazonium chloride is unstable and reacts readily with water to produce phenol, N_2 and HCl.

Stage 2: coupling reaction with a phenol to form the azo dye

Reagent: phenol

Conditions: alkaline solution ($OH^-(aq)$)

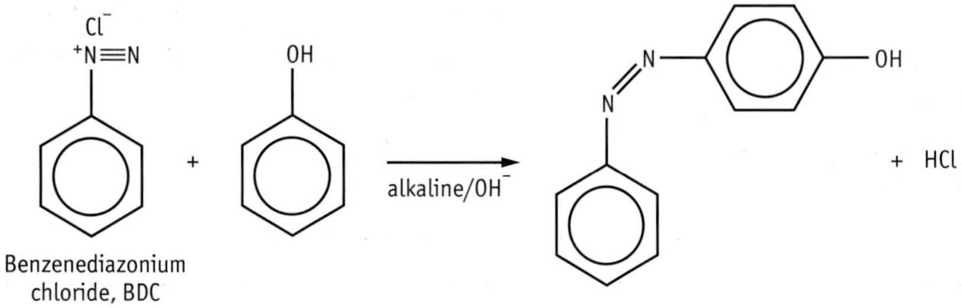

$$C_6H_5N^+{\equiv}N\ Cl^- + C_6H_5OH \xrightarrow[\text{alkaline}/OH^-]{} C_6H_5N{=}NC_6H_4OH + HCl$$

Benzenediazonium chloride, BDC

This reaction introduces two new functional groups; the diazonium group, $N{\equiv}N^+$, and the azo group, $-N{=}N-$.

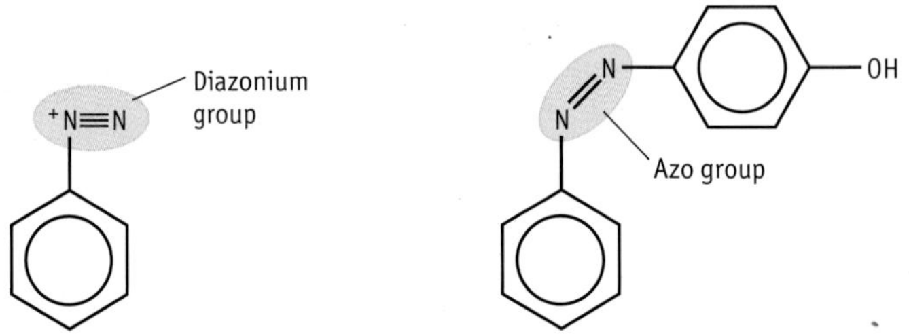

Diazonium group

Azo group

The $-N{=}N-$ group absorbs light in the visible region of the spectrum. Therefore, azo compounds are brightly coloured. They are used in the formation of dyestuffs. The exact colour depends on the substituents on the aromatic rings around the $-N{=}N-$ group. Many of the E numbers, used as food colorants, are azo compounds. The structure of E105, a yellow food dye is shown below.

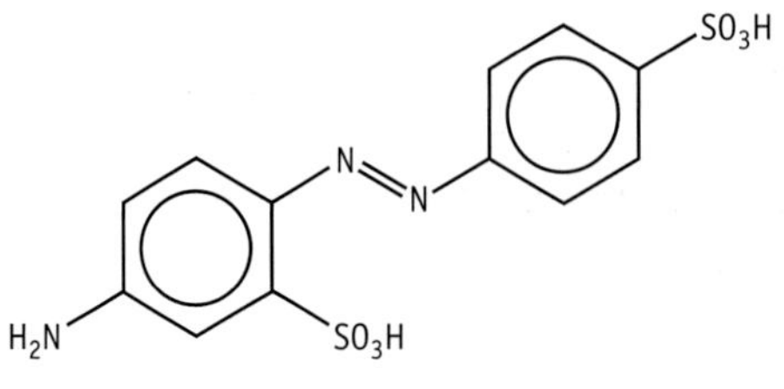

SO_3H

H_2N SO_3H

For exam purposes, a base is defined as a proton acceptor. The nitrogen uses its lone pair of electrons to form a dative covalent bond.

Questions

1 a Name each of the following compounds:

A CH_3 branch: $H_3C-CH(CH_3)-CH_2-NH_2$

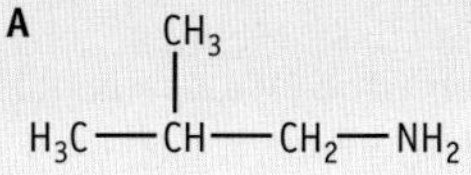

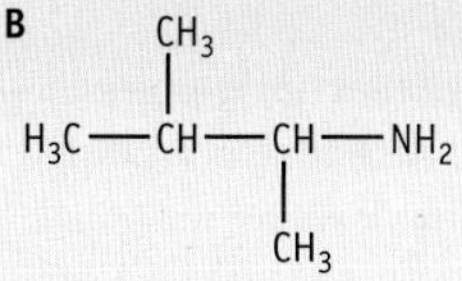

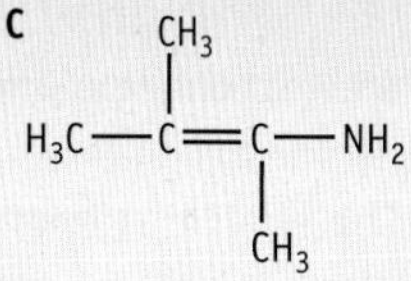

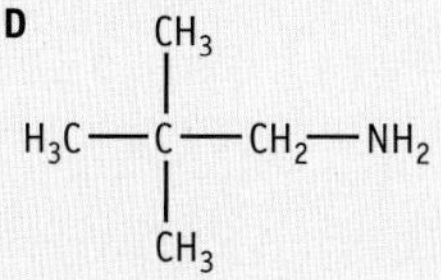

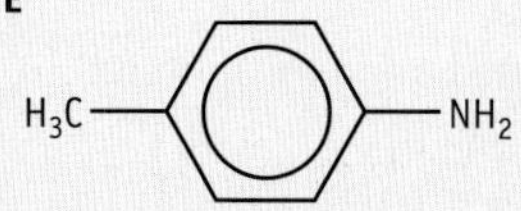

B $H_3C-CH(CH_3)-CH(CH_3)-NH_2$

C $H_3C-C(CH_3)=C(CH_3)-NH_2$

D $H_3C-C(CH_3)_2-CH_2-NH_2$

E $H_3C-C_6H_4-NH_2$

F $C_6H_5-CH_2-CH_2-NH_2$

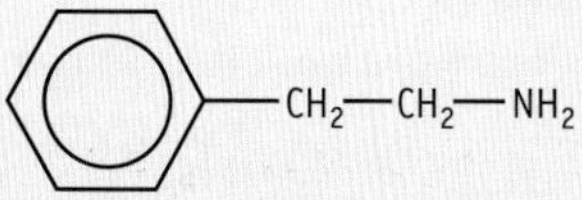

b Draw the skeletal formulae of **A**, **B** and **C**.

c What is the molecular formula of compound **E**?

d Starting from 1-chloro-2-methylpropane, show, with the aid of an equation, how compound **A** could be prepared.

e Starting from benzene, explain how compound **E** could be prepared. Give reagents, conditions and equations for each step.

2 Diazonium compounds can be formed from aromatic amines. Azo dyes can be formed by the coupling reaction between a diazonium compound and a phenol.

Complete the table below by drawing the structures of the diazonium compounds and the azo dyes.

Amine	Diazonium	Phenol	Azo dye
CH_3, NH_2 (2-methylaniline) $\xrightarrow[\text{Temp.} < 10°C]{HNO_2 + HCl}$		+ OH (naphthol) $\longrightarrow$	
NH_2, $S(=O)(=O)OH$ $\xrightarrow[\text{Temp.} < 10°C]{HNO_2 + HCl}$		+ OH, Cl, Cl (dichlorophenol) $\longrightarrow$	

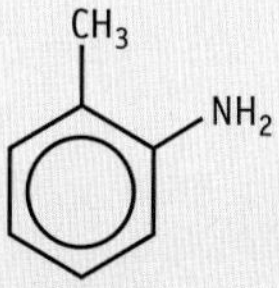

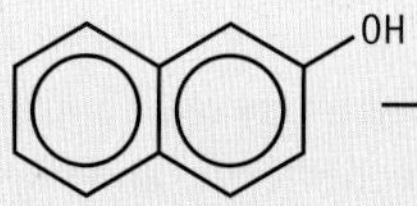

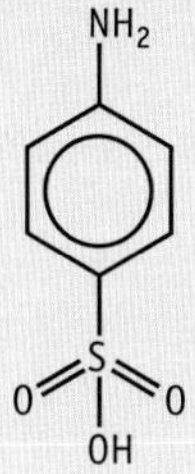

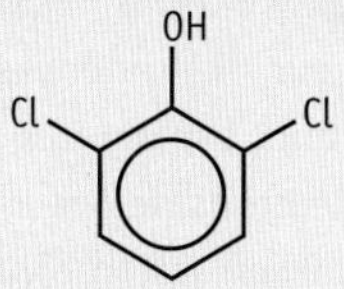

Summary

You should now be able to:

- explain the basicity of amines and their reactions with acids to produce salts
- describe the preparation of aliphatic amines and of aromatic amines
- describe the synthesis of diazonium compounds and azo dyes

Additional reading

Basicity of amines

All amines contain a nitrogen atom that has a lone pair of electrons. The lone pair enables the amine to behave as a base by accepting a proton, H^+, and, hence, form a salt.

The relative strength of amines can be measured by using the base dissociation constant, K_b. This is equivalent to the K_a values for carboxylic acids discussed at the end of Chapter 3.

The pH of a weak base can be calculated using the formula:

$$pH = -\log\left(\frac{K_w}{\sqrt{cK_b}}\right)$$

where c = concentration of the base and $K_w = 1.0 \times 10^{-14}\,mol^2\,dm^{-6}$

The K_b values and the pH of $1.00\,mol\,dm^{-3}$ solutions of ammonia and some simple amines are given in Table 4.1.

Table 4.1
K_b values and the pH of 1.00 mol dm^{-3} solutions of ammonia and amines

Amine		K_b/mol dm^{-3}	pH	[H$^+$(aq)]/mol dm^{-3}
Ammonia	NH_3	1.8×10^{-5}	11.6	2.51×10^{-12}
Methylamine	CH_3NH_2	4.4×10^{-4}	12.3	5.01×10^{-13}
Phenylamine	$C_6H_5NH_2$	4.2×10^{-10}	9.3	4.90×10^{-10}
Phenylmethylamine	$C_6H_5CH_2NH_2$	2.2×10^{-5}	11.7	2.00×10^{-12}

As you know from previous discussion, alkyl groups have a positive inductive effect and are electron releasing. This has the effect of increasing the electron density on the amine group and makes it more likely to accept a proton, i.e. more likely to behave as a base. Alkyls group increase the relative basicity of an amine.

The pH difference between methylamine and phenylamine solutions does not at first sight seem to be very great but, because pH is measured on a logarithmic scale, the difference is in fact 10 raised to the power $(12.3 - 9.3) = 10^{3.0} = 1000$. Hence methylamine is approximately 1000 times more basic than phenylamine.

Comparing the [H$^+$(aq)] values in Table 4.1, shows that the difference in basicity between methylamine and phenylamine is:

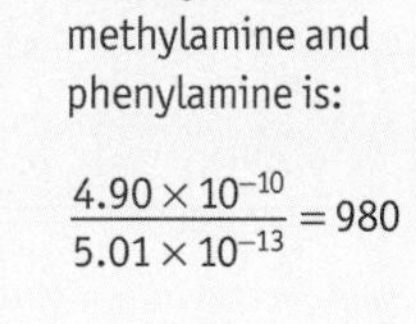

$$\frac{4.90 \times 10^{-10}}{5.01 \times 10^{-13}} = 980$$

Simple inductive effects alone cannot explain this large difference in pH. It can only be accounted for by the lone pair of electrons on the nitrogen being incorporated into the ring and becoming part of the delocalised system. The electron density on the nitrogen is greatly reduced, as is its ability to accept a proton. Therefore, it follows that the basicity of aromatic amines is much lower than that of aliphatic amines.

Overlap of lone pair from nitrogen into the ring

Stretch and challenge

1 Aliphatic amines can be prepared by the reaction between chloroalkanes and ammonia. The ammonia must be dissolved in ethanol.

a Explain why the ammonia must not be dissolved in water.

b Suggest why ammonia dissolved in a saturated aqueous solution of NH_4Cl might be a suitable reagent.

c Chloromethane reacts with ammonia to produce methylamine. Complete the following mechanism by adding curly arrows, relevant dipoles and lone pairs of electrons.

Step 1:

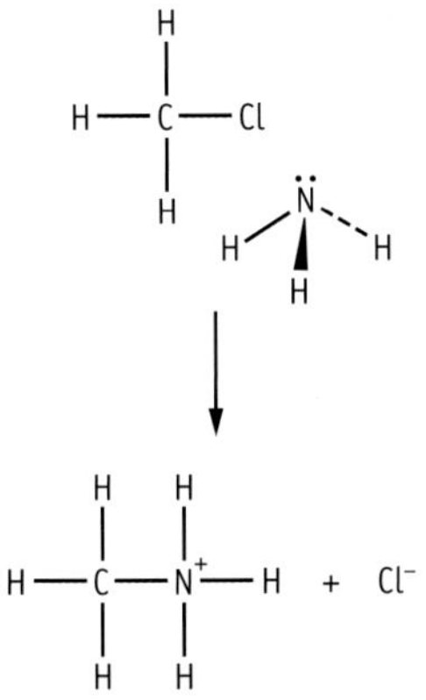

Step 2:

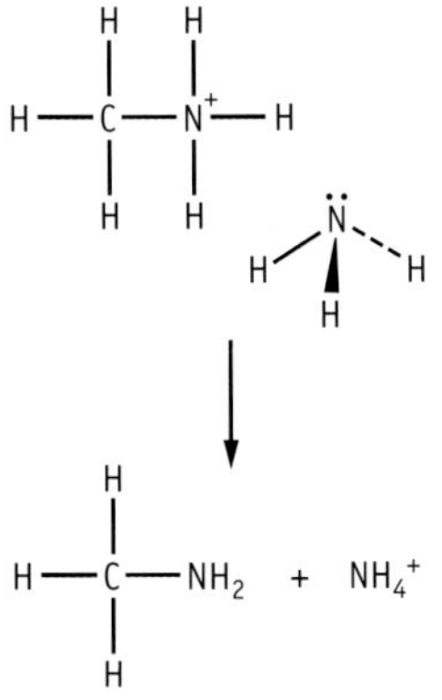

d In the reaction between chloromethane and ammonia various different organic products are formed including:

- $CH_3CH_2NH_2$
- $(CH_3CH_2)_2NH$
- $(CH_3CH_2)_3N$
- $(CH_3CH_2)_4N^+$

Suggest how you might manipulate the reaction conditions to ensure that you obtain mainly:

(i) $CH_3CH_2NH_2$

(ii) $(CH_3CH_2)_4N^+$

2 Put the following in order of increasing basicity. Justify the order.

NH_3

$CH_3CH_2NH_2$

$(CH_3CH_2)_3N$

$C_6H_5NH_2$

$(C_6H_5)_3N$

$C_6H_5CH_2NH_2$

Amino acids and chirality

Chapter 5

Amino acids

Amino acids are the building blocks of proteins. There are about 20 naturally occurring amino acids, some of which are shown in Table 5.1. The general formula is $H_2NCH(R)COOH$ where R represents the side chain.

***Table 5.1** Properties of amino acids*

Amino acid	Abbreviation	Side chain	pH of isoelectric point	Polarity	pH of side chain
Alanine	Ala	$-CH_3$	6.0	Non-polar	
Arginine	Arg	$-(CH_2)_3NHC(NH)NH_2$	10.8	Polar	Strongly basic
Asparagine	Asn	$-CH_2CONH_2$		Polar	
Aspartic acid	Asp	$-CH_2COOH$		Polar	Acidic
Cysteine	Cys	$-CH_2SH$		Non-polar	Acidic
Glutamic acid	Glu	$-CH_2CH_2COOH$	3.1	Polar	Acidic
Glutamine	Gln	$-CH_2CH_2CONH_2$		Polar	
Glycine	Gly	$-H$	5.8	Non-polar	
Isoleucine	Ile	$-CH(CH_3)CH_2CH_3$		Non-polar	
Leucine	Leu	$-CH_2CH(CH_3)_2$	6.0	Non-polar	
Lysine	Lys	$-(CH_2)_4NH_2$	9.7	Polar	Basic
Methionine	Met	$-CH_2CH_2SCH_3$		Non-polar	
Phenylalanine	Phe	$-CH_2C_6H_5$	5.5	Non-polar	
Serine	Ser	$-CH_2OH$		Polar	
Threonine	Thr	$-CH(OH)CH_3$		Polar	Weakly acidic
Tyrosine	Tyr	$-CH_2C_6H_4OH$		Polar	
Valine	Val	$-CH(CH_3)_2$		Non-polar	

All naturally occurring amino acids are α-amino acids in which the amine group, $-NH_2$, and the carboxyl group, $-COOH$, are attached to the same carbon.

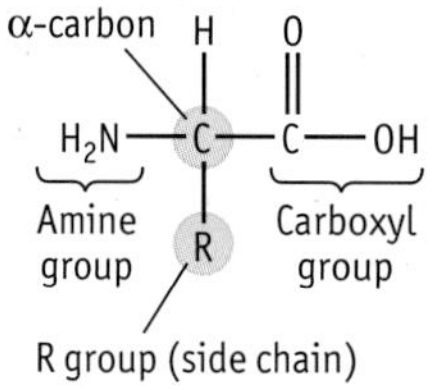

α-carbon
H_2N—C—C(=O)—OH
Amine group
Carboxyl group
R group (side chain)

All amino acids have trivial names — for example, glycine, alanine, valine and phenylalanine. These names are often abbreviated — for example, Gly, Ala, Val and Phe.

The simplest amino acid is glycine, which has the systematic name aminoethanoic acid. The R group in glycine is H. In 2-aminopropanoic acid (alanine), the R group is CH_3. The two amino acids are shown below.

$$H_2N{-}CH_2{-}C(=O){-}OH \qquad H_2N{-}CH(CH_3){-}C(=O){-}OH$$

Glycine   Alanine

Glycine is the only α-amino acid that does not have a stereoisomer.

You will recall from Unit F322 that stereoisomers are compounds that have the same molecular formula, the same structural formula but different three-dimensional shapes. There are two kinds of stereoisomerism: At AS you studied *E/Z* (*cis–trans*) isomerism, which occurs in alkenes. The key features to look for are compounds that contain a C=C double bond in which each carbon atom is bonded to two different atoms or groups.

There is restricted rotation about the C=C double bond and the different atoms or groups attached to each carbon atom ensure that there is no symmetry around the carbons in the C=C double bond.

$$H_3CH_2C(H)C{=}CH_2 \qquad H_3C(H)C{=}C(H)CH_3 \qquad H_3C(H)C{=}C(CH_3)H$$

But-1-ene   *Z*-but-2-ene (*cis*)   *E*-but-2-ene (*trans*)

But-1-ene and but-2-ene both contain a C=C double bond. However, the right-hand carbon in the C=C double bond in but-1-ene is bonded to two hydrogen atoms and, therefore, but-1-ene does not exhibit *E/Z* (*cis–trans*) isomerism. But-2-ene possesses both essential key features and hence has a *Z* (*cis*) and an *E* (*trans*) isomer.

Chirality and optical isomerism

There is a second type of stereoisomerism known as **optical isomerism**. This occurs in all amino acids apart from glycine. The key feature to look for is a **chiral** centre. A chiral compound contains an asymmetric atom. The mirror images of the molecule are, therefore, not identical and are **non-superimposable**. Most chiral compounds have a carbon atom attached to four different atoms or groups. The asymmetric carbon atom is usually shown by an asterisk, *.

$$H_2N{-}CH_2{-}C(=O){-}OH \qquad H_2N{-}C^{*}H(CH_3){-}C(=O){-}OH$$

Glycine — The α-carbon is not bonded to four different groups

Alanine — The α-carbon is bonded to four different groups

An asymmetric carbon atom is a carbon atom that is bonded to four different atoms or groups.

Alanine has an asymmetric carbon atom but glycine does not. Glycine is not chiral because the carbon atom is bonded to two hydrogen atoms. The optical isomers of alanine are shown below.

H, C*, H_2N, COOH, H_3C | H, *C, HOOC, NH_2, CH_3

Optical isomers

> **e** When drawing optical isomers it is essential that you represent them as 3-dimensional structures that are mirror images of each other.

The two optical isomers look the same, but they are non-superimposable.

Optical isomers are so called because they behave differently when plane-polarised light is passed through them. One isomer (the D-isomer) rotates the plane of plane-polarised light to the right (clockwise) and the other (the L-isomer) rotates it to the left (anti-clockwise). Both isomers are optically active.

Light is made up of electromagnetic radiation that has vibrations all directions. Plane-polarised light is created by restricting the vibrations to one plane only.

Other properties of amino acids

An amino acid contains two functional groups; the carboxylic acid group and the amine group. They are described as 'bi-functional'. One functional group is acidic and the other is basic so they can react with each other. The –COOH donates a H^+ to the $-NH_2$ group forming an 'internal salt' known as a **zwitterion.**

Proton is transferred

Zwitterion

For each amino acid, the zwitterion exists at a particular pH known as the **isoelectric point.**

The isoelectric point of an amino acid depends on the nature of the R group (Table 5.1 on page 58) — for example, that of alanine is pH 6.0. If the pH changes, the ion also changes. At a pH lower than the isoelectric point, the COO^-

is protonated, so a positive ion (cation) is formed; at a higher pH than the isoelectric point, the NH_3^+ is de-protonated, so a negative ion (anion) is formed.

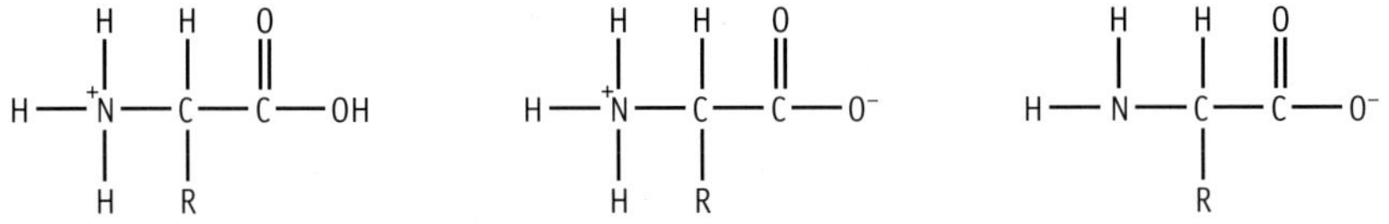

Alanine at pH = 2.0
Cation

Alanine at pH = 6.0
Zwitterion

Alanine at pH = 10.0
Anion

If the amino acid is in an acidic solution it forms a cation; in an alkaline solution it forms an anion.

An amino acid can behave as both an acid and a base. The carboxylic acid group can react with a base to form a salt and with an alcohol to form an ester:

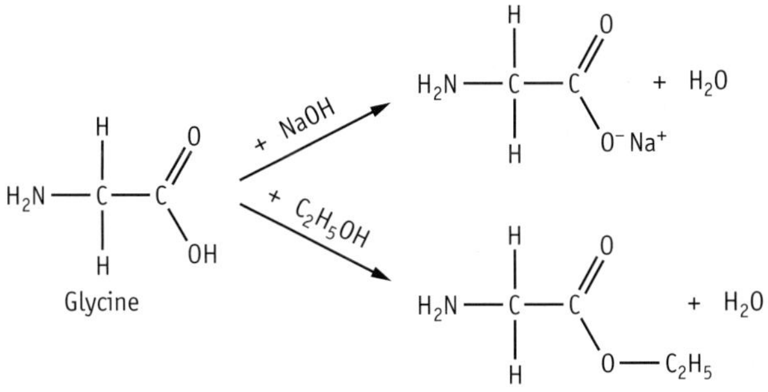

Glycine + NaOH → $H_2N{-}CH_2{-}COO^-Na^+$ + H_2O

Glycine + C_2H_5OH → $H_2N{-}CH_2{-}COO{-}C_2H_5$ + H_2O

The amine group is a base and can react with an acid, such as HCl, to produce a salt.

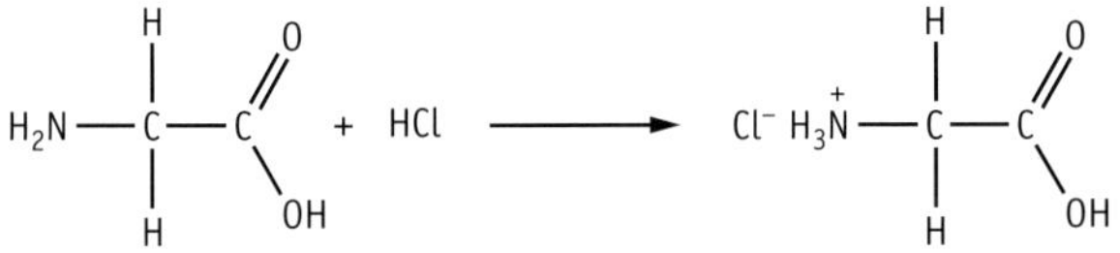

Glycine + HCl → $Cl^-\ H_3N^+{-}CH_2{-}COOH$

Peptide formation

Amino acids also display properties that depend on *both* functional groups. They react to form peptides by a condensation reaction.

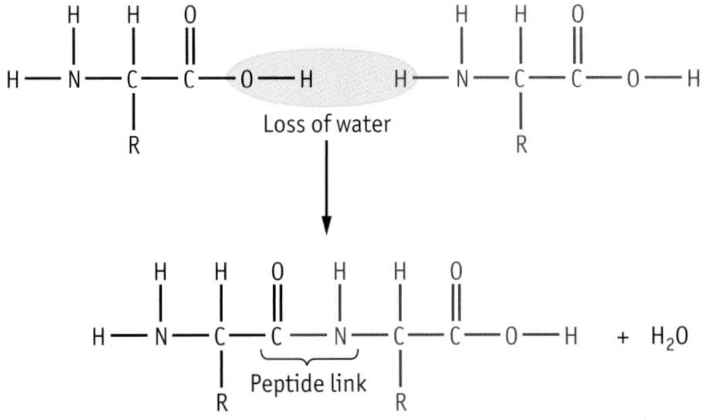

Loss of water

Peptide link

Dipeptide + H_2O

A dipeptide contains one peptide link (–CONH–) and two side chains (R groups).

If two different amino acids are mixed and allowed to react, it is possible to form four different dipeptides. The sequence in which the amino acids link gives rise to the different dipeptides. If glycine (Gly) reacts with alanine (Ala) it is possible to form a dipeptide designated Gly–Ala (this indicates the sequence of the amino acids).

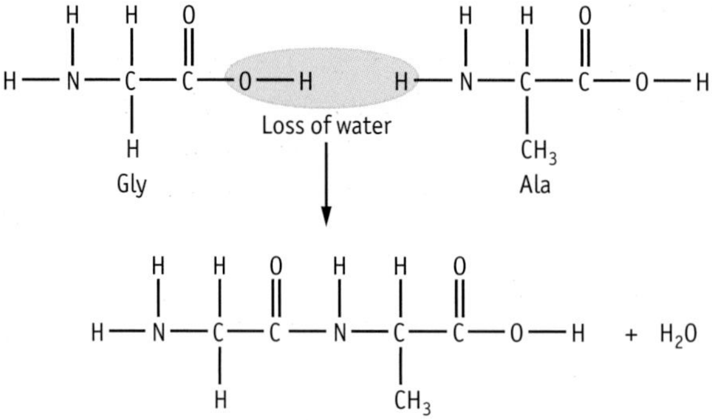

Gly Loss of water Ala

+ H_2O

Gly-Ala

It is also possible to form Ala–Gly:

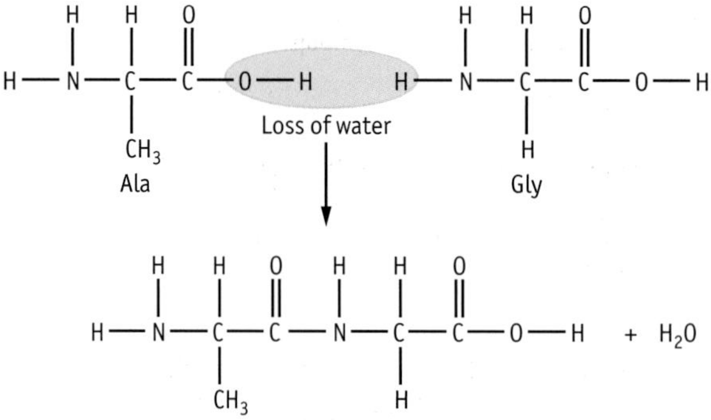

Ala Loss of water Gly

+ H_2O

Ala-Gly

Starting with glycine and alanine the four possible dipeptides are Gly–Gly, Gly–Ala, Ala–Gly and Ala–Ala.

Dipeptides can react further with additional amino acids, thus extending the chain length. This leads to the formation of tripeptides, polypeptides and proteins:

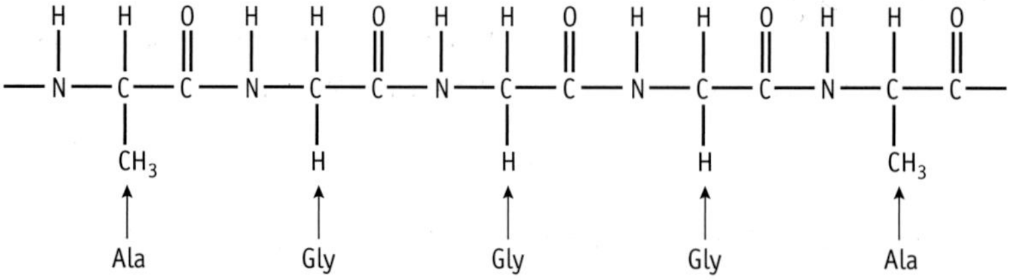

Ala Gly Gly Gly Ala

The sequence of amino acids is determined by the order of the R groups. The sequence in the section above is –Ala–Gly–Gly–Gly–Ala–.

Remember that all amino acids have the general formula $H_2NCH(R)COOH$.

Hydrolysis of peptides and proteins

The peptide link is polar and is subject to attack:

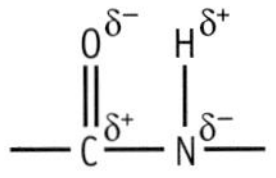

$$-\overset{\overset{O^{\delta-}}{\|}}{C^{\delta+}}-\overset{\overset{H^{\delta+}}{|}}{N^{\delta-}}-$$

It can be hydrolysed in both acidic and alkaline conditions.

Peptides and proteins can be hydrolysed to amino acids by refluxing with 6.0 mol dm^{-3} hydrochloric acid. Acid hydrolysis results in the formation of α-amino acids. The section of a polypeptide shown below contains three different amino acids. It is possible to identify the amino acids in a polypeptide by the R groups.

Since hydrolysis is either acid-catalysed or base-catalysed, the amino acids form the corresponding ions.

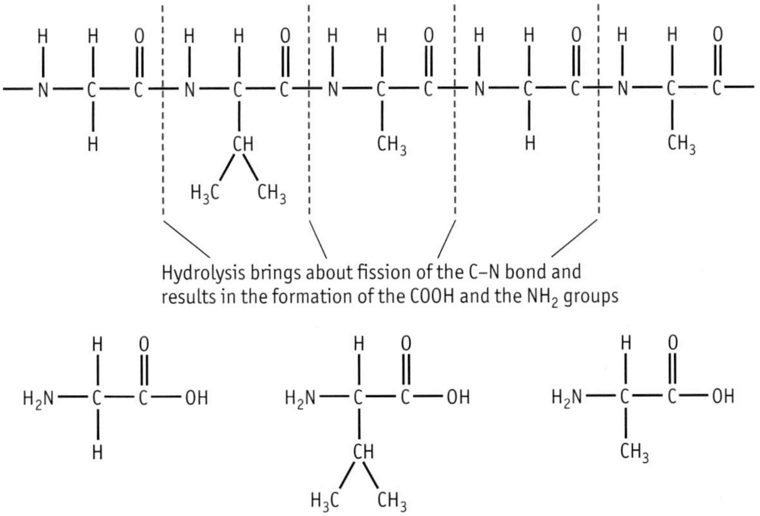

$$-NH-CH(H)-CO-NH-CH(CH(CH_3)_2)-CO-NH-CH(CH_3)-CO-NH-CH(H)-CO-NH-CH(CH_3)-CO-$$

Hydrolysis brings about fission of the C–N bond and results in the formation of the COOH and the NH_2 groups

$$H_2N-CH(H)-COOH \qquad H_2N-CH(CH(CH_3)_2)-COOH \qquad H_2N-CH(CH_3)-COOH$$

Using the general formula, $H_2NCH(R)COOH$, the three amino acids are identified as $H_2NCH(H)COOH$, $H_2NCH(CH(CH_3)_2)COOH$ and $H_2NCH(CH_3)COOH$.

Questions

1 a Give the systematic chemical names of amino acids **A**, **B**, **C** and **D**.

Use Table 5.1 on page 58 to look up their common names and three-letter abbreviations.

b Draw the two optical isomers of compound **A**.

c Draw the skeletal formulae of compounds **A** and **B**.

d Draw the zwitterions formed by compounds **A** and **C**.

e Draw the ions formed by compounds **B** and **D** at pH 2 and at pH 12.

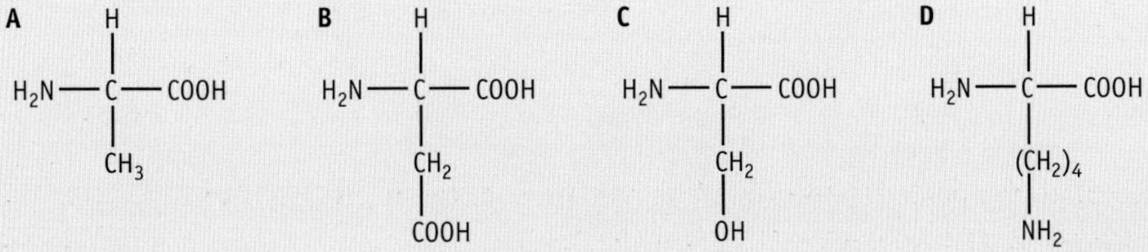

A $H_2N-CH(CH_3)-COOH$

B $H_2N-CH(CH_2COOH)-COOH$

C $H_2N-CH(CH_2OH)-COOH$

D $H_2N-CH((CH_2)_4NH_2)-COOH$

f Using compounds **A** and **C**, a mixture of dipeptides can be formed.
- Identify how many different dipeptides can be formed.
- Use the three-letter abbreviation of each compound to identify the different dipeptides.
- Draw two of the possible dipeptides.

g Draw the polypeptide sequence:
–A–B–B–A–C–A–D–

2 Explain how hydrolysis of proteins is achieved.

Summary

You should now be able to:

- describe the general formula of an amino acid
- draw the formula of a zwitterion and deduce the effect of changing pH on the zwitterion
- describe the acid–base reactions of amino acids
- explain the formation of peptide links, dipeptides, polypeptides and proteins
- describe the hydrolysis of proteins and polypeptides
- describe and explain optical isomerism, identify chiral centres and draw optical isomers

Additional reading

Optical isomerism

Compounds that contain a carbon atom bonded to four different atoms or groups are described as **asymmetric** or **chiral**. Asymmetric refers to the lack of symmetry around the carbon atom; chiral is derived from the Greek *cheir*, meaning hand. Individual optical isomers are known as **enantiomers** and are said to be optically active; they rotate the plane of plane-polarised light. One of the two mirror images rotates the plane of plane-polarised light to the right (**dextrorotatory** or D-enantiomer) and the other (**laevorotatory** or L-enantiomer) rotates it to the left. Samples prepared in the laboratory often contain a 50:50 mixture of the two enantiomers. This is called a **racemic mixture** or **racemate.** A racemic mixture is optically inactive because the optical activities of the enantiomers cancel. Enantiomers have almost identical physical and chemical properties. However, the chemical properties do vary when the enantiomers react with other optically active compounds.

Some compounds contain more than one asymmetric carbon atom. For example, a molecule of 2-hydroxy-3-methylbutanedioic acid contains two asymmetric carbon atoms:

$$HOOC-\overset{OH}{\underset{H}{C^*}}-\overset{H}{\underset{CH_3}{C^*}}-COOH$$

There are four optical isomers:

$$\begin{array}{c} COOH \\ | \\ H-C^*-OH \\ | \\ H_3C-C^*-H \\ | \\ COOH \end{array}$$

Non-superimposable mirror-images = one pair of enantiomers

Non-superimposable mirror-images = one pair of enantiomers

The differences in the isomers are more apparent if you build models.

In general, if there are *n* asymmetric carbons in a monomer, there will be 2^n optical isomers. However, this can vary if the asymmetric carbons form an internal mirror image as in 2,3-dihydroxybutanedioic acid or tartaric acid.

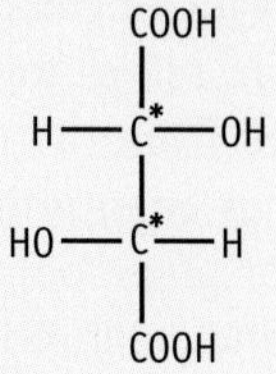

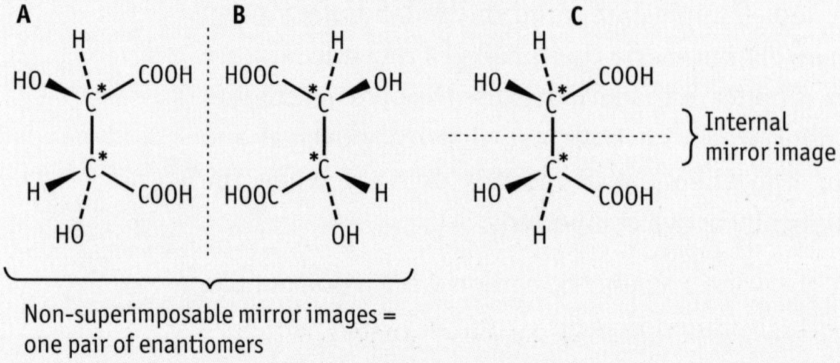

A B C

Internal mirror image

Non-superimposable mirror images = one pair of enantiomers

Structures **A** and **B** are mirror images of each other and are not superimposable. Therefore, both **A** and **B** are optically active. One is the D(+) enantiomer and the other is the L(–) enantiomer. The third structure (**C**) is not superimposable on either **A** or **B**, but it is *not* a mirror image of either, so it is not an enantiomer of either **A** or **B**. Structure **C** is optically inactive because the two asymmetric carbon atoms are mirror images of each other. This is described as being *internally* compensated and is known as the **meso**-form. An equimolar mixture of structures **A** and **B** is also optically inactive; this is described as being *externally* compensated. The D(+) and the L(–) isomers have identical physical properties, unlike the **meso**-isomer. Properties of the optical isomers of tartaric acid are shown in Table 5.2.

***Table 5.2** Properties of the optical isomers of tartaric acid*

Property	D(+)	L(–)	(meso)
Density/g cm^{-3}	1.76	1.76	1.67
Solubility in H_2O/g per 100 cm^3	139	139	125
Melting point/°C	170	170	140

Stretch and challenge

1 Draw 3-dimensional diagrams of the optical isomers of each of the following:

a **(i)** valine, $H_2NCH(CH(CH_3)_2)COOH$

(ii) threonine, $H_2NCH(CH(OH)CH_3)COOH$

b 2,3-dimethyl-2,3-diphenylbutanedioic acid

2 Polymers such as poly(propene) can exist in different forms:

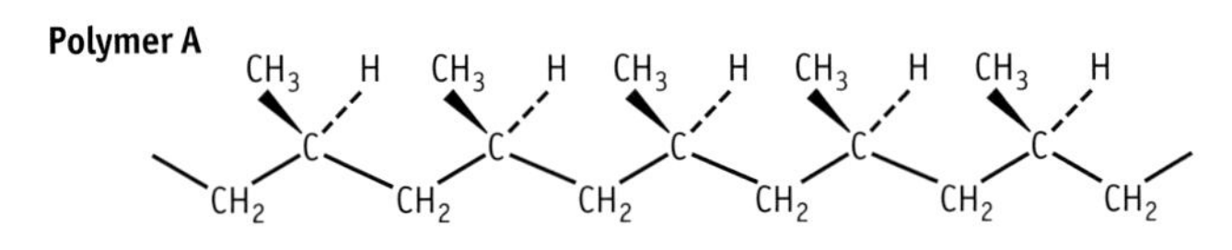

Polymer B

CH3 H H CH3 CH3 H H CH3 CH3 H

C C C C C

CH2 CH2 CH2 CH2 CH2 CH2

State the difference between the two structures and explain why one cannot be rotated and superimposed on the other.

Polyesters and polyamides

Chapter 6

Introduction

Synthetic polymers are chain-like molecules made from small individual units called monomers. Many polymers are plastics that play an important part in everyday life. The amount of plastic we use is enormous and growing. Plastics are easy to make, easy to colour and can be chemically tailored to have specific properties. They are often cheaper to produce than the materials that they replace. There are many instances where plastic is used instead of wood, metal, stone, glass, leather or a natural fabric such as cotton or silk. However, there are problems with their disposal and there is growing concern over the environmental aspects associated with this.

In the specification, there are two categories of polymer:

- addition polymers
- condensation polymers

Addition polymers are formed when alkenes undergo a reaction in which one alkene molecule joins to another and a long molecular chain is built up. The individual alkene molecule is referred to as a **monomer**; the long chain molecule is known as the **polymer**.

e Addition polymerisation was covered in Unit 2 (see Chapter 9 in the AS textbook).

The polymerisation can be initiated in a number of ways. The initiator may be incorporated at either end of the long molecular chain. However, if the initiator is disregarded the empirical formulae of the monomer and the polymer are unaffected.

Common monomers include:

- ethene, $CH_2{=}CH_2$
- propene, $CH_3CH{=}CH_2$
- phenylethene (also known as styrene), $C_6H_5CH{=}CH_2$

It is worth recalling that the bonds in addition polymers are strong covalent bonds and the molecules are generally non-polar. This makes most of the polymers resistant to chemical attack and they are also not broken down by bacteria — they are **non-biodegradable**. The widespread use of these polymers has created a major problem of disposal and much work is being carried out to rectify this.

Condensation polymers are formed when monomers link together to form a long chain in which each link is accompanied by the loss, or condensation, of a small molecule such as H_2O or HCl. The two main types of condensation polymer are polyesters and polyamides.

Polyesters

Terylene® (PET, **p**oly**e**thylene **t**erephthalate) is a common polyester used in synthetic fibres and in containers for food and beverages. It is one of the most important raw materials of man-made fibres. It is made by reacting the monomers ethane-1,2-diol and benzene-1,4-dicarboxylic acid:

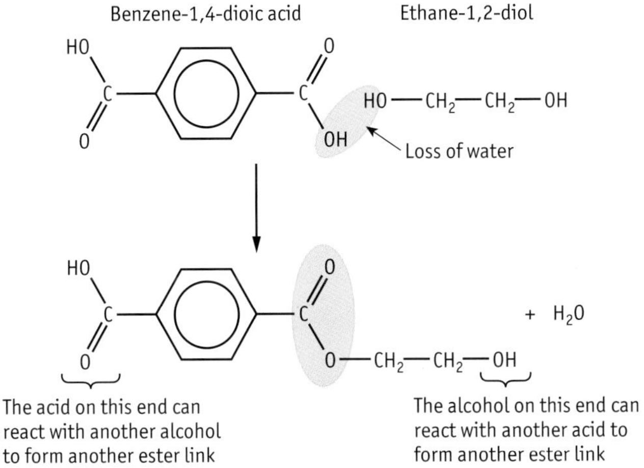

Benzene-1,4-dioic acid

Ethane-1,2-diol

Loss of water

$$HOOC-C_6H_4-COOH + HO-CH_2-CH_2-OH \rightarrow HOOC-C_6H_4-COO-CH_2-CH_2-OH + H_2O$$

The acid on this end can react with another alcohol to form another ester link

The alcohol on this end can react with another acid to form another ester link

Both monomers react at each end, building up a long-chain molecule held together by a large number of ester linkages — hence producing a polyester.

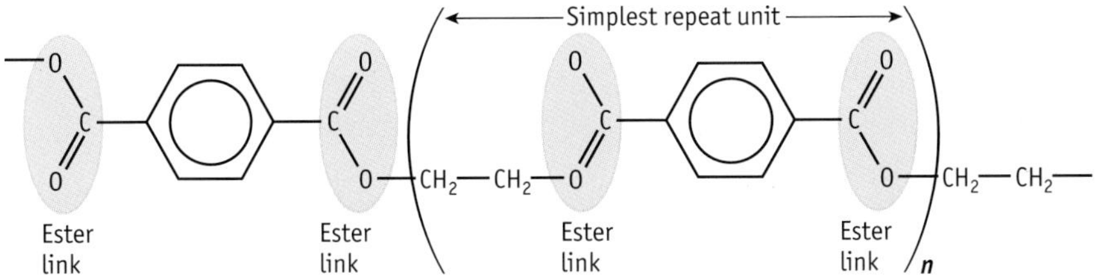

Simplest repeat unit

Ester link

Ester link

Ester link

Ester link

n

It is important to be able to identify the simplest repeat unit.

The resulting polymer is a regular, approximately linear, structure. The polymer chains pack closely together, so there are strong intermolecular forces.

Polyesters can be formed by:

- the reaction between any dioic acid and any diol
- the polymerisation of hydroxycarboxylic acids — for example 2-hydroxy-propanoic acid (lactic acid), $CH_3CH(OH)COOH$

A molecule of lactic acid has a chiral centre, $CH_3{}^*CH(OH)COOH$, and contains an alcohol group (–OH) and a carboxylic acid group (–COOH). These can react to produce ester linkages.

Polylactic acid, PLA, is not made by a simple esterification reaction or directly from lactic acid. It is made via the formation of a lactide (a cyclic dimer) followed by the ring opening polymerisation of the lactide. This ensures that no water is formed during the polymerisation and prevents hydrolysis of the polymer.

The reaction scheme below shows lactic acid as the monomer and polylactic acid as the resultant polymer. The cyclic dimers are not included here but detail of this can be found in the stretch-and-challenge questions at the end of this chapter.

Monomer = lactic acid (2-hydroxypropanoic acid)

Loss of water by condensation

$$HO-CH(CH_3)-CO-OH + HO-CH(CH_3)-CO-OH + HO-CH(CH_3)-CO-OH + HO-CH(CH_3)-CO-OH$$

Polymer = poly(lactic acid), PLA

$$-CH(CH_3)-CO-O-CH(CH_3)-CO-O-CH(CH_3)-CO-O-CH(CH_3)-CO-O-$$

Ester links

The introduction of polymers such as polylactic acid reflects the role of chemists in minimising the impact on the environment through the use of renewable sources and by the development of degradable polymers. Polylactic acid is particularly attractive as a sustainable alternative to products derived from petrochemicals, since the monomer can be produced by the bacterial fermentation of agricultural by-products, such as corn starch or sugar cane, which are renewable feedstocks.

Polylactic acid is fully compostable and degrades in a relative short period of time to form CO_2 and H_2O. Under controlled conditions it is also possible for polylactic acid to be converted back to the monomer. Polymers that contain groups such as the carbonyl group, $-C=O$, may also be photodegradable because the $C=O$ bond absorbs radiation.

Polylactic acid currently has a number of biomedical applications — for example internal stitches and drug release systems. Being biodegradable, it can also be employed in the preparation of 'bioplastic' used in food packaging, female hygiene products and disposable nappies. Polylactic acid also has potential uses in fibres and non-woven textiles, and it is being evaluated as a possible material for tissue engineering.

PLA is more expensive than many petroleum-derived commodity plastics, but its price has been falling as more production comes online. The degree to which the price will fall and the degree to which PLA will be able to compete with non-sustainable petroleum-derived polymers is uncertain.

Polyamides

Polyamides can be prepared from two monomers, one with an amine group at each end and the other with a carboxylic acid group at each end.

Nylon

Nylon is the general name for a family of condensation polymers first produced by Wallace Carothers in 1935.

Today, nylon is one of the most common polymers used as a fibre. Nylons are formed by reacting equal parts of a diamine and a dicarboxylic acid so that an amide (or peptide) bond is formed at both ends of each monomer. Diacyl dichlorides, $ClOC(CH_2)_nCOCl$, can be used in place of the dioic acids.

Different nylons have a different numerical suffix which specifies the numbers of carbons in each monomer; the diamine is first and the diacid second. The most common variant is nylon-6,6, the name of which indicates that the diamine (1,6-diaminohexane) and the dicarboxylic acid (hexanedioic acid) each donate six carbons to each repeat unit of the polymer chain.

CHARLES D. WINTERS/SCIENCE PHOTO LIBRARY

In the nylon rope experiment, a strand of nylon can be pulled from the interface of 1,6-diaminohexane and decanedioic acid

Loss of water

$$\mathrm{H{-}N(H){-}(CH_2)_6{-}N(H){-}H + HO{-}C({=}O){-}(CH_2)_4{-}C({=}O){-}OH + H{-}N(H){-}(CH_2)_6{-}N(H){-}H + HO{-}C({=}O){-}(CH_2)_4{-}C({=}O){-}OH}$$

$$\downarrow$$

Simplest repeat unit

$$\mathrm{{-}N(H){-}(CH_2)_6{-}N(H){-}C({=}O){-}\left[(CH_2)_4{-}C({=}O){-}N(H){-}(CH_2)_6{-}N(H){-}C({=}O)\right]_n{-}(CH_2)_4{-}C({=}O){-}} + \mathrm{H_2O}$$

Amide link

Other nylons include nylon-6,10, which is made from the monomers 1,6-diaminohexane and decane-dicarboxylic acid.

Nylon forms a strong flexible fibre when it is melt-spun. Nylon has a range of uses including clothing, ropes and fishing lines, toothbrush bristles, parachutes, flexible tubing and pantyhose.

> **e** You will not be expected to recall the names or formulae of the monomers.

Nylon products

MARTYN F. CHILLMAID/SCIENCE PHOTO LIBRARY

Kevlar®

Kevlar® is another polyamide. It is stronger than steel and is fire-resistant. It is used for making bulletproof vests, crash helmets and protective clothing used by firefighters. It is also used for sports equipment and loud-speaker cones.

Kevlar® is made from the monomers benzene-1,4-diamine and benzene-1,4-dicarboxylic acid:

Loss of water

$$H-N(H)-C_6H_4-N(H)-H + HO-C(=O)-C_6H_4-C(=O)-OH + H-N(H)-C_6H_4-N(H)-H + HO-C(=O)-C_6H_4-C(=O)-OH$$

$$\downarrow$$

Simplest repeat unit

$$-N(H)-C_6H_4-N(H)\left(-C(=O)-C_6H_4-C(=O)-N(H)-C_6H_4-N(H)-\right)_n C(=O)-C_6H_4-C(=O)- \quad + \quad H_2O$$

Amide link

Fibres of Kevlar® consist of long molecular chains and there are many inter-chain bonds, which makes the material extremely strong. Kevlar® derives its high strength partly from intermolecular hydrogen bonds formed between the

JAMES KING-HOLMES/SCIENCE PHOTO LIBRARY

Products made from Kevlar® include these Concorde fuel tank liners

carbonyl groups and protons on neighbouring polymer chains and partly from the partial π-stacking of the aromatic rings. An aromatic, or π–π, interaction is a non-covalent interaction caused by the intermolecular overlapping of *p*-orbitals in adjacent π-delocalised systems. These interactions become stronger as the number of π-electrons increases.

The extensive hydrogen bonding is shown in the diagram below.

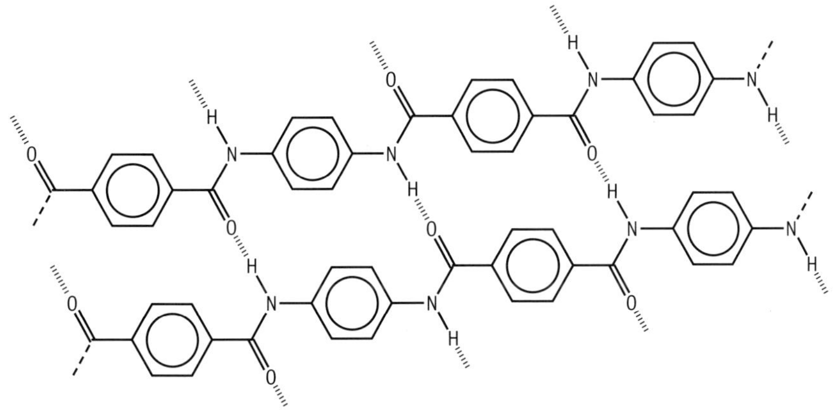

Hydrolysis of condensation polymers

The ester link in a polyester and the amide link in a polyamide are both polar and are subject to acid-catalysed and base-catalysed hydrolysis.

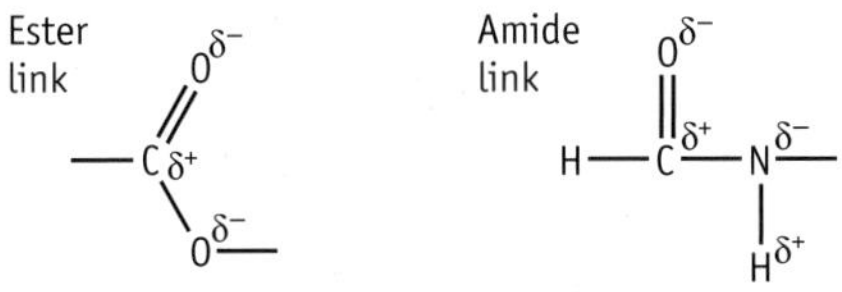

Ester link

Amide link

Acid hydrolysis of a polyester results in the formation of a diol and a dioic acid:

$$\text{Polyester} + H_2O \xrightarrow{H^+(aq)\text{ catalyst}} \text{Diol} + \text{Dioic acid}$$

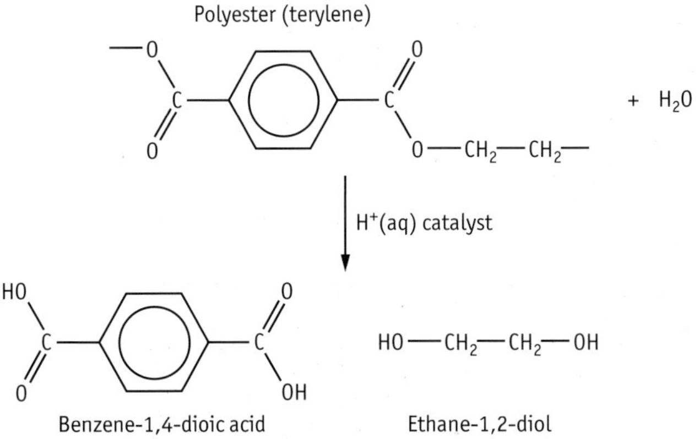

Polyester (terylene)

H⁺(aq) catalyst

Benzene-1,4-dioic acid

Ethane-1,2-diol

Base hydrolysis of a polyester also forms a diol, but the dioic acid formed then reacts with the base catalyst to form the dioate salt. The products of refluxing Terylene® with an aqueous solution of NaOH(aq) are:

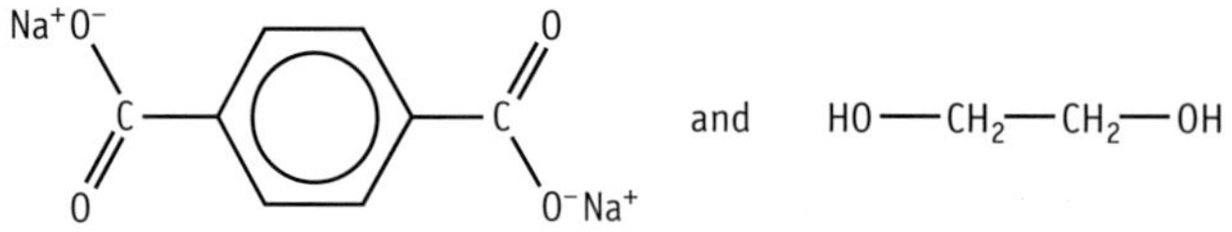

$Na^+\,{}^-O\text{–}CO\text{–}C_6H_4\text{–}CO\text{–}O^-\,Na^+$ and $HO\text{—}CH_2\text{—}CH_2\text{—}OH$

Polyamides can also be hydrolysed:

- Acid-catalysed hydrolysis results in the formation of the dioic acid and the di-salt of the diamine.
- Base-catalysed hydrolysis results in the formation of the diamine and the di-salt of the dioic acid.

Hydrolysis of polyamides is summarised in the reaction scheme below.

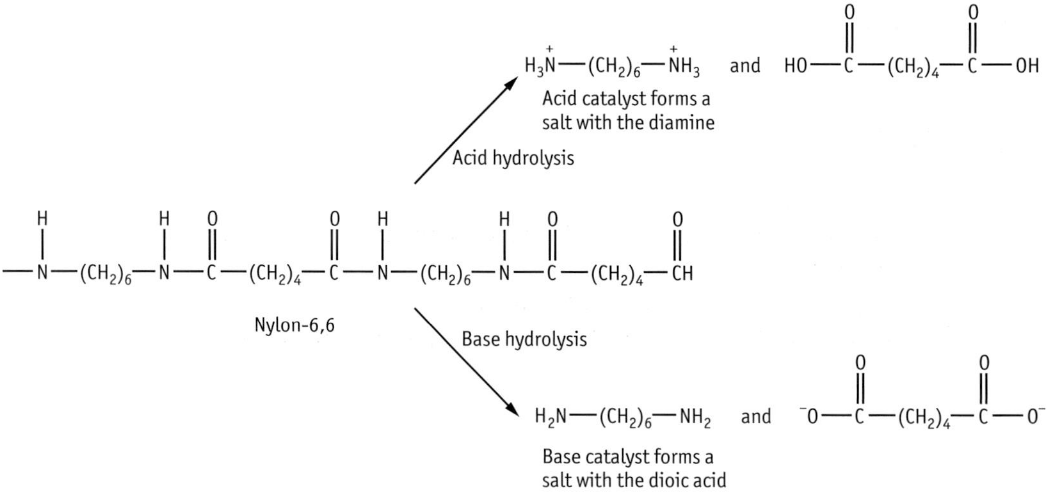

Nylon-6,6:

$$\text{—NH—(CH}_2)_6\text{—NH—CO—(CH}_2)_4\text{—CO—NH—(CH}_2)_6\text{—NH—CO—(CH}_2)_4\text{—CH}$$

Acid hydrolysis → $H_3\overset{+}{N}\text{—(CH}_2)_6\text{—}\overset{+}{N}H_3$ and $HO\text{—CO—(CH}_2)_4\text{—CO—OH}$

Acid catalyst forms a salt with the diamine

Base hydrolysis → $H_2N\text{—(CH}_2)_6\text{—NH}_2$ and ${}^-O\text{—CO—(CH}_2)_4\text{—CO—O}^-$

Base catalyst forms a salt with the dioic acid

Questions

1 Draw a section of the polymer, showing two repeat units, which could be formed from each of the following pairs of monomers:

a 1,6-diaminohexane and decanedioic acid

b benzene-1,4-dioic acid and 1,3-diaminobenzene

c ethane-1,2-diol and benzene-1,3-dioic acid

2 Draw two repeat units of the polymer that could be formed from:

a

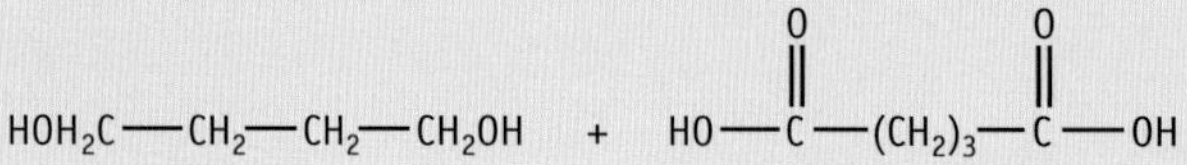

$$HOH_2C\text{—}CH_2\text{—}CH_2\text{—}CH_2OH \quad + \quad HO\text{—CO—(CH}_2)_3\text{—CO—OH}$$

b

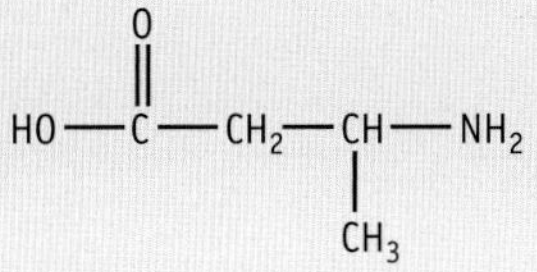

$$HO{-}\overset{\overset{\displaystyle O}{\|}}{C}{-}CH_2{-}\underset{\underset{\displaystyle CH_3}{|}}{CH}{-}NH_2$$

3 Identify the organic products formed when the polymer

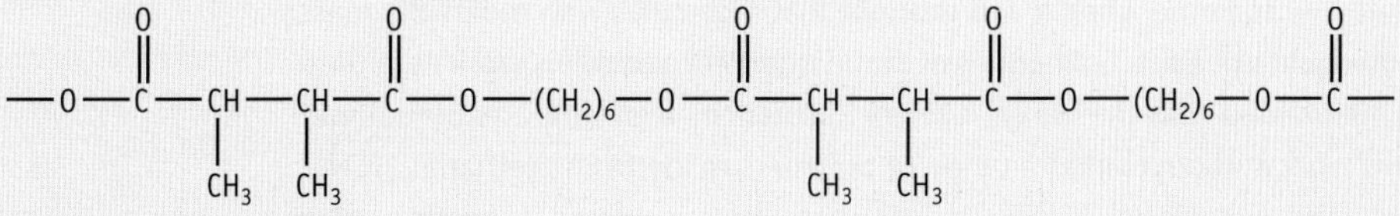

$$-O-\overset{\overset{\displaystyle O}{\|}}{C}-\underset{\underset{\displaystyle CH_3}{|}}{CH}-\underset{\underset{\displaystyle CH_3}{|}}{CH}-\overset{\overset{\displaystyle O}{\|}}{C}-O-(CH_2)_6-O-\overset{\overset{\displaystyle O}{\|}}{C}-\underset{\underset{\displaystyle CH_3}{|}}{CH}-\underset{\underset{\displaystyle CH_3}{|}}{CH}-\overset{\overset{\displaystyle O}{\|}}{C}-O-(CH_2)_6-O-\overset{\overset{\displaystyle O}{\|}}{C}-$$

is hydrolysed by heating with:

a an acid catalyst

b a base catalyst

Summary

You should now be able to:

- describe condensation polymerisation and deduce the structure of a polymer from its given monomer(s)
- describe the hydrolysis of polyesters and polyamides
- describe the efforts made by chemists to develop biodegradable polymers

Additional reading

The age of polymers

Polymers come in all shapes and sizes and extend far beyond the addition polymers, polyesters and polyamides included in the specification.

Polymer science is relatively new and was developed during the twentieth century. The father of polymer science was Hermann Staudinger, who in the 1920s investigated a series of 'macromolecules' — at a time when it was thought that molecules with molecular mass greater than 5000 could not exist. Colleagues of Staudinger referred to his work as 'grease chemistry' and maintained that the waxy, rubbery substances were impure. Staudinger persisted with his work and developed a range of polymers made from methanal, HCHO. In 1953, he received the Nobel Prize. The diagram below shows a section of a poly(methanal) pioneered by Staudinger.

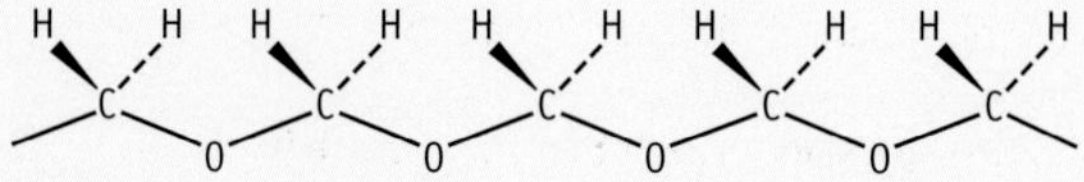

One of the most important polymers, poly(ethene) was discovered almost by accident. A group of research chemists working for ICI (Imperial Chemical Industries) in Cheshire were attempting to react ethene, $H_2C{=}CH_2$, with benzaldehyde, C_6H_5CHO, when they noticed a white waxy solid on the inside of the reaction vessel. After about another 5 or 6 years of experimentation, a material with an average relative molecular mass of 10 000 had been isolated. A few years later they were able to manufacture materials with relative molecular masses up to about 30 000. The initial research was difficult and expensive; ethene was not readily available and had to be prepared by the dehydration of ethanol. Today, ethene is produced in bulk from oil by the petrochemical industry.

Some polymers have now been developed that conduct electricity; others are light emitting.

Polymers are used universally and are essential to modern life.

Stretch and challenge

1 Copy the diagram below and add curly arrows to show the movement of electrons in the formation of poly(methanal).

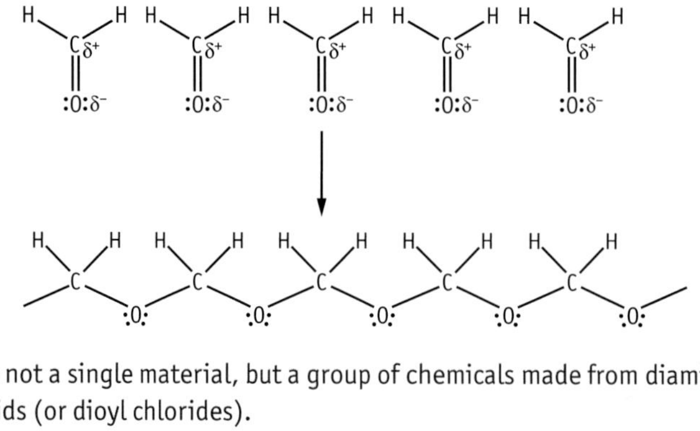

2 Nylon is not a single material, but a group of chemicals made from diamines and dioic acids (or dioyl chlorides).

Identify the monomers used to make

a nylon-4,6

b nylon-6,4

c nylon-6,10

3 Nylon-6 can be made from caprolactam:

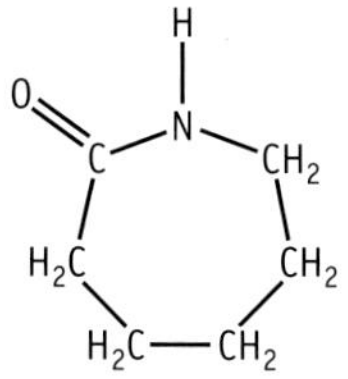

H
O C N CH_2
H_2C CH_2
H_2C — CH_2

Caprolactam

Draw two repeat units of nylon-6 and explain how it differs from nylon-6,6.

4 Polylactic acid, PLA, is a polymer of lactic acid, $CH_3CH(OH)COOH$. It is made by first forming a cyclic dimer from lactic acid (2-hydoxypropanoic acid), rather than directly from $CH_3CH(OH)COOH$.

a Write an equation that would represent the formation of PLA from $CH_3CH(OH)COOH$.

Suggest why PLA is not manufactured directly from the monomer $CH_3CH(OH)COOH$.

b Write a balanced equation for the dimerisation and draw the displayed formula of the cyclic dimer.

c Write an equation for the formation of PLA from the cyclic dimer. Suggest why PLA is manufactured directly from the cyclic dimer.

d Draw a section of PLA showing two repeat units.

Synthesis

Chapter 7

You have now studied a range of topics in organic chemistry. In the AS Unit 2, the functional groups covered were:

- alkanes and alkenes
- alcohols and halogenoalkanes

In this A2 Unit 4, the functional groups covered are:

- arenes and phenols
- aldehydes and ketones
- carboxylic acids and esters
- amines and amino acids
- diazonium compounds and azo dyes

All of these compounds are interrelated.

Synthetic routes

An ideal reaction involves one step, produces one product and requires no purification or separation. Some organic reactions can be accomplished by such a single reaction — for example, the conversions:

- ethanol → ethanoic acid
- ethanoic acid → ethyl ethanoate

However, many other reactions cannot be accomplished in a single step and involve a multi-stage synthetic route.

Few, if any, chemical conversions are 100% efficient and it is extremely unusual for a reaction to give a 100% yield in line with the stoichiometry of a reaction. For example, the formation of ethanoic acid, CH_3COOH, by the oxidation of ethanol, CH_3CH_2OH, using acidified dichromate is unlikely to be 100% efficient.

$$CH_3CH_2OH + 2[O] \rightarrow CH_3COOH + H_2O$$

The equation tells us that the molar ratio of ethanol and ethanoic acid is 1:1. Therefore, 1 mol (46.0 g) of ethanol should produce 1 mol (60.0 g) of ethanoic acid, but this is unlikely to happen. There are a number of factors that influence the percentage yield of a reaction. In this case it may be that the oxidation is only partial and some ethanal is formed. It is also possible that as the ethanoic acid is formed, some of it reacts with unreacted ethanol to form ethyl ethanoate. It follows that, in this single-stage process, the reaction products are likely to be a complex mixture of chemicals.

The more stages there are in a synthesis, the lower the yield is likely to be. The number of possible impurities increases with the number of stages in a synthesis.

The flow charts shown in Figures 7.1 and 7.2 cover the functional groups listed above. They show the relationships between them and give the essential reagents and conditions for their reactions. Figure 7.1 covers the aliphatic chemistry.

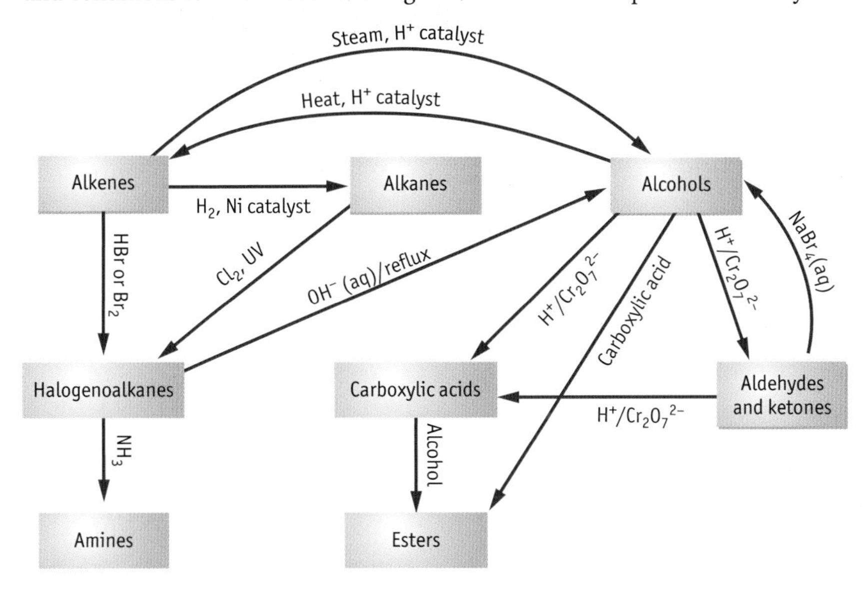

Figure 7.1
A flow chart covering aliphatic chemistry

e You could use the flow chart to work out other multi-stage conversions.

From Figure 7.1, it can be seen that an alkene can be converted into an alkane by a single step. However, the reverse conversion cannot be achieved in a single step. To convert an alkane into an alkene, the alkane is first converted into a halogenoalkane, then into an alcohol and finally into an alkene. This multi-stage conversion is less efficient than a single-stage conversion.

The flow chart in Figure 7.2 is simpler and covers aromatic chemistry.

Figure 7.2
A flow chart covering aromatic chemistry

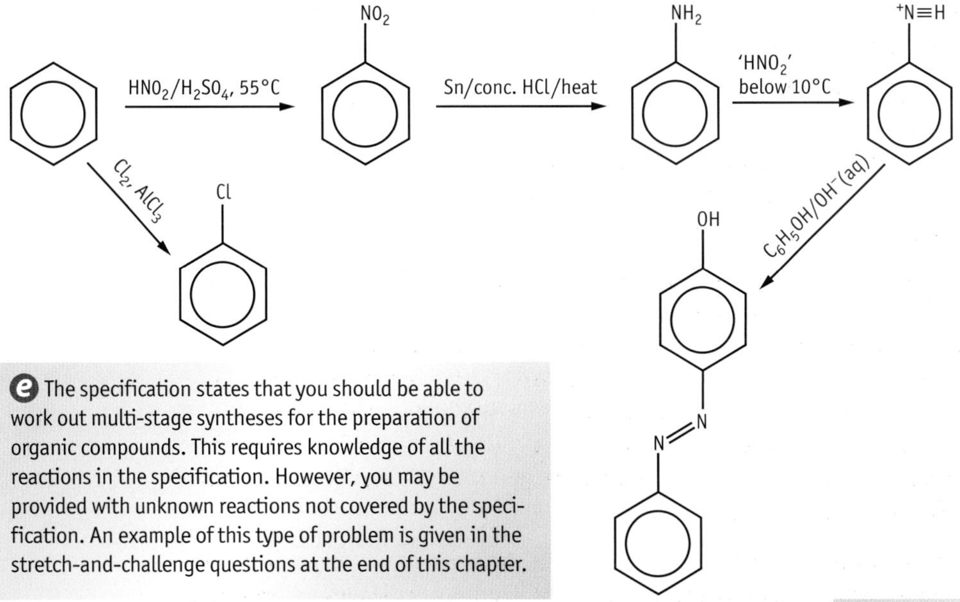

NO_2 NH_2 $^{+}N{\equiv}H$

HNO_2/H_2SO_4, 55°C   Sn/conc. HCl/heat   'HNO_2' below 10°C

Cl_2, $AlCl_3$ Cl

OH $C_6H_5OH/OH^-(aq)$

N=N

e The specification states that you should be able to work out multi-stage syntheses for the preparation of organic compounds. This requires knowledge of all the reactions in the specification. However, you may be provided with unknown reactions not covered by the specification. An example of this type of problem is given in the stretch-and-challenge questions at the end of this chapter.

Chirality in pharmaceutical synthesis

Chiral comes from the Greek 'cheir' — meaning hand. Our right and left hands are mirror images of each other.

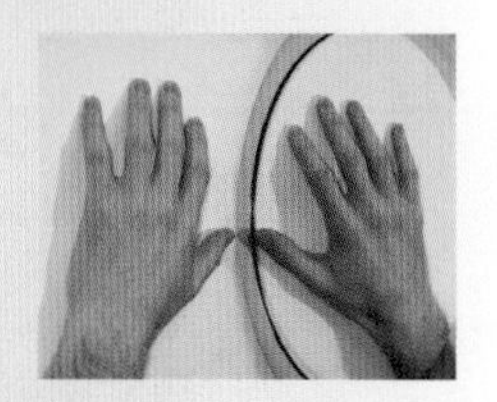

Molecules with a chiral centre (see page 65) exist as one of two possible optical isomers. Optical isomers are so-called because they behave differently in the presence of plane-polarised light. One isomer rotates the plane of plane-polarised light to the right (the D-isomer); the other rotates it to the left (the L-isomer). Each isomer is optically active. The D- and the L-forms have different shapes. Since many biochemical processes require molecules of a specific shape it is easy to see why one isomer is naturally predominant. All naturally occurring α-amino acids occur in the L-form only.

Phenylalanine is a naturally occurring α-amino acid. The derivative 3,4-dihydroxy-L-phenylalanine, L-dopa, is used in the management of Parkinson's disease. The correct orientation of functional groups must be present in order to make a successful interaction with a biological receptor, i.e. L-dopa must have a specific shape.

Models of phenylalanine, L-dopa and dopamine

Dopamine is a chemical messenger (similar to adrenaline). It affects the brain processes that control movement and the ability to experience pain and pleasure. Dopamine plays a crucial role in our mental and physical health. Patients with Parkinson's disease are deficient in dopamine. To help relieve their symptoms they can be treated with L-dopa, which is a neurotransmitter. It is extremely useful as it can cross the blood–brain barrier, which dopamine itself cannot.

The preparation of a single chiral compound in the laboratory is difficult. If an attempt were made to synthesise L-dopa in the laboratory it would probably result in the formation of an equal amount of each optical isomer. The product would probably contain 50% of the isomer that rotates the plane of plane-polarised light to the right (D-dopa) and 50% of the isomer that rotates the plane of plane-polarised light to the left (L-dopa).

Only L-dopa has the correct shape to be active pharmaceutically. It would have to be separated from D-dopa because it is possible that D-dopa could have adverse side effects or make the correct isomer less effective. Therefore, it is vital that the effective isomer is isolated. Separation of optical isomers can be achieved but it is tedious and expensive. Techniques such as chiral chromatography can be used.

e Chromatography is discussed in more detail in Chapter 8.

The usual chromatographic techniques do not distinguish between optical isomers. However, if the column is packed with a solid that contains an active site in the form of either an enzyme or a chiral stationary phase, then one optical isomer will be adsorbed on to the column more effectively than the other and separation can be achieved.

By contrast, synthesis using naturally occurring enzymes or bacteria results in the formation of a single optical isomer. This occurs because active sites have specific shapes that promote specific reactions. Only the pharmaceutically active isomer is produced, which eliminates possible adverse side effects from the other optical isomer and also eliminates the need to separate the pharmacologically active isomer from the inactive isomer.

Single optical isomers can also be produced from starting materials such as L-amino acids or L-sugars or by using a chiral catalyst. Chiral catalysts were first developed in the 1960s by William S. Knowles. In 2001, Knowles, Ryoji Noyori and K. Barry Sharpless were jointly awarded the Nobel Prize for Chemistry for their work on developing chiral catalysts for hydrogenation and oxidation reactions. Their work is now being applied to the development of new drugs, the production of flavourings, sweetening agents and insecticides and to other aspects of material and medical science.

A chiral catalyst is a catalyst that has the correct shape to promote a reaction involving one optical isomer, rather than the other.

Questions

1 Each of the following conversions involves a multi-stage synthesis. In each case, explain how the synthesis could be achieved. Write equations and state the conditions (if any) for each stage in the synthesis.

a propene → propanone

b 3-chloropropan-1-ol → 3-hydroxypropene

c phenylethanone → poly(phenylethene)

2 Compound **X** exhibits stereoisomerism.

$$HO-C_6H_4-CH=CH-CH(CH_3)-CHO$$

X

a (i) Define stereoisomerism.

(ii) State the types of stereoisomerism shown by compound **X** and draw the stereoisomers.

b Compound **X** reacts with a number of different chemicals. Identify the organic product formed when compound **X** reacts with each of the following and state the type of reaction(s) involved in each:

(i) hot $NaBH_4(aq)$

(ii) hot $Na_2Cr_2O_7$ in the presence of H_2SO_4

(iii) Tollens' reagent

(iv) Br_2

3 During the late 1950s and early 1960s, thalidomide, shown below, was given to pregnant women to help combat morning sickness.

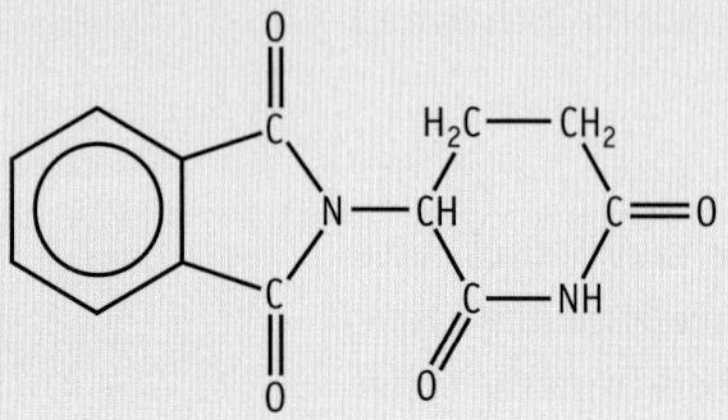

a What is the molecular formula of thalidomide?

b Draw the organic product when thalidomide is treated with:

(i) an excess of hot $NaBH_4(aq)$

(ii) $HCl(aq)$

c Copy the structure of thalidomide and include an asterisk, *, beside the chiral carbon.

d Explain why it is important that pharmaceutical companies ensure that drugs contain only a single optical isomer.

e Outline how pharmaceutical companies manufacture drugs that contain only a single optical isomer.

Summary

You should now be able to:

- identify functional groups and predict the chemistry of unfamiliar molecules
- devise multi-stage synthetic routes for the preparation of organic compounds
- explain the importance of chirality in pharmaceuticals
- explain how modern synthetic techniques can result in the production of a single optical isomer

Additional reading

The conversion of benzene into 4-methylphenylamine might at first sight seem to be straightforward.

$$C_6H_6 \xrightarrow{?} CH_3C_6H_4NH_2$$

From the specification, the synthesis would appear to be achievable in three separate stages:

Stage 1: nitration of benzene

$$C_6H_6 + HNO_3 \xrightarrow[\text{Heat}]{\text{Conc. } H_2SO_4} C_6H_5NO_2 + H_2O$$

Stage 2: reduction of nitrobenzene

$$C_6H_5NO_2 + 6[H] \xrightarrow[\text{Heat}]{\text{Conc. HCl/Sn}} C_6H_5NH_2 + 2H_2O$$

Stage 3: alkylation of phenylamine

$$C_6H_6 + CH_3Cl \xrightarrow[\text{(Halogen carrier)}]{AlCl_3} CH_3C_6H_4NH_2 + HCl$$

Stages 1 and 2 are taken directly from the specification; stage 3 is an adaptation of chlorination of an arene using a halogen carrier. Halogen carriers can also be used to substitute alkyl groups such as $-CH_3$ or $-C_2H_5$ into the ring. If CH_3COCl and benzene are reacted in the presence of a halogen carrier it is also possible to substitute the acyl group, $-COCH_3$ into the ring.

Stages 1 and 2 should present no problems. However, stage 3 will result in the formation of a number of isomers. It is possible to place the $-CH_3$ group in the 2, 3 or 4 position in the ring, so the product should be a mixture of the following compounds:

2-methylphenylamine 3-methylphenylamine 4-methylphenylamine

Logic dictates that the mixture should contain:

2-methylphenylamine : 3-methylphenylamine : 4-methylphenylamine

2 : 2 : 1

In percentage terms:

40 : 40 : 20

Analysis of the product, however, reveals that the mixture contains 59% 2-methylphenylamine, 4% 3-methylphenylamine and 37% 2-methylphenylamine, i.e. the ratio is 59:4:37.

This surprising variation can be explained, but requires a little thought and application. The major difference is at the 3-position on the ring. Statistically, the mixture should contain 40% 3-methylphenylamine, but this has been reduced to 4%. You may recall from Chapter 1, page 15 that one of the lone pairs of electrons on the phenoxide ion can increase the charge density at certain positions on the ring. The same is true for phenylamine —the lone pair of electrons on the nitrogen increases the electron density in the ring, but only at certain positions:

Reactions involving arenes are electrophilic substitution reactions. It is clear from the structures above that an electrophile will be attracted to positions 2, 4 and 6 but is unlikely to be attracted to the 3 or 5 positions. This is why the amount of 3-methylphenylamine in the mixture is lower than expected.

In a monosubstitution reaction the 2- and 6-positions are equivalent, so the mixture ought to contain twice as much 2-methylphenylamine as 4-methylphenylamine. However, this is not the case. It seems that the 4-position is favoured.

The reason for this is subtle and is related to the size of both the electrophile, CH_3^+, and the amine group, $-NH_2$.

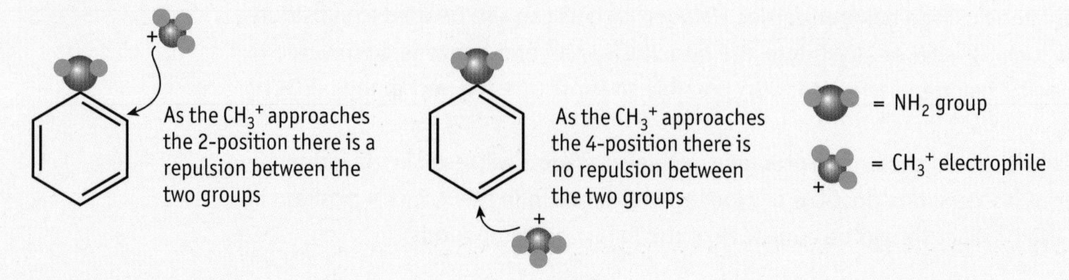

The size of the $-NH_2$ group hinders the approach of the methyl electrophile, CH_3^+, as it approaches the 2- or the 6-positions. This is not the case at the 4-position. This is called the 'steric hindrance effect' and it can be used to protect certain positions from attack.

It follows that if the group on the ring were even larger it would protect the 2- and the 6-positions even more, so that the major product would be 4-methylphenylamine.

This can be achieved by reacting phenylamine with ethanoyl chloride, CH_3COCl, such that:

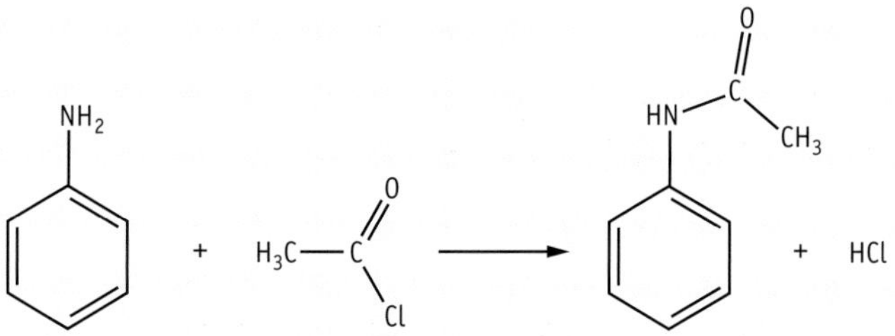

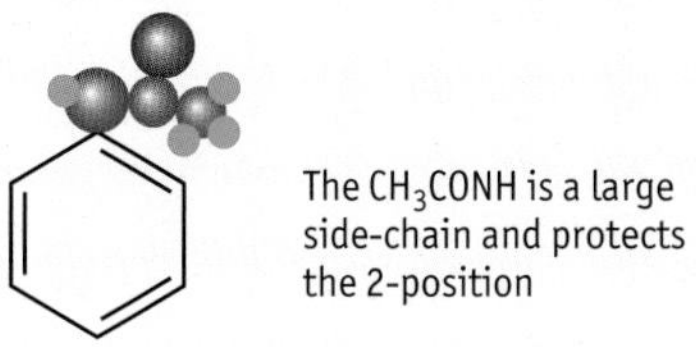

The CH_3CONH is a large side-chain and protects the 2-position

The product, phenylethanamide ($C_6H_5NHCOCH_3$), is then reacted with CH_3Cl in the presence of a halogen carrier and the major product is:

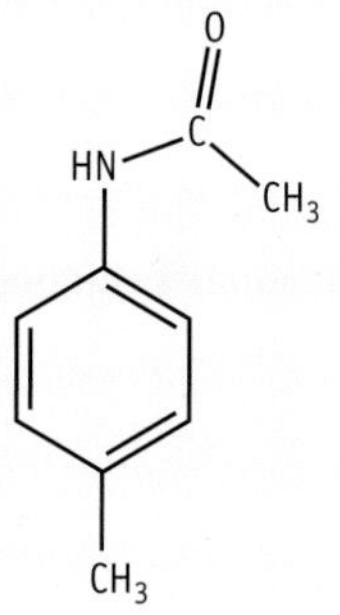

This is then hydrolysed with NaOH(aq) to produce the desired product:

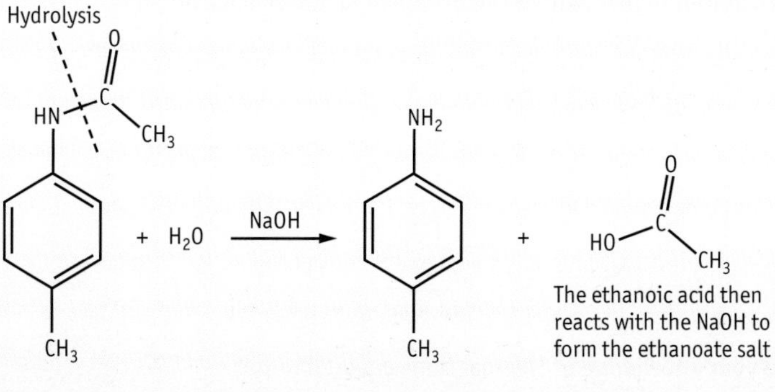

The ethanoic acid then reacts with the NaOH to form the ethanoate salt

In multi-stage syntheses, there are often unexpected complications.

About 100 years ago it was discovered by chance that phenylethanamide ($C_6H_5NHCOCH_3$) (also known as acetanilide) had painkilling properties. Unfortunately, phenylethanamide is toxic and is harmful in even moderate doses. Drug companies recognised the potential and after much research developed the related compound paracetamol.

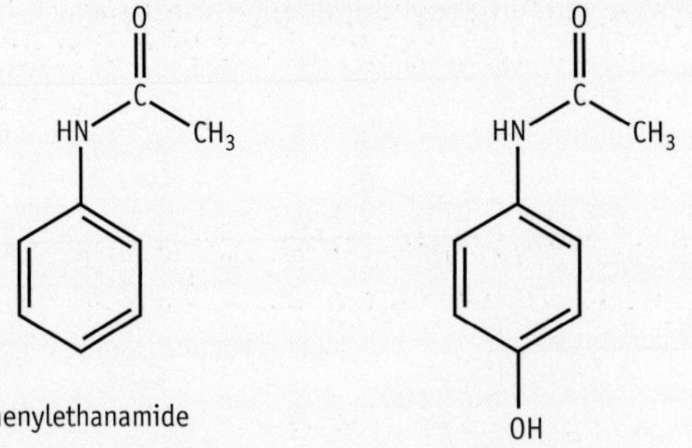

Phenylethanamide

Paracetamol

The conversion of phenylethanamide into paracetamol looks as if it would be easy, but this is not the case. Information on the manufacture of paracetamol can be obtained by a search on the internet. There are many sites that chart its development.

Stretch and challenge

1 Diazonium compounds are useful synthetic chemicals that can be used to prepare a variety of compounds. During the preparation of diazonium compounds the temperature has to be controlled carefully. This is because in the presence of water, if the temperature rises, the diazonium compound decomposes into a phenol, nitrogen and hydrogen chloride.

Use this information and your knowledge of the reactions covered in the specification to suggest how benzene might be converted to 4-methylphenol. Identify any likely impurities or by-products.

2 Carbonyl compounds that contain at least one hydrogen on the carbon atom adjacent to the carbonyl group (CHC=O) can, in the presence of a base such as NaOH, undergo a condensation reaction — for example:

propanone → 4-hydroxy-4-methylpentan-2-one

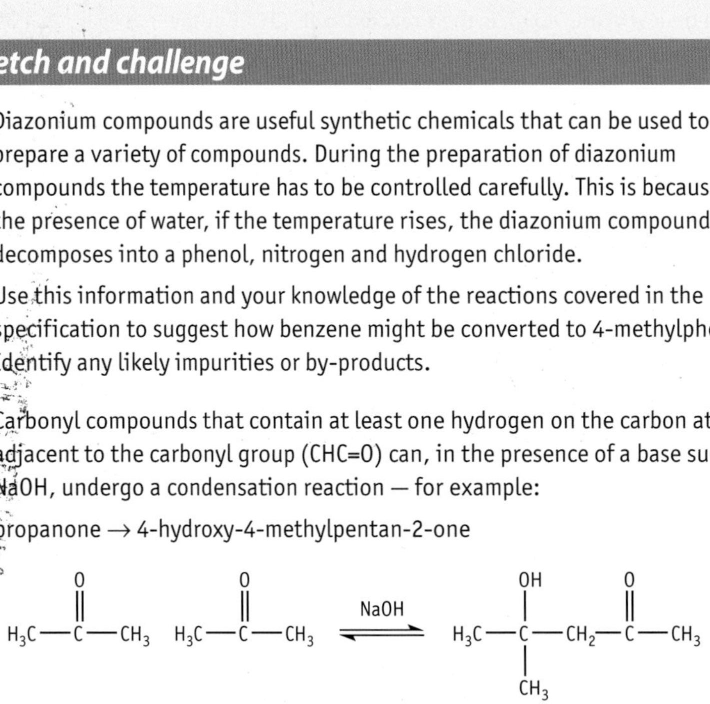

$$H_3C{-}C({=}O){-}CH_3 \quad H_3C{-}C({=}O){-}CH_3 \xrightleftharpoons{NaOH} H_3C{-}C(OH)(CH_3){-}CH_2{-}C({=}O){-}CH_3$$

4-hydroxy-4-methylpentan-2-one

Use this information and your knowledge of the reactions covered in the specification to suggest how but-2-enal, $CH_3CHCHCHO$, could be prepared from ethanal.

Analysis

Chapter 8

Chromatography

You will have met chromatography while studying science at GCSE and you may have used paper chromatography to separate out the components of black ink (Figure 8.1).

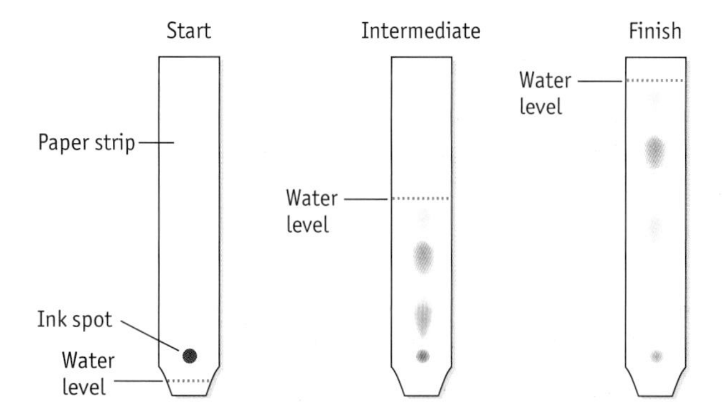

Figure 8.1 *Chromatographic separation of black ink*

Black is a mixture of several coloured components, as can be seen from the colour chart below.

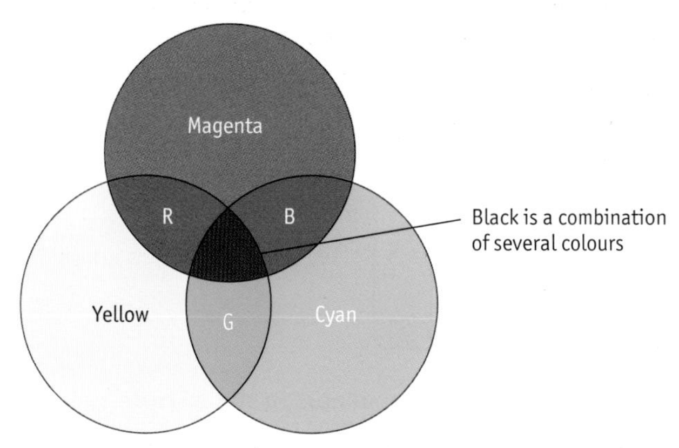

Chromatography involves the small-scale separation of components within a mixture. All types of chromatography contain a **stationary phase** and a **mobile phase**. In the paper chromatography of black ink, the mobile phase is water that moves over the paper and the stationary phase is the water trapped by the cellulose fibres in the paper. Different types of chromatography separate the components of a mixture by either **adsorption** or by **partition**.

e Thin-layer chromatography (TLC) and gas chromatography (GC) are covered in this unit.

In paper chromatography, the separation is achieved by partition. The coloured components of the black ink are not equally soluble in the mobile and stationary phases and an equilibrium is set up between the solvents in the two phases.

$$\frac{\text{concentration of solute in mobile phase}}{\text{concentration of solute in stationary phase}} = \text{constant}$$

In the paper chromatography shown in Figure 8.1, the cyan (light blue) component is more soluble in the mobile phase and the yellow component is more soluble in the stationary phase. This allows separation to be achieved.

Thin-layer chromatography (TLC)

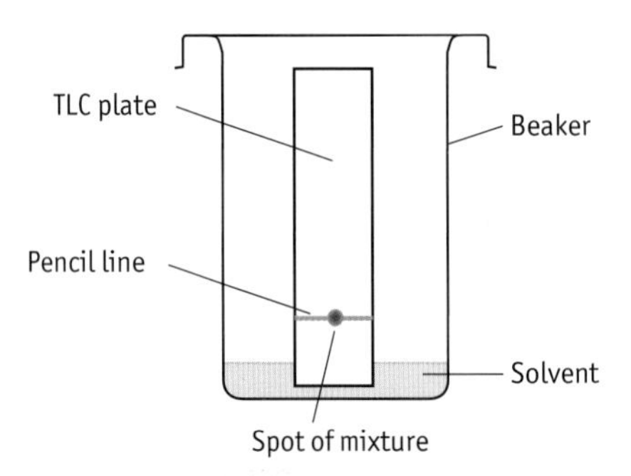

Figure 8.2 *Apparatus for thin-layer chromatography*

The TLC plate is a glass, metal or plastic plate coated with a uniform thin layer of either silica gel, SiO_2, or alumina, Al_2O_3. The mixture to be separated is spotted onto the base line (drawn in pencil) and is allowed to dry. The plate is then placed in a beaker containing the solvent. The beaker is then covered with a watch glass to ensure that the air inside the beaker is saturated with solvent vapour. This stops the solvent evaporating as it rises up the plate.

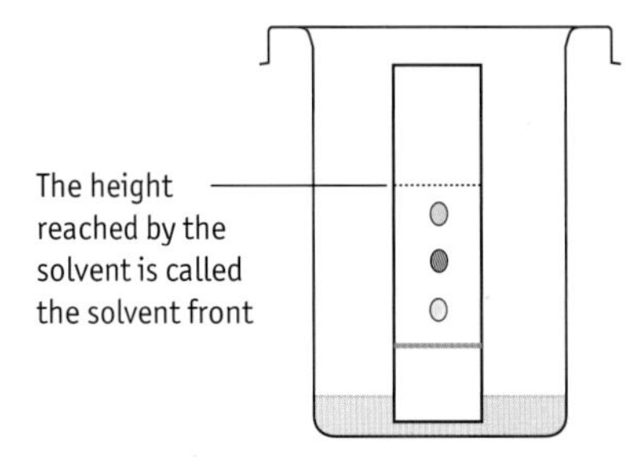

As the solvent travels up the plate, the components in the mixture travel at different rates. The solvent is allowed to rise until it almost reaches the top of the plate, which is then removed and the height of the solvent front is marked in pencil.

The separated components may be identified by using R_f values.

R_f stands for **retardation factor**. It is measured by using the equation:

$$R_f = \frac{\text{distance moved by spot/solute}}{\text{distance moved by solvent}}$$

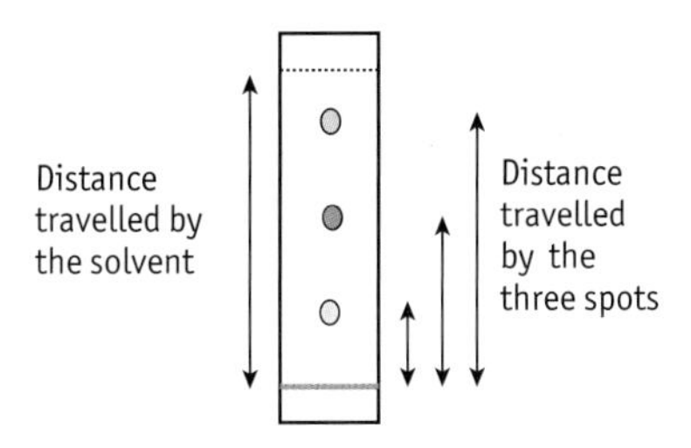

Sometimes the components of mixtures are not visible and the separation cannot be seen until the TLC plate has been developed. There are two simple ways of achieving this:

- Spots can be made visible by adding substances, such as ninhydrin and iodine, that react with the chemicals in the spots to produce coloured compounds.
- An alternative method is to add a substance to the stationary phase that is fluorescent when exposed to ultraviolet light. An unused TLC plate will glow uniformly when exposed to UV light, but the presence of separated substances on the plate prevents those parts of the plate from glowing and they appear as dark spots.

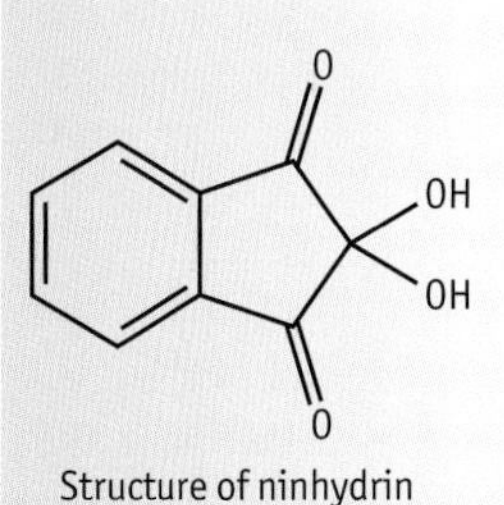

Structure of ninhydrin

The stationary phase in TLC is either silica gel (SiO_2) or alumina (Al_2O_3), both of which have a giant lattice structure. Silica gel is a giant covalent lattice with each silicon atom bonded to four oxygen atoms and each oxygen atom bonded to two silicon atoms. At the surface, the silicon atoms are bonded to –OH groups:

```
   OH        OH        OH        OH    Surface
   |         |         |         |
 —Si———O———Si———O———Si———O———Si—
   |         |         |         |     Giant
   O         O         O         O     covalent
   |         |         |         |     lattice
 —Si———O———Si———O———Si———O———Si—
   |         |         |         |
   O         O         O         O
   |         |         |         |
 —Si———O———Si———O———Si———O———Si—
   |         |         |         |
```

The –OH groups at the surface of the silica gel make it polar. This means that there are hydrogen bonds, dipole–dipole interactions and van der Waals forces between the silica gel and components in the mixture. The aluminium atoms in alumina are also bonded to –OH groups at the surface.

As the solvent rise up the plate, the components in the mixture rise up the plate with it. How fast each component rises depends on:

- how soluble the component is in the solvent
- how much the component 'sticks' or adsorbs to the stationary phase

The latter depends on the polarity of the component and its ability to form hydrogen bonds or van der Waals forces with the surface of either the silica gel or the alumina. TLC achieves separation by adsorption.

Gas chromatography

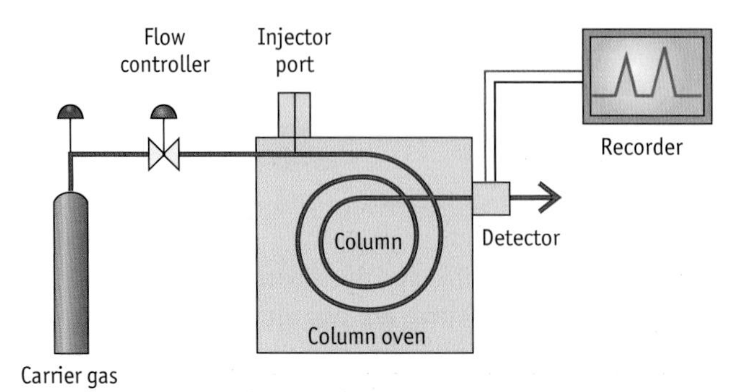

Figure 8.3
Diagram representing the apparatus for gas chromatography

Gas chromatography, GC, involves a sample being vaporised and injected onto the head of the chromatographic column. Like TLC, GC has a mobile phase and a stationary phase. However, GC is different from other forms of chromatography because the sample travels through the column in the gaseous state.

The mobile phase is an unreactive or an inert gas such as nitrogen or one of the noble gases. The gas is known as the carrier gas and flows under pressure through the column. The column itself contains either a liquid or a solid adsorbed onto the surface of an inert solid that makes up the stationary phase. The way in which separation is achieved depends on whether the stationary phase is a liquid or a solid. If the stationary phase is a liquid on the surface of the inert solid, separation depends on the relative solubility of the component in the stationary phase and in the mobile phase. If the stationary phase is a solid on the surface of the inert solid, separation depends on the adsorption of the component onto the stationary phase.

The time between injection of the sample and the emergence of a component from the column is called the **retention time**. For examination purposes, retention time is defined as the time taken from the injection of the sample for each component to leave the column. Retention time depends on volatility of the solute and the relative solubility of the solutes in the mobile and stationary phases.

In a GC analysis, a known volume of gas or liquid is injected into the 'entrance' (head) of the column and the carrier gas sweeps the molecules in the mixture through the column. The rate at which the molecules move through the column depends on either the relative solubility or the adsorption of each component; it also depends on the pressure of the carrier gas and the temperature of the column.

A detector is used to monitor the outlet stream from the column in order to determine the time at which each component reaches the outlet and the amount of each component. The recorder produces a chromatogram showing each component as a separate peak. The substances are identified qualitatively by the order in which they emerge from the column and quantitatively by the area of the peak of each component.

If a mixture of methanal, ethanal, propanal and butanal were passed through a GC the resulting spectrum would look something like this:

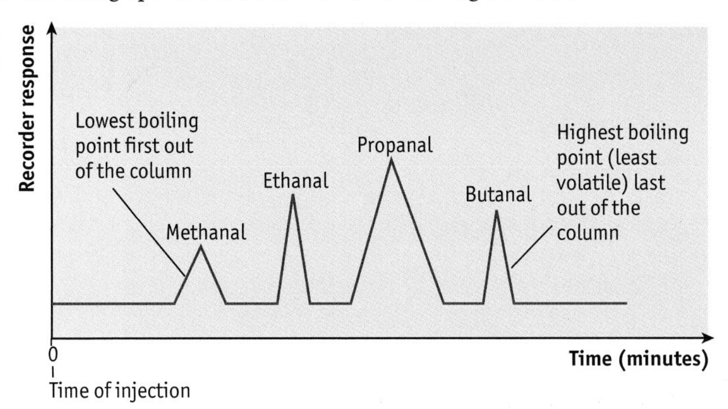

The area under each peak is proportional to the amount of that component. On the graph above, propanal is the most abundant. The relative amounts of the components can be estimated by assuming that the peak is a triangle and by using the equation:

$$\text{area} = \text{base} \times \tfrac{1}{2}\text{height}$$

Analysis by GC has its limitations, in that similar compounds often have similar retention times. It is likely that in a mixture of methanal, ethanal, propanal and butanal some of the peaks would overlap. Identification of unknown compounds is difficult because reference times vary depending on flow rate of the carrier gas and on the temperature of the column — indeed, retention times vary from one GC machine to another. These limitations have been largely overcome by coupling GC with mass spectrometry.

The chromatogram below shows the breath alcohol levels in a sample taken from a suspected drink-driver. This is a complex mixture containing many different chemicals, but ethanol is clearly the most abundant.

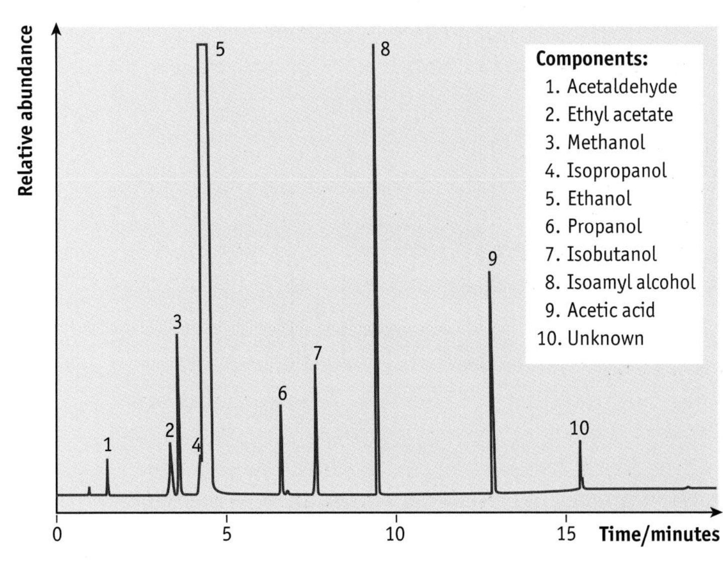

Combining chromatography with mass spectrometry (GC–MS)

The combination of gas chromatography and mass spectrometry provides a powerful analytical tool that is used widely in areas such as forensics, environmental analysis and airport security.

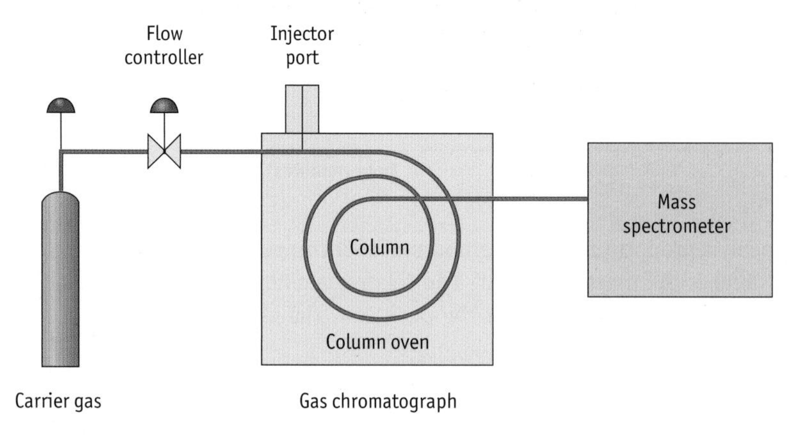

***Figure 8.4** Diagram representing apparatus that combines gas chromatography with mass spectrometry*

In Chapter 10 of the AS textbook, the process of fragmentation was outlined and a number of simple fragment ions were identified in order to determine the relative isotopic masses of elements. For example, if a sample of propane were analysed, its mass spectrum would always look like that in the diagram below:

The mass spectrometer was discussed in Chapters 1 and 10 of the AS textbook. It is important that you understand the basic principles and you are advised to revise these sections.

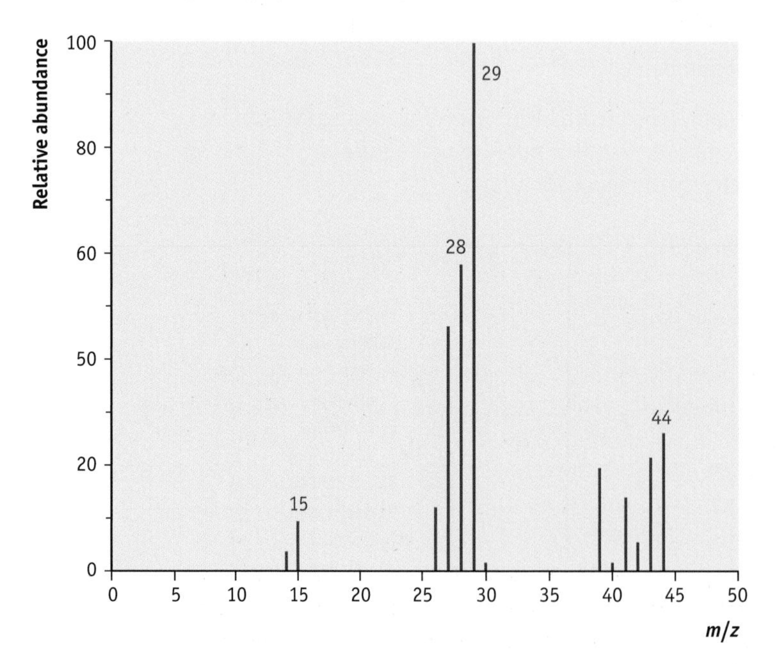

The relative molecular mass of propane is 44 and the peak furthest to the right of the spectrum represents the ion $C_3H_8^+$. This is called the **molecular ion peak**.

$$e^- + H_3C{-}CH_2{-}CH_3(g) \longrightarrow (H_3C{-}CH_2{-}CH_3)^+(g) + 2e^-$$

An electron has been removed from the molecule to produce the molecular ion

Always remember that it is a positive ion that is detected in a mass spectrometer

However, the bombardment of electrons also creates ions of fragments of the molecule that also register on the printout. For example, the peak at 29 on the mass spectrum of propane occurs because a CH_3 unit has been broken from the $CH_3CH_2CH_3$ chain and the residual ion $CH_3CH_2^+$ has been detected.

$$(H_3C{-}CH_2{-}{\vdots}{-}CH_3)^+(g) \longrightarrow (H_3C{-}CH_2)^+(g) + CH_3(g)$$

The C—C bond breaks and produces two fragments

One fragment has + charge and the other is neutral

Similarly, the peak at 15 is due to a CH_3^+ ion:

$$(H_3C{-}CH_2{-}{\vdots}{-}CH_3)^+(g) \longrightarrow CH_3^+(g) + (H_3C{-}CH_2)(g)$$

The C—C bond breaks and produces two fragments

One fragment has + charge and the other is neutral

It is possible to suggest the identity of all the other peaks in the spectrum.

Even in molecules as simple as propane there are a large number of lines in the mass spectrum. The combination of the lines and their size is specific to the individual compound — the fragmentation peaks are said to be a **fingerprint** of the molecule. This fingerprint can be cross-matched against a computer database to identify the compound.

Coupling together GC and MS enables separation of components in a mixture (GC) and identification of each component (MS), which provides a strong analytical tool.

Uses of GC–MS

- **Environmental analysis** — GC–MS can be used to identify most organic compounds. The detection of pesticides and herbicides is both sensitive and effective.
- **Forensic science** — GC–MS can be used to analyse particles from a human body to help link a criminal to a crime and to detect and identify small amounts of narcotics. GC–MS has proved particularly useful in the analysis of fire debris linked to arson.
- **Airport security** — GC–MS is used in airports to detect explosives.
- **Food and drink analysis** — GC–MS is used extensively in the analysis of compounds that occur in food and drink. These include esters, fatty acids and alcohols. It is also used to measure contamination by pesticides and herbicides.
- **Medicine** — GC–MS is used widely in medicine for the analysis of pharmaceuticals and of metabolic compounds labelled with ^{13}C.
- **Astrochemistry** — GC–MS has provided much information from the planets, including Mars and Saturn.

Nuclear magnetic resonance spectroscopy (NMR)

NMR is a powerful tool used to determine the structure of a compound. In this unit you are expected to be able to predict and recognise the high-resolution NMR spectra of simple organic compounds.

If the nucleus of an atom contains an odd number of protons and/or neutrons, the nucleus has a net nuclear spin that can be detected by using radio frequency — for example, ^{1}H and ^{13}C can both be detected.

If the nucleus of an atom contains an even number of protons and an even number of neutrons the nucleus does *not* have a net nuclear spin and *cannot* be detected by using radio frequency — for example, ^{12}C and ^{16}O cannot be detected.

The net spin on a nucleus that has an odd number of protons or neutrons means that the nucleus behaves like a tiny bar magnet and as it spins it generates a magnetic moment. Adjacent nuclei also have magnetic moments. Therefore, each nucleus is affected by neighbouring nuclei. The frequency at which each nucleus absorbs radio waves depends on its environment.

The radio waves are at the low energy end of the electromagnetic spectrum. ^{1}H and ^{13}C both absorb energy in the radio-wave part of the spectrum. However, the frequency of the radio waves absorbed depends on the surrounding atoms, i.e. the exact frequency absorbed depends on the chemical environment. This variation in the frequency absorbed is the key to the determination of structure. It is known as the chemical shift, δ. All absorptions are measured relative to tetramethylsilane (TMS):

CH_3

H_3C—Si—CH_3

CH_3 Chemical shift, δ = 0 ppm

The chemical shift, *δ*, of TMS is set at zero.

TMS is used as a standard because:

- All 12 hydrogens are equivalent and the chemical shift is standardised at δ = 0 ppm. All ^{1}H chemical shifts occur between 0 ppm and 12 ppm.
- All four carbons are equivalent and the chemical shift is standardised at δ = 0 ppm. All ^{13}C chemical shifts occur between 0 ppm and 220 ppm.
- It is chemically inert and does not react with the sample.
- It is volatile and easy to remove at the end of the procedure.
- It absorbs at a higher frequency than other organic compounds. Therefore, its mass spectrum does not overlap with that of the sample.

e All other peaks are measured relative to TMS. You do not have to learn these.

e All relevant absorptions for ^{1}H and ^{13}C NMR are listed in the data booklet, which you will be given in the examination. However, it is essential that you are able to recognise different chemical environments.

^{1}H (proton)-NMR spectroscopy

Consider a molecule of ethanol, C_2H_5OH:

```
      H   H
      |   |
  H — C — C — OH
      |   |
      H   H
```

The six hydrogen atoms are not identical:

- The three hydrogen atoms in the CH_3 group are in the same environment and can be labelled H_a.
- The two hydrogen atoms in the CH_2 group are in the same environment (labelled H_b).
- The hydrogen in the OH group is different from all of the rest (labelled H_c).

```
        Ha   Hb
        |    |
  Ha — C  — C — OHc
        |    |
        Ha   Hb
```

The six hydrogens in ethanol are, therefore, in three different environments. In ^1H (proton) NMR spectroscopy this leads to three different absorptions and hence three different peaks (H_a, H_b and H_c) in the spectrum at different chemical shifts.

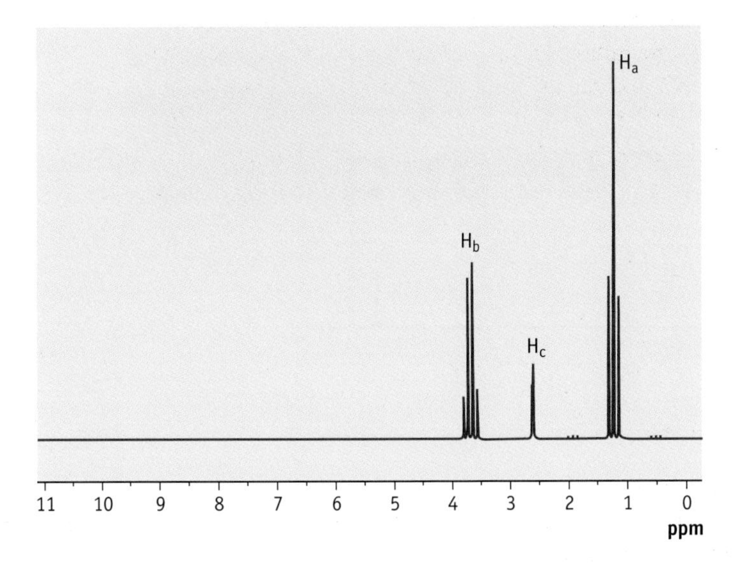

In the NMR spectrum of ethanol there are three peaks. The peaks are of different sizes and may be split.

The relative size of each peak reflects the number of hydrogens in each environment:

- H_a — there are three hydrogens in this environment
- H_b — there are two hydrogens in this environment
- H_c — there is one hydrogen in this environment

It follows that the relative intensity of the peaks H_a, H_b and H_c is 3:2:1.

In the examination, the relative numbers of each type of proton present will either be given as a ratio or presented in the form of an integration trace.

Worked example

Propane, C_3H_8, has two different proton environments, H_a and H_b:

```
     Ha  Hb  Ha
     |   |   |
 Ha—C———C———C—Ha
     |   |   |
     Ha  Hb  Ha
```

How many peaks are there in its NMR spectrum and what is the relative intensity of the peaks?

Answer

There are two different proton environments so there are two peaks.

There are six protons in the environment H_a and two protons in the environment H_b. The simplest ratio is therefore 3:1.

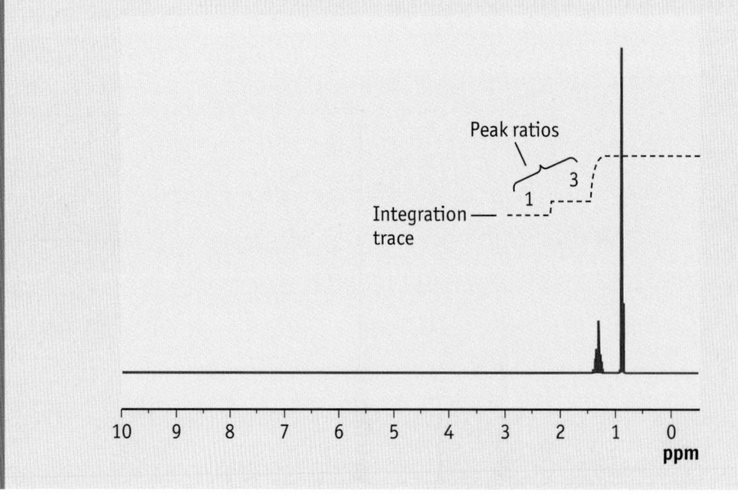

The hydrogens attached to one carbon atom influence the hydrogens on adjacent carbon atoms. This is called **spin–spin coupling**. The easiest way to predict the splitting pattern is to count the number of hydrogens on the adjacent carbon atoms and then use the **'n + 1'** rule, where n is the number of hydrogens on the adjacent carbon atoms.

In the NMR spectrum of ethanol, each of the peaks will be split differently.

- H_a is next to two hydrogens in CH_2 and hence the peak is split into (2 + 1) — a **triplet**.
- H_b is next to three hydrogens in CH_3 and hence the peak is split into (3 + 1) — a **quartet**.
- H_c is not attached to a carbon atom and hence does not undergo spin–spin coupling. It is, therefore, a **singlet**.

We would, therefore, expect the high-resolution NMR spectrum of ethanol to have three peaks of relative intensity 3:2:1 and split into a triplet, a quartet and a singlet.

The exact position (the chemical shift) of each peak can be obtained from the data sheet shown in Figure 8.5.

Type of proton	Chemical shift, δ/ppm
$R-CH_3$	0.7–1.6
$N-H$ NH_2 $R-OH$	1.0–5.5*
$R-CH_2-R$	1.2–1.4
R_3CH	1.6–2.0
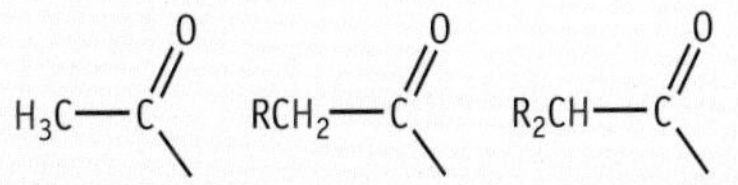 $H_3C-C(=O)-$ $RCH_2-C(=O)-$ $R_2CH-C(=O)-$	2.0–2.9
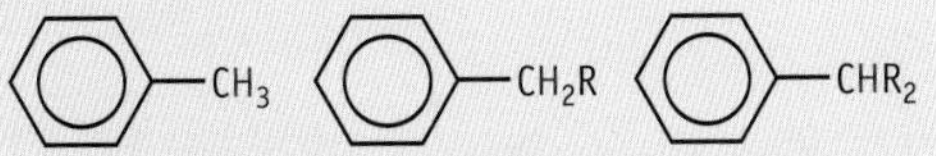 $C_6H_5-CH_3$ $C_6H_5-CH_2R$ $C_6H_5-CHR_2$	2.3–2.7
$N-CH_3$ $N-CH_2R$ $N-CHR_2$	2.3–2.9
$O-CH_3$ $O-CH_2R$ $O-CHR_2$	3.3–4 .3
Br or $Cl-CH_3$ Br or $Cl-CH_2R$ Br or $Cl-CHR_2$	3.0–4 .2
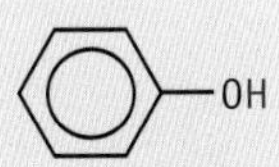 C_6H_5-OH	4.5–10.0*
$-CH=CH-$	4.5–6.0
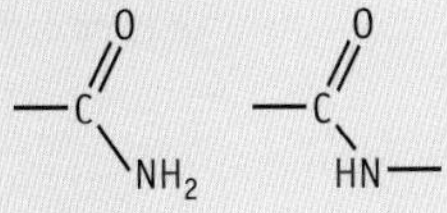 $-C(=O)-NH_2$ $-C(=O)-HN-$	5.0–12.0*
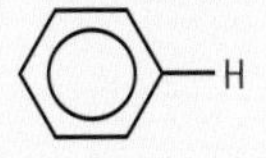 C_6H_5-H	6.5–8.0
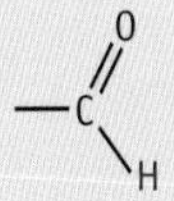 $-C(=O)-H$	9.0–10.0
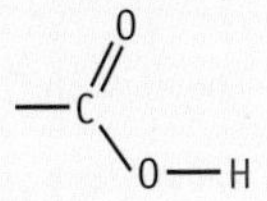 $-C(=O)-O-H$	11.0–12.0*

* OH and NH chemical shifts are very variable and may be sometimes outside these limits. Signals for OH and NH are usually singlets and are not split.

***Figure 8.5** Data sheet. Chemical shifts shown are based on average values and can vary slightly depending on the conditions used.*

In the ^1H (proton) NMR of ethanol:

- The CH_3 (H_a) has a chemical shift between 0.7 ppm and 1.6 ppm.
- The CH_2 (H_b) has a chemical shift between 2.0 ppm and 4.3 ppm.
- The OH (H_c) has a chemical shift between 1.0 ppm and 5.5 ppm.

Use of D_2O

The O–**H** and the N–**H** peaks have chemical shifts that differ between compounds and sometimes lie outside the range given in the data sheet. Therefore, they are difficult to assign. When alcohols, carboxylic acids, amines or amides are dissolved in water there is a rapid exchange between the protons in the functional groups (**labile protons**) and the protons in the water — for example:

$$H_a{-}C(H_a)_2{-}C(H_b)_2{-}OH_c + H{-}O{-}H \rightleftharpoons H_a{-}C(H_a)_2{-}C(H_b)_2{-}OH + H_c{-}O{-}H$$

If water is replaced by deuterated water, ^2H_2O, the peak at H_c disappears. The H_c proton is replaced by deuterium, ^2H, which does not absorb in this region of the spectrum.

Deuterated water, ^2H_2O can be written as D_2O. Ethanol dissolved in deuterated water can be represented as:

$$H_a{-}C(H_a)_2{-}C(H_b)_2{-}OH_c + D{-}O{-}D \rightleftharpoons H_a{-}C(H_a)_2{-}C(H_b)_2{-}OD + H_c{-}O{-}D$$

This is no longer detected and the peak for H_c disappears

The use of ^2H_2O to identify labile protons is a valuable technique in proton (^1H) NMR.

When samples are prepared for NMR it may be necessary to dissolve them in a suitable solvent. Solvents containing protons are unsuitable because the protons would be detected and interfere with the spectrum. This is overcome by using a deuterated solvent, such as $CDCl_3$.

e You will be expected to be able to predict the ^1H NMR of simple organic compounds.

Worked example 1

Draw the structure of propan-1-ol and determine:

- the number of different H environments
- the relative ratio of the peaks
- the splitting of each peak
- the chemical shift of each peak

Answer

	Number of protons	Ratio	Splitting	Chemical shift, δ ppm
$H_a{-}C(H_a)_2{-}C(H_b)_2{-}C(H_c)_2{-}OH_d$	Four peaks H_a, H_b, H_c, H_d	H_a, H_b, H_c, H_d 3 : 2 : 2 : 1	H_a — triplet H_b — sextet H_c — triplet H_d — singlet	H_a 0.7–1.6 H_b 1.2–1.4 H_c 3.3–4.3 H_d 1.0–5.5* *OH peaks are variable and should be confirmed by using D_2O as the solvent

Worked example 2

Compound **A** has the empirical formula C_2H_4O and a molar mass of 88. The ^{1}H-NMR spectrum of compound **A** is shown below.

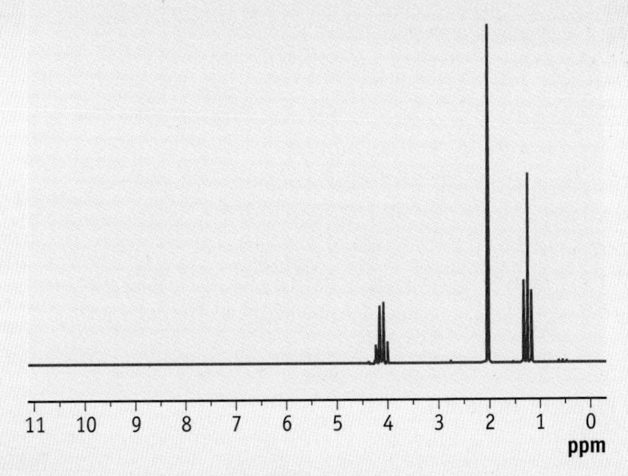

Deduce the identity of compound **A**.

Answer

Molar mass of $C_2H_4O = 24 + 4 + 16 = 44\,g\,mol^{-1}$

Therefore the molecular formula is $C_4H_8O_2$.

The NMR spectrum shows three peaks, so there are three different proton environments.

The ratio between the peaks is 3:3:2. This adds up to 8, which indicates that the molecule is likely to contain two CH_3 groups and a CH_2.

The peak at 1.2 is a triplet, indicating that the proton is next to a CH_2.The peak at 4.2 is a quartet, indicating that the proton is next to a CH_3.This indicates the presence of a $CH_3–CH_2–$ grouping.

The peak at 2.1 is a singlet suggesting that it is a $CH_3–$ and that the adjacent carbon has no hydrogen. The data sheet confirms that $CH_3–$ bonded to a carbonyl (C=O) absorbs in the region 2.0–2.9 ppm.

Compound **A** is known to contain $CH_3–CH_2–$ and

$$H_3C—C(=O)—$$

The molecular formula is $C_4H_8O_2$, so the extra oxygen is likely to be part of an ester group:

$$H_3C—C(=O)—O—$$

Compound A is ethyl ethanoate:

$$H_3C—C(=O)—O—CH_2—CH_3$$

ⓔ There are a number of ways to approach this; one method is shown here.

Carbon-13, ^{13}C, NMR spectroscopy

Carbon-12, ^{12}C, is the most abundant isotope of carbon. It does not have spin because it has an even number of protons and an even number of neutrons. The second isotope of carbon, carbon-13, ^{13}C, can be detected using low-energy radio waves and it is possible to generate ^{13}C-NMR spectra. The ^{13}C atom is about 6000 times more difficult to detect than ^{1}H atoms because of its low abundance (only about 1.1% of naturally occurring carbon is ^{13}C) and its low magnetic moment. Interaction between adjacent ^{13}C atoms is unlikely because of their low abundance. ^{13}C atoms do interact with adjacent protons but these interactions are removed by decoupling and all absorptions appear as singlets.

All ^{13}C-NMR spectra used will be decoupled and all peaks will appear as singlets.

The information obtained from ^{13}C-NMR is entirely based on the chemical shift values.

Some chemical shift values are given in Table 8.1.

Chemical shift values are provided on the data sheet that is supplied in the examination. You do not need to learn them.

***Table 8.1** Some ^{13}C chemical shift values. These are average values*

Type of carbon	Chemical shift, δ/ppm
C—C (alkanes)	5–55
C—C(=O)—	20–30
C—Cl or C—Br	30–70
C—N (amines)	35–60
C—OH	50–70
C=C (alkenes)	115–140
Aromatic (benzene ring of six C)	110–165
Carbonyl (ester, carboxylic acid, amide) —C(=O)O, —C(=O)N	160–185
Carbonyl (aldehyde, ketone) —C(=O)H, —C(=O)C	190–220

Ethanol has two carbon atoms, C_1 and C_2:

$$H-\underset{H}{\overset{H}{C_1}}-\underset{H}{\overset{H}{C_2}}-OH$$

- C_1 is joined to a CH_2 and is part of an alkyl group. The chemical shift should be between 5 ppm and 55 ppm.
- C_2 is joined to an alcohol OH. The chemical shift should be between 50 ppm and 70 ppm.

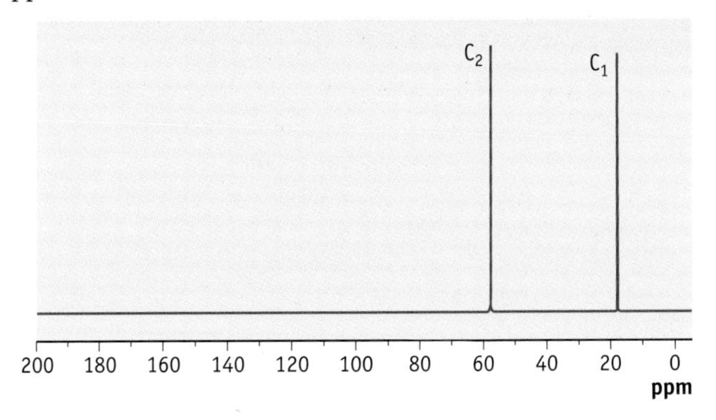

The key to interpreting ^{13}C-NMR spectra is to identify the number of different carbon environments and then to match them with the groups in the data sheet.

Worked example 1

Determine the number of carbon environments in propanoic acid. For each environment, predict the chemical shift (the δ value).

Answer

There are three different C environments – C_1, C_2 and C_3

The ^{13}C-NMR spectrum should therefore contain three peaks:

- C_1 is part of an alkyl group and should therefore have a δ value between 5 ppm and 55 ppm.
- C_2 is next to a carbonyl group and should therefore have a δ value between 20 ppm and 30 ppm.
- C_2 is part of a carboxylic acid group and should therefore have a δ value between 160 ppm and 185 ppm.

The ^{13}C spectrum of propanoic acid is shown below.

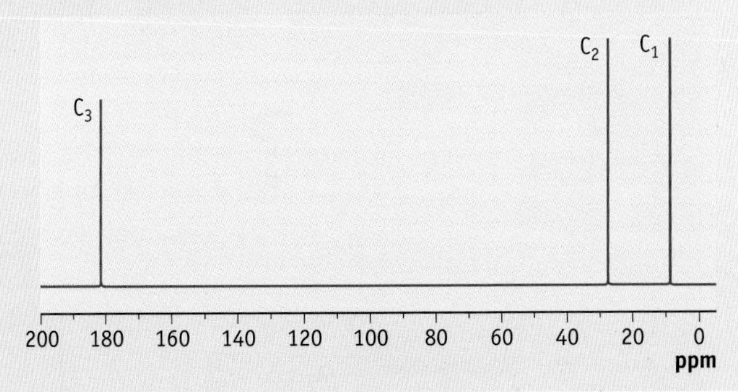

Worked example 2

Determine the number of carbon environments in propan-2-ol. For each environment, predict the chemical shift (the δ value).

Answer

```
      H  OH  H                                  H  OH  H
      |   |  |                                  |   |   |
  H — C — C — C — H     There are           H — C1 — C2 — C1 — H
      |   |  |          two different           |   |   |
      H   H  H          C environments —        H   H   H
                        C1 and C2
```

The ^{13}C-NMR spectrum should therefore contain two peaks:

- C_2 is next to an OH group and should therefore have a δ value between 50 ppm and 70 ppm.
- C_1 is part of an alkyl group and should therefore have δ value between 5 ppm and 55 ppm. However, because the adjacent carbon is bonded to an OH the *δ* value will be towards the high end of the range.

The ^{13}C spectrum of propan-2-ol is shown below.

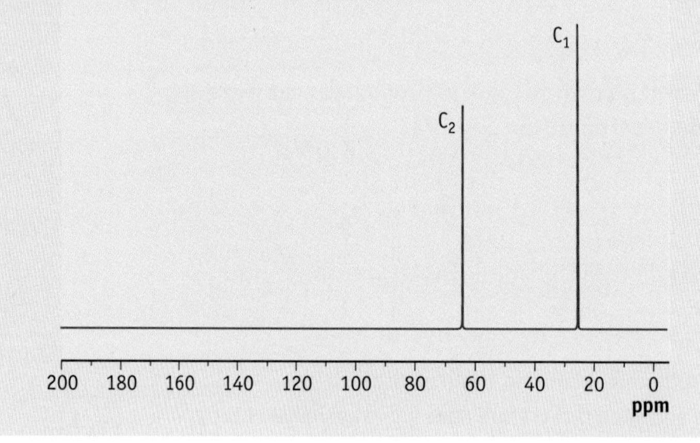

Combined techniques

Analytical chemistry is rather like detective work — pieces of evidence are gathered from different places. Usually no one piece of evidence is conclusive but when the pieces of evidence are slotted together, the combination is definitive. Gathering together the evidence is a bit like doing a jigsaw.

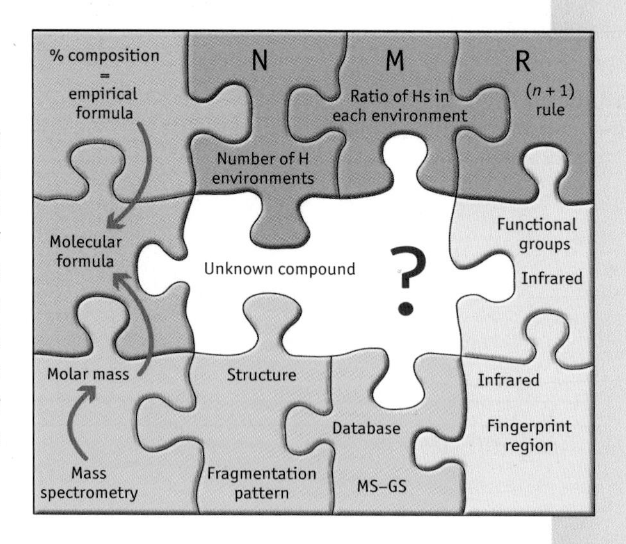

Worked example 1

Compound **X** has the following percentage composition by mass: carbon 54.5%; hydrogen 9.1%; oxygen 36.4%.

Use this information and the spectra below to identify compound **X**.

Mass spectrum

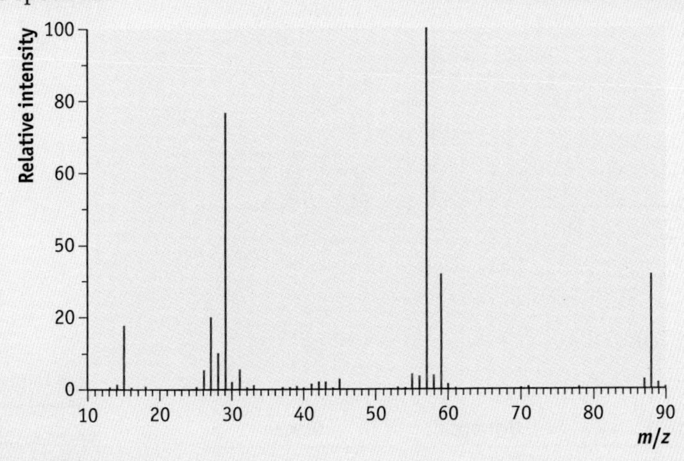

Infrared spectrum

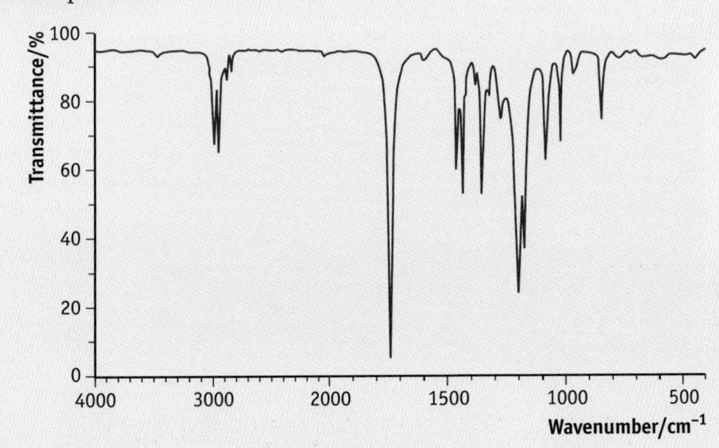

^{13}C-NMR spectrum

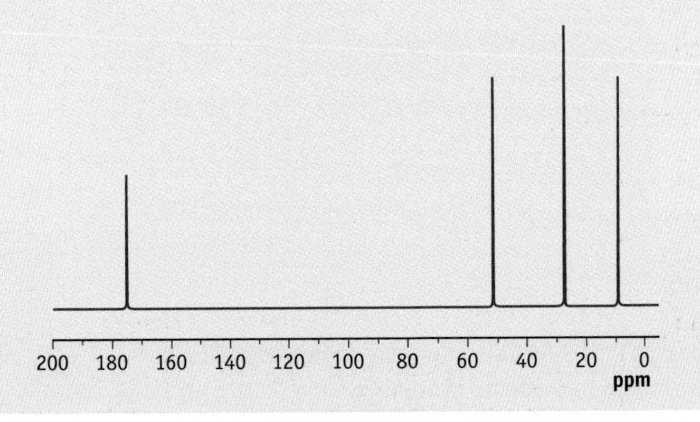

^{1}H-NMR spectrum

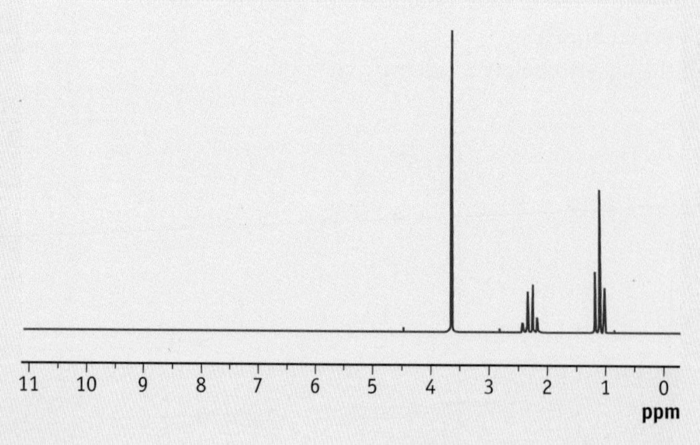

Answer

Empirical formula:

Method	Carbon	Hydrogen	Oxygen
Percentage	54.5	9.1	36.4
Divide by relative molecular mass	54.5/12.0 = 4.54	9.1/1.0 = 9.1	36.4/16.0 = 2.28
Divide by smallest	4.54/2.28 = 1.99 ≈ 2	9.1/2.28 = 3.99 ≈ 4	2.28/2.28 = 1

The simplest ratio of C:H:O is 2;4:1. The empirical formula is C_2H_4O.
empirical formula mass = 24.0 + 4.0 + 16.0 = 44.0

Molecular formula:

mass peak in the mass spectrum = 88
empirical formula = C_2H_4O, empirical formula mass = 44.0
Hence, the molecular formula is $C_4H_8O_2$.

Information from the spectra:

(1) The peak at 175 in the ^{13}C-NMR spectrum is due to

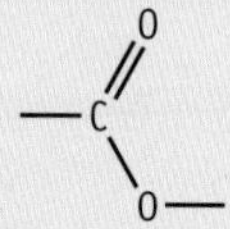

This suggests that compound **X** is either an ester or a carboxylic acid.

(2) The infrared spectrum has a peak at about 1700 cm^{-1} confirming a C=O group. However, there is no broad absorption in the range 2500–3300 cm^{-1}. This indicates that compound **X** is not a carboxylic acid. Hence, it is probably an ester.

(3) The molecular formula is $C_4H_8O_2$ and the ^{13}C-NMR shows four peaks. This confirms that all four carbons are in different environments.

(4) The ^{1}H-NMR spectrum has only three peaks. This means that one of the carbon atoms has no hydrogens attached. There is a familiar pattern of a triplet and a quartet indicating a CH_3 next to a CH_2. This accounts for five of the eight hydrogens. The third peak in the ^{1}H-NMR spectrum is a singlet, which suggests that there is a CH_3 next to a carbon with no hydrogens.

e There is no set way of solving this problem — there are a number of different approaches. However, it makes sense to use the percentage composition data to calculate the empirical formula, and then to link this with the mass spectrum to find the molecular formula.

There is now enough information to conclude that compound **X** is methyl propanoate:

$$H_3C-CH_2-C(=O)-O-CH_3$$

- In the mass spectrum, the fragment ions at 29 and 57 confirm the $CH_3CH_2^+$ and the $CH_3CH_2CO^+$ ions respectively.
- The absorption at about 3000 cm^{-1} in the infrared spectrum confirms C–H.
- The chemical shifts in the ^{13}C-NMR spectrum can be used to confirm the identity of the carbons.

e This is not the only way of deducing the identity of a compound — there are many others

Worked example 2

a An unknown compound **Y** reacted with 2,4-dinitrophenylhydrazine to give an orange precipitate. When treated with Tollens' reagent, compound **Y** did not produce a silver mirror. The mass spectrum of compound **Y** is shown below.

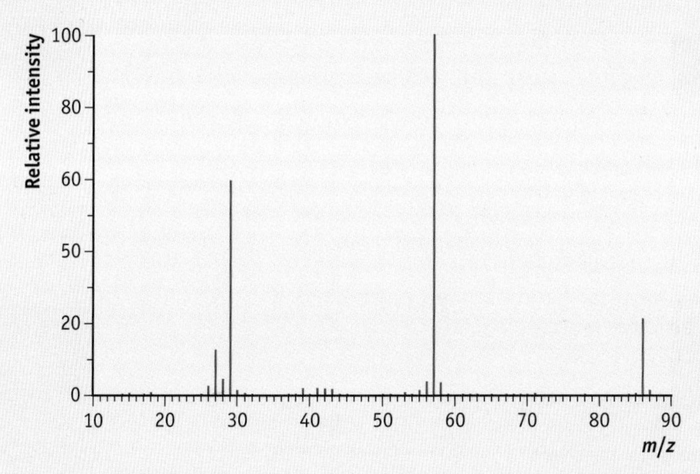

(i) Identify the functional group present in compound **Y**. Explain your reasoning.

(ii) Deduce the molecular formula of compound **Y**.

(iii) Draw, and name, all possible isomers of compound **Y**.

b (i) The ^{13}C-NMR of compound Y contains three peaks. Identify compound **Y**.

(ii) Explain how ^{1}H-NMR spectroscopy could be used to confirm the identity of compound **Y**.

Answer

a (i) The reaction with 2,4-DNPH indicates that compound **Y** contains a carbonyl group, C=O. It is, therefore, either an aldehyde or a ketone. The lack of reaction with Tollens' reagent shows that compound **Y** is not an aldehyde; it is therefore a ketone.

(ii) The mass spectrum of compound **Y** has a molecular ion with $m/z = 86$.

mass of C=O = 12 + 16 = 28

mass of the rest of the molecule = (86 – 28) = 58

mass of four carbon atoms = 48

Hence, there are ten hydrogen atoms.

Compound **Y** has the formula $C_4H_{10}C{=}O$; its molecular formula is $C_5H_{10}O$.

(iii)

Pentan-2-one: $H_3C{-}CH_2{-}CH_2{-}C({=}O){-}CH_3$

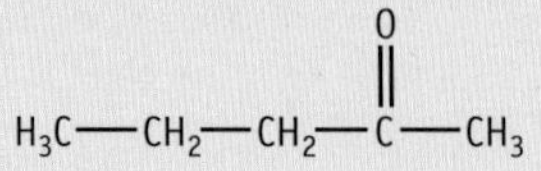

Pentan-3-one: $H_3C{-}CH_2{-}C({=}O){-}CH_2{-}CH_3$

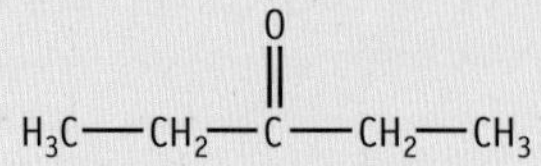

3-methylbutan-2-one: $H_3C{-}CH(CH_3){-}C({=}O){-}CH_3$

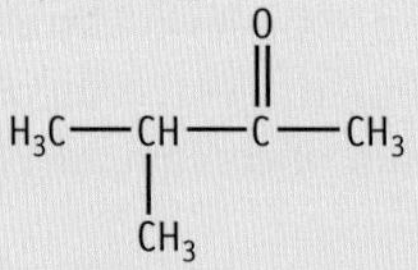

b (i) Compound **Y** has three peaks in the ^{13}C-NMR spectrum. Therefore, although it has five carbons, only three are in different environments:

Pentan-2-one has five different C environments: $H_3C{-}CH_2{-}CH_2{-}C({=}O){-}CH_3$ (carbons numbered 5, 4, 3, 2, 1)

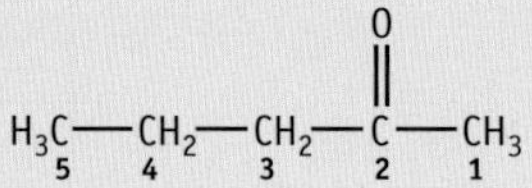

Pentan-3-one is symmetrical and only has three different C environments: $H_3C{-}CH_2{-}C({=}O){-}CH_2{-}CH_3$ (carbons numbered 1, 2, 3, 2, 1)

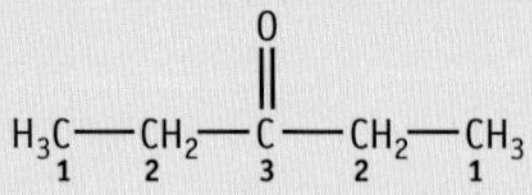

The two methyls attached to the CH are in the same environment and therefore 3-methylbutan-2-one has four different C environments: $H_3C{-}CH({}_1CH_2){-}C({=}O){-}CH_3$ (carbons numbered 1, 2, 3, 4; branch carbon numbered 1)

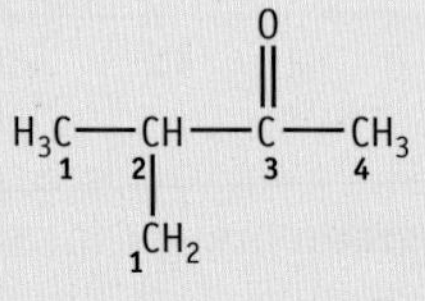

(ii) The ^1H-NMR spectrum of pentan-3-one would contain two peaks only:

Group	Chemical shift, δ/ppm	Peak ratio	Splitting
CH_3	0.7–1.6	3	Triplet
CH_2	2.0–2.9	2	Quartet

(Pentan-2-one would have four peaks in the ^1H-NMR spectrum, comprising a singlet, two triplets and a sextet. 3-methylbutan-2-one would have three peaks — a singlet, a doublet and a septet.)

e You should be able to assign chemical shift values and peak ratio numbers to each peak in an NMR spectrum.

Questions

1 Define each of the following terms:

a mobile phase

b stationary phase

c adsorption

d partition

e R_f value

f retention time

2 Explain how TLC could be used to identify the individual amino acids present in a mixture of amino acids.

3 State four uses of GC–MS.

4 The structure of cyclopentanone, C_5H_8O, is shown below:

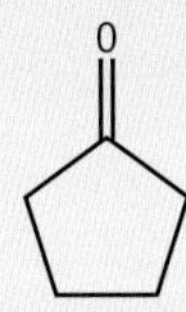

a Explain why the ^{1}H-NMR spectrum has only two peaks whereas the ^{13}C-NMR spectrum has three peaks.

b Predict the chemical shift (δ value) for each peak in the ^{13}C-NMR spectrum.

c Predict the chemical shift (δ values) the splitting pattern and the relative peak area for each peak in the ^{1}H-NMR spectrum.

5 Lactic acid (2-hydroxypropanoic acid, $CH_3CH(OH)COOH$) was analysed using a combination of analytical techniques.

a Calculate the percentage composition by mass of lactic acid:

Carbon

Hydrogen

Oxygen

b What would you expect the following techniques to reveal about lactic acid?

(i) infrared spectroscopy

(ii) mass spectrometry

6 Identify compound **X** from the spectra.

Mass spectrum

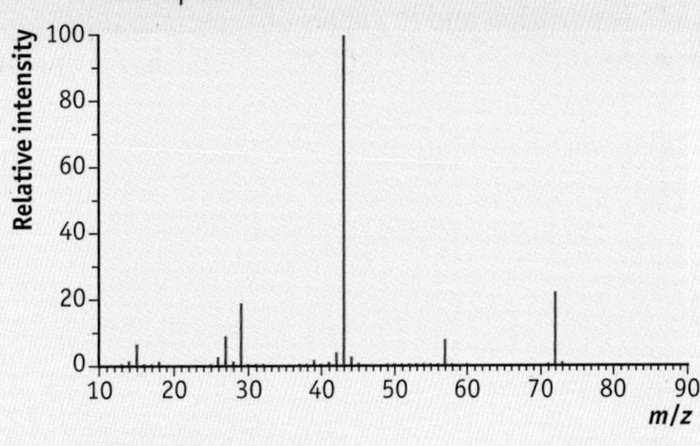

^{13}C-NMR spectrum

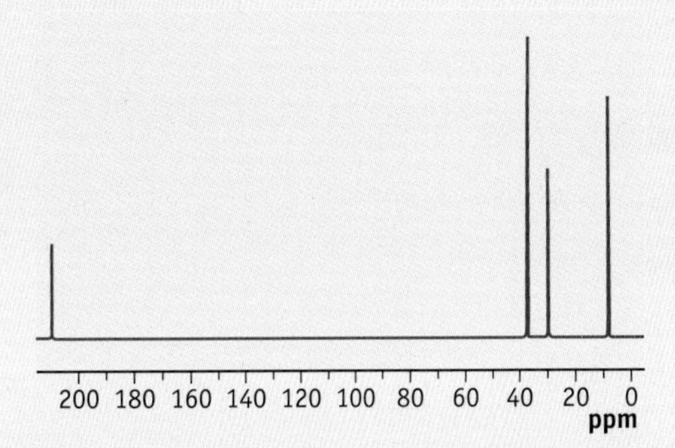

Infrared spectrum

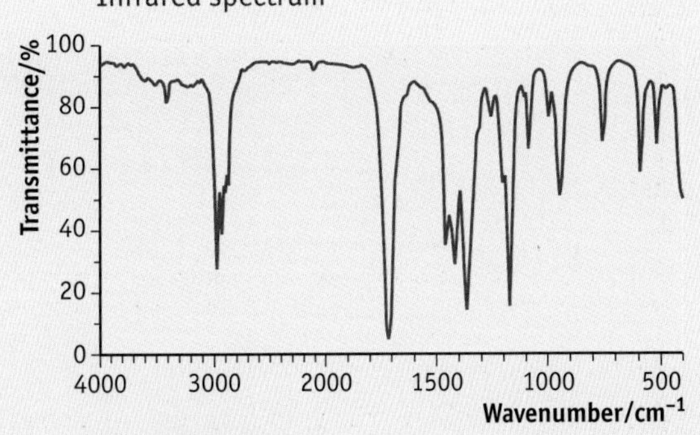

^{1}H-NMR spectrum

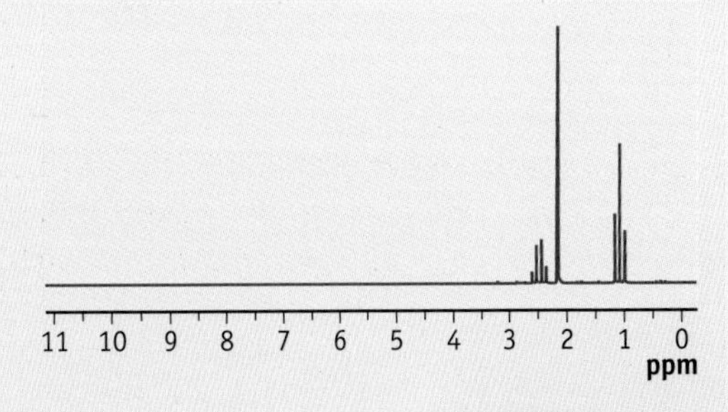

Summary

You should now be able to:

- describe chromatography in terms of mobile and stationary phases, adsorption or partition and R_f values or retention times
- describe the use of GC–MS as a powerful analytical tool
- describe and interpret ^{1}H-NMR spectra
- describe and interpret ^{13}C-NMR spectra
- use a combination of analytical techniques to determine the identity of an unknown compound.

Additional reading

Nuclear magnetic resonance

Today NMR can be used to identify the network of atoms and bonds in complex molecules. It is not limited to a single state of matter; NMR is used on solids, liquid crystals, liquids and solutions. Nuclear magnetic imaging is becoming increasingly important as a diagnostic tool in medicine.

Protons and neutrons (and electrons) all spin about an axis. As they spin, they induce magnetic moments. A nucleus contains both protons and neutrons. If it has an odd number of protons and/or neutrons it has a net spin and behaves like a tiny bar magnet. ^{1}H, ^{13}C, ^{19}F and ^{31}P each has a net spin value, but ^{12}C and ^{16}O do not.

If nuclei are subjected to a strong external magnetic field, those nuclei that behave as tiny bar magnets line up in parallel with the external field:

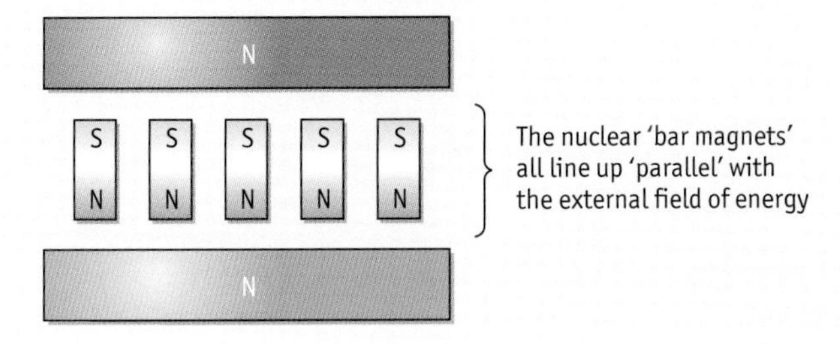

If electromagnetic energy, in the radio frequency, is applied to the sample some of the nuclei 'flip' from the parallel to anti-parallel position:

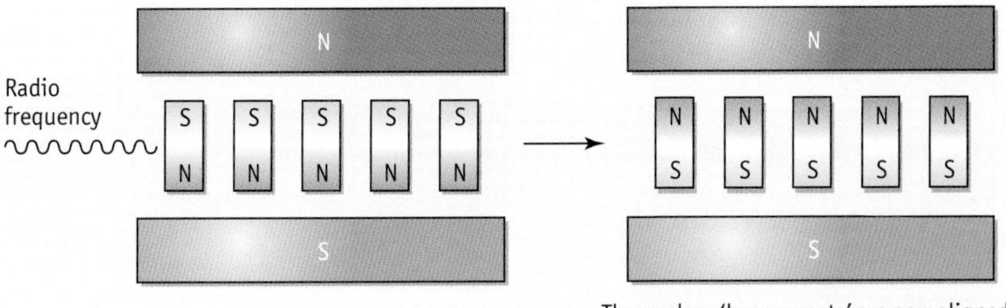

The nuclear 'bar magnets' are now aligned against the external feld, which is at higher energy, and are now anti-parallel

The frequency that brings about this 'flipping' is called the **resonance frequency**. Different nuclei flip at different radio frequencies because the nuclei are **shielded** by the electrons, which also have spin and therefore magnetic moments. The distribution of electrons around the nucleus affects the radio frequency required for resonance; hence the resonance frequency of any particular nucleus is affected by its chemical environment. This can be used to identify different ^{1}H and ^{13}C environments. The extent of shielding by the electrons depends on the electron density around the proton. The electron density of the hydroxyl proton in the OH group is decreased because the oxygen atom has high electronegativity and draws the electron cloud towards itself, which decreases the electron density around the proton. The oxygen atom is said to be 'de-shielding'. Molecules that contain π-bonding, particularly if delocalisation is possible, are also de-shielded.

Spin–spin coupling: the $(n + 1)$ rule

If we look more closely at ethanol we see that the CH_3 (H_a) and the CH_2 (H_b) signals are not single peaks but are split. For simple molecules the splitting can be deduced by the so-called $(n + 1)$ rule — H_a is split into a triplet by the two adjacent hydrogens in the CH_2.

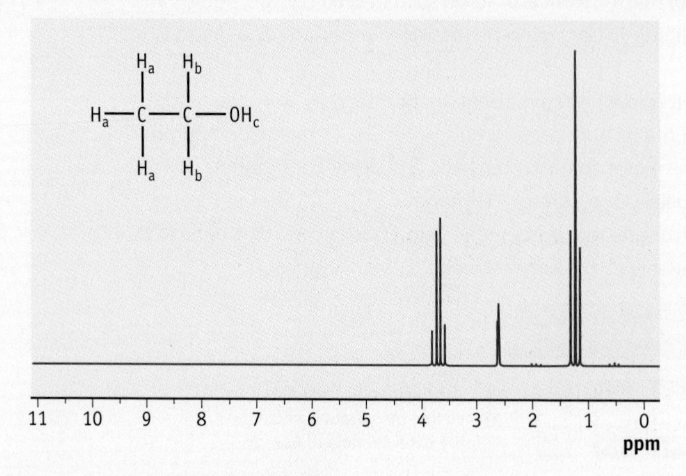

The three protons (H_a) in the CH_3 are in the same environment. Each one 'feels' three possible effects for the two (H_b) protons in the CH_2 group. The CH_2 protons can be either parallel or anti-parallel, so that the (H_a) protons feel three net fields from the CH_2. The three fields are:

- both (H_b) protons parallel with the external field
- one (H_b) proton parallel with and one (H_b) proton anti-parallel with the external field
- both (H_b) protons anti-parallel with the external field

There are two possible ways of achieving the second arrangement. This makes it twice as likely as the other two possibilities, so this sub-peak has twice the intensity of the other two sub-peaks.

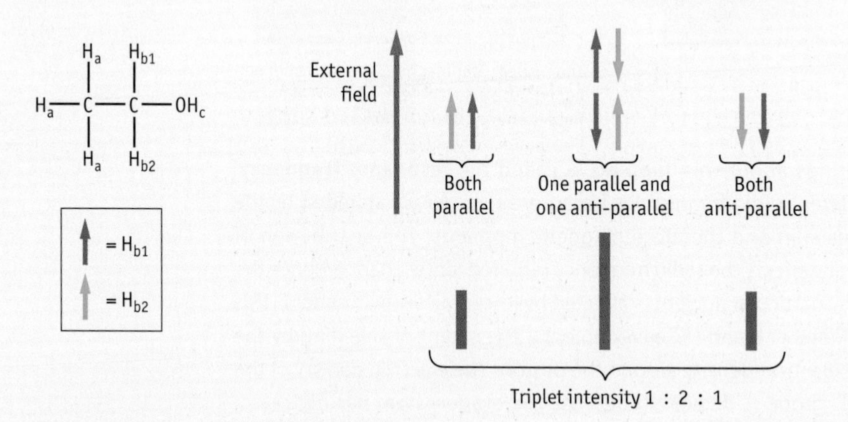

The CH_2 peak is split into a quartet because of the adjacent three hydrogens in the CH_3 group.

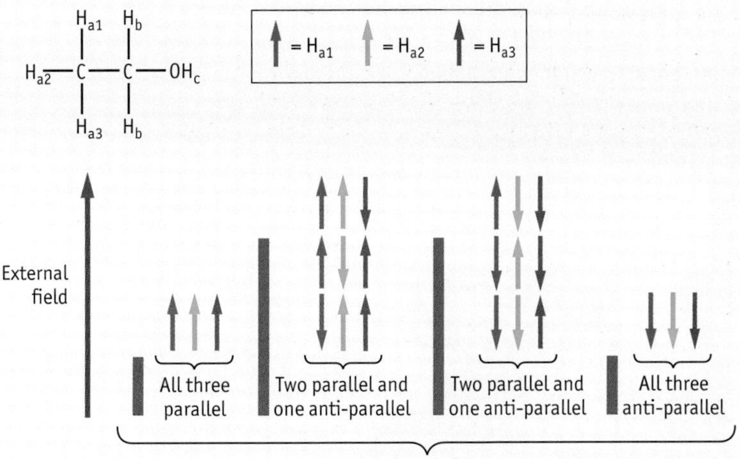

H_{a1} H_b — H_{a2}—C—C—OH_c — H_{a3} H_b

↑ = H_{a1} ↑ = H_{a2} ↑ = H_{a3}

External field

All three parallel | Two parallel and one anti-parallel | Two parallel and one anti-parallel | All three anti-parallel

The splitting and the intensity (relative size) can be predicted using Pascal's triangle:

Number of adjacent hydrogens, *n*	Splitting, *n* + 1	
0	1	1
1	2	1 1
2	3	1 2 1
3	4	1 3 3 1
4	5	1 4 6 4 1
5	6	1 5 10 10 5 1

When $n = 2$ we get 3 peaks $(n + 1)$ rule intensity = 1 : 2 : 1

When $n = 3$ we get 4 peaks $(n + 1)$ rule intensity = 1 : 3 : 3 : 1

With more complex molecules, splitting becomes more complicated.

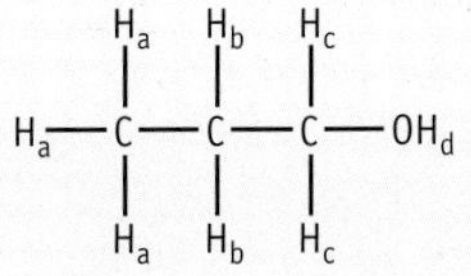

H_a H_b H_c — H_a—C—C—C—OH_d — H_a H_b H_c

Propan-1-ol

In propan-1-ol, the H_b proton peak is split by both H_a and H_c protons. It is split into a quartet (four sub-peaks) by the three H_a protons; each of these four sub-peaks is split into three by the two adjacent H_c protons. This gives a total of 12 sub-peaks, which is quite a complex spectrum.

It is possible to simplify the spectrum by saturating the sample with the resonance frequency of one of the adjacent groups, either H_a or H_c. This causes these protons to flip rapidly. If the sample is flooded with the resonance frequency for H_a, the spin–spin coupling for H_a disappears and we are left with a triplet due to coupling with H_c only.

Magnetic resonance imaging (MRI)

The body is made up of about 70% water. Since NMR can detect the protons in water it should be possible to detect the water in the body. Water in the body is in a very different environment from free water and these differences can be detected by NMR.

NMR is a common diagnostic tool used in medicine, where it is known as magnetic resonance imaging. Use is made of a magnetic field that is varied across the object being examined. It is able to detect differences in water content and other soft tissue deep inside the object. It is non-invasive and unlike X-rays does not damage any tissue. It is used extensively in the study of tissues, muscles and blood flow.

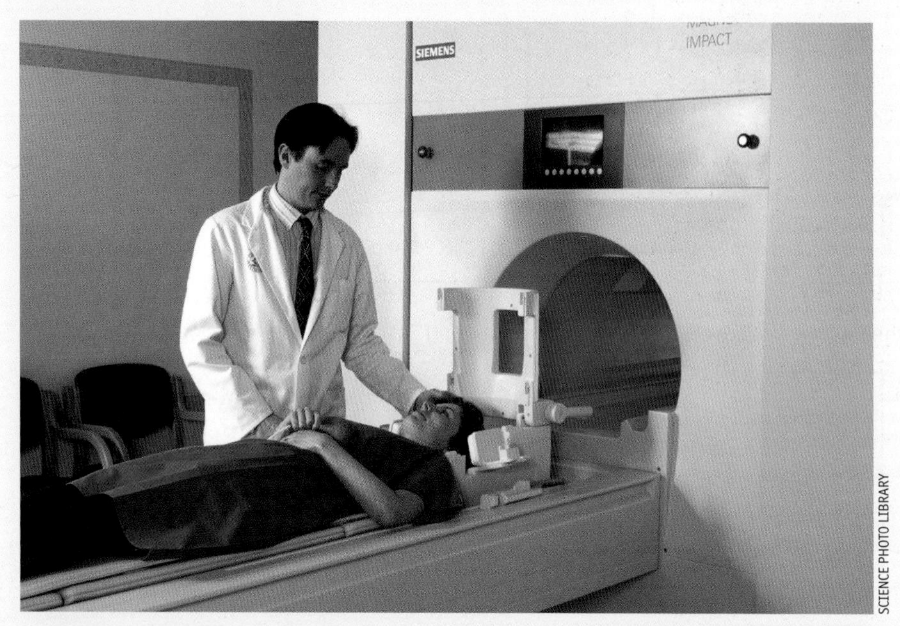

An MRI scan in progress

SCIENCE PHOTO LIBRARY

Hydrocephalus means water (hydro) in the head (cephalus). It is sometimes called 'water on the brain'. In the photograph, the MRI scan on the left shows the build up of extra fluid in the brain compared with the normal MRI scan on the right. MRI scans are now often coloured.

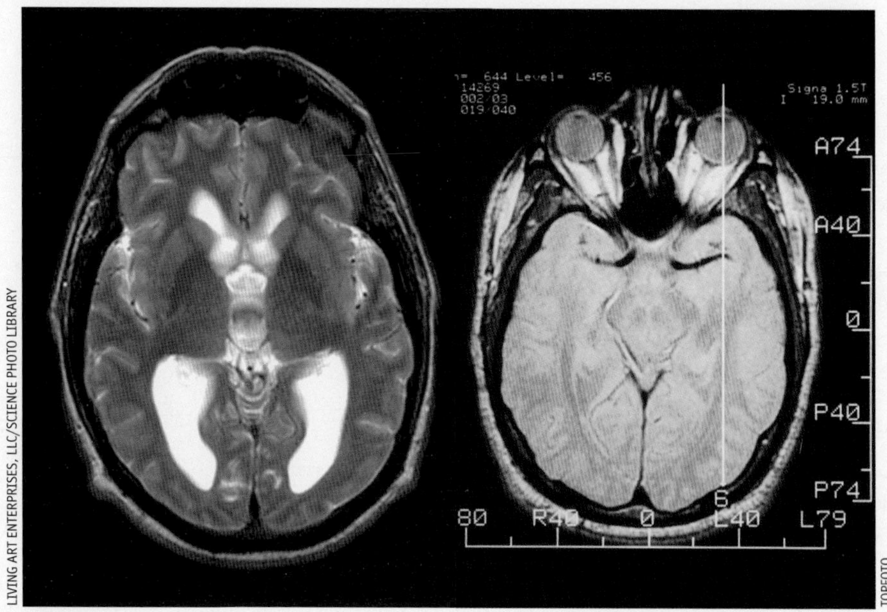

MRI scans of a brain with hydrocephalus (left) and a normal brain (right)

LIVING ART ENTERPRISES, LLC/SCIENCE PHOTO LIBRARY

TOPFOTO

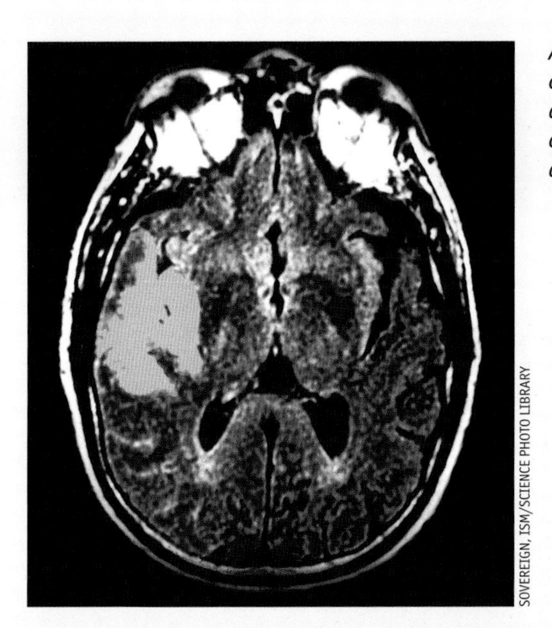

A coloured MRI scan of the brain of a patient after a stroke. A large area of dead tissue can be seen on the left of the image (orange area)

Stretch and challenge

1 Consider the proton environments in H_2O, CH_4 and CH_3OH. Consider the extent of shielding and de-shielding. State in which molecule the protons are:

- most shielded
- least de-shielded

Explain your reasoning.

2 Predict the spin–spin splitting in propane, $CH_3CH_2CH_3$.

Deduce the intensity of each peak.

3 Predict the number of peaks in the ^{13}C-NMR spectra of compounds **A**–**F**.

A CH_3 CH_3

B CH_3 CH_3

C CH_3 CH_3

D

E OH

F OH HO

Unit 5

Equilibria, energetics and elements

Introduction to Unit 5

Unit 5 (F325) builds on some of the ideas met in Units 1 and 2 of the AS chemistry course. There is little that depends on knowledge of the organic chemistry in the specification, but there are a number of essential ideas that come from the other modules.

The main areas of the unit are:

- Module 1: Rates, equilibrium and pH
  - how fast?
  - how far?
  - acids, bases and buffers
- Module 2: Energy
  - lattice enthalpy
  - enthalpy and entropy
  - electrode eotentials and fuel cells
- Module 3: Analysis
  - transition elements

As with the earlier chapters in this book, the main body of the text covers the basic principles and facts required to complete the course to a high standard. There are also sections of additional reading and 'stretch-and-challenge' questions that are designed to stimulate those who are aiming for an A^* grade or who simply wish to extend their knowledge further.

Summary of essential AS chemistry required for Unit 5

From Unit 1: Atoms, bonds and groups

The ability to write correct formulae and balanced equations is essential for anyone pursuing chemistry to a higher level. It is a requirement on which much depends. Calculations based on an equation that is not balanced are likely to be wrong. Such calculations include those based on titrations and energy cycles; both are expanded in this unit. If you have any uncertainty, then at least the standard reactions of acids with metals, bases, alkalis and carbonates should be practised before attempting Module 1.

Considerable time was spent at AS on calculations of mass, volumes of gases and concentration of solutions. It is the latter type of calculation that features most in this unit and, although some further examples are given, it is worth making sure that the principles are well understood.

Much of the work covered in the section on electrode potentials and the transition metals concerns redox reactions, so you should check the ideas of electron transfer and oxidation number. You should be able to recognise oxidation and reduction reactions, using the displacement reactions of the halogens as examples. For example, in the reaction of chlorine with potassium bromide, the chlorine is reduced while the bromide ion is oxidised:

$$Cl_2 + 2Br^- \rightarrow 2Cl^- + Br_2$$

Module 3 extends your knowledge of electron arrangement to include the transition elements and it is helpful to remember the order of filling of the orbitals. The triangle below should help you to remember this order.

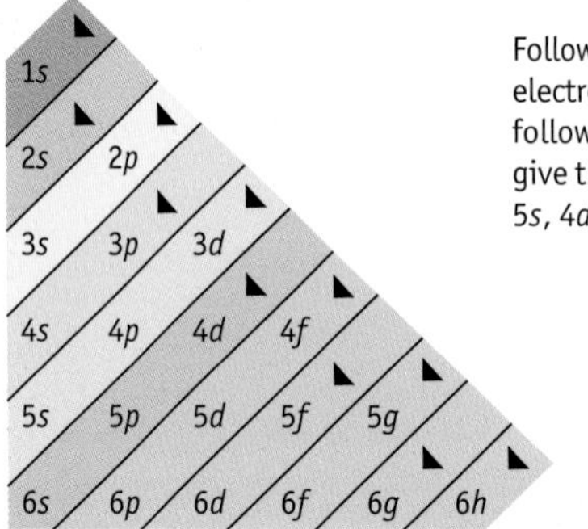

Follow the arrows to obtain the order that the electrons fill the orbitals. Start with the red and follow with the orange, yellow, green etc. to give the sequence, 1*s*, 2*s*, 2*p*, 3*s*, 3*p*, 4*s*, 3*d*, 4*p*, 5*s*, 4*d*, 5*p* etc.

Electron arrangement leads to an understanding of how elements combine and, although remembering all the details of types of bonding are unnecessary, you need to appreciate that dative bonding occurs when one of the elements forming a covalent bond provides both electrons. It is also helpful to understand the principles governing the shapes of molecules.

From Unit 2: Chains, energy and resources

The important knowledge from this unit comes from Module 3: Energy. It is probably true to say that revision of all of this module is essential before starting Unit 5, although the key points about rates of reaction and equilibria can be picked up while studying Chapters 9 and 10.

The ability to construct energy cycles and to use Hess's law is perhaps the single most important thing to revise and an understanding of this is assumed in Chapters 12 and 13.

The following example is a reminder of what is required.

Worked example

Use the data in the table below to calculate the standard enthalpy change of combustion for ethane:

Substance	Formula	$\Delta H^\ominus_f$ /kJ mol^{-1}
Ethane	$C_2H_6(g)$	−85
Carbon dioxide	$CO_2(g)$	−394
Water	$H_2O(l)$	−286

Answer

The equation for the reaction is:

$$C_2H_6(g) + 3\tfrac{1}{2}O_2(g) \rightarrow 2CO_2(g) + 3H_2O(l)$$

The enthalpy cycle required is:

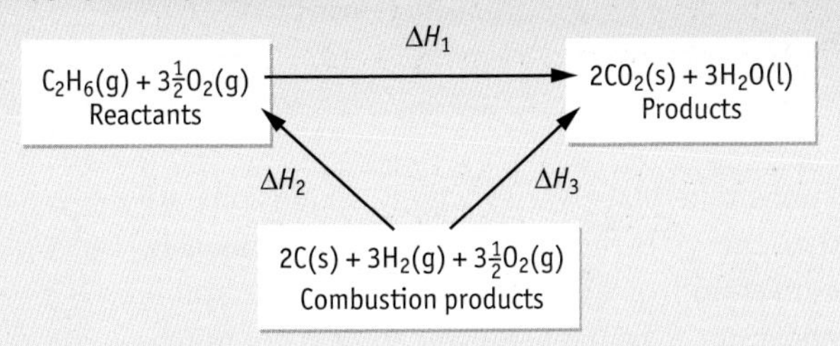

Using Hess's law $\Delta H_1 + \Delta H_2 = \Delta H_3$,

Therefore, $\Delta H_1 = \Delta H_3 - \Delta H_2$

ΔH_2 for formation of $C_2H_6(g) = -85$ kJ mol^{-1}

$\Delta H_3 = 2 \times \Delta H_f(CO_2(g)) + 3 \times \Delta H_f(H_2O(l))$

$= 2 \times (-394) + 3 \times (-286)$

$= -1646$ kJ mol^{-1}

Therefore, $\Delta H_1 = -1646 - (-85) = -1561$ kJ mol^{-1}

$$C_2H_6(g) + 3\tfrac{1}{2}O_2(g) \rightarrow 2CO_2(g) + 3H_2O(l)$$

Therefore, the standard enthalpy of combustion of ethane, $\Delta H^{\circ}{}_c(C_2H_6(g)) = -1561$ kJ mol^{-1}

How fast?

Chapter 9

The rate at which chemical reactions take place varies widely. Explosions are reactions that occur so rapidly that energy is released almost instantaneously. Other processes, such as the conversion of diamond into the more stable graphite, are so slow that they appear not to be happening at all. It is not unreasonable to say that diamonds are forever! You should already appreciate that the controlling factor is the activation energy of a reaction and you should be familiar with 'energy-profile' diagrams, which provide a simple explanation. This chapter considers this in more detail and focuses on the route by which the final products are created. It is possible to provide a more quantitative view of the role of the reactants and this can sometimes provide an insight into the way a reaction takes place.

Ⓔ Before proceeding, it would be wise to check that you are familiar with the work on rates of reaction that was covered at AS. In particular, you need to appreciate that reactions occur through the collisions of particles that have sufficient energy to overcome the activation energy barrier.

Reaction profiles

A balanced chemical equation gives important information about the exact amounts of substances required for a chemical reaction. However, it gives no indication of how efficient the reaction is, and no detail of how the products are formed. For example, the nucleophilic substitution of a halogenoalkane can be summarised by the equation:

$$OH^- + C_4H_9Cl \rightarrow C_4H_9OH + Cl^-$$

The reaction, however, proceeds by a specific mechanism.

Ⓔ Use the additional reading pages 192–194 in the AS textbook and try to explain why 1-chlorobutane reacts by an S_N2 mechanism whereas 2-chloro-2-methylpropane reacts by an S_N1 mechanism.

Although some reactions do occur simply by collision of the reactant particles, many involve a number of steps before the products are generated. This means that the energy profile of a reaction shown in Figure 9.1 is really a simplification of a profile that might look like that in Figure 9.2.

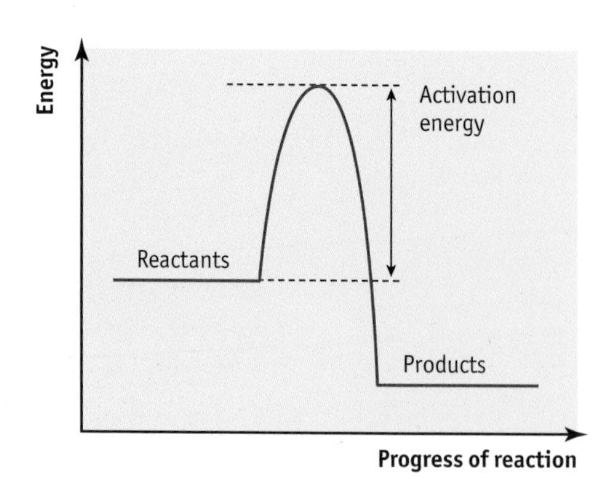

Figure 9.1
A simplified energy-profile diagram

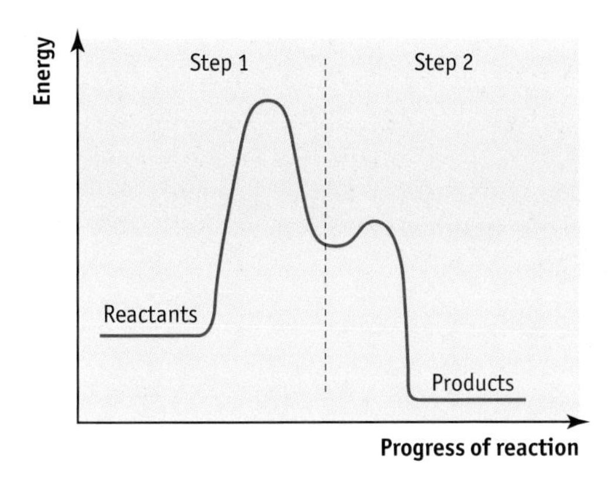

Figure 9.2
A complete energy-profile diagram

Figure 9.2 represents a reaction that has two steps, each with its own activation energy. Step 2 would almost certainly be instantaneous because the energy released as step 1 took place would be sufficient to allow conversion to the final products. Step 1 might be difficult to achieve because its activation energy is so large. Therefore, it is this step that effectively controls the overall rate of the reaction because the time taken for it to occur will be much greater than that for the other step.

The rate-determining step

Although it might supposed that a reaction could proceed via several slow steps, in practice it is found that in almost all cases just one step controls the overall rate. This step is called the **rate-determining step**.

It is important to appreciate that:

- the numbers of particles involved in the rate-determining step are not necessarily related to the balanced equation — quite often they are different
- all the reactants are not necessarily involved in the rate-determining step

The second point is significant because, contrary to what you may have assumed to be the case, increasing the concentration of a reagent does *not* always result in an increase in the rate of the reaction.

For example, the reaction between propanone, CH_3COCH_3, and iodine, when catalysed by hydrogen ions, forms iodopropane, CH_3COCH_2I, hydrogen ions and iodide ions as products. The equation for the reaction is:

$$I_2 + H^+ + CH_3COCH_3 \longrightarrow CH_3COCH_2I + H^+ + I^-$$

The equation is usually written as:

$$I_2 + CH_3COCH_3 \xrightarrow{H^+} CH_3COCH_2I + HI$$

From experimental data (obtained by methods described later) it is found that the rate-determining step involves only propanone and hydrogen ions. Iodine reacts only in a subsequent fast step. Therefore, increasing the concentration of acid or propanone results in a faster reaction whereas changing the concentration of iodine does not.

The rate-determining step may be summarised by the equation:

$$H^+ + CH_3COCH_3 \rightarrow CH_3C(OH){=}CH_2 + H^+$$

This reaction occurs slowly because a C–H bond must be broken, which is difficult. A subsequent step is the electrophilic addition of an iodine atom, which occurs rapidly.

Orders of reaction

Where one particle of a reagent is involved in the rate-determining step, increasing the concentration of that reagent increases the rate proportionately. The reaction is said to be **first order** with respect to that reagent. When increasing the concentration of a reagent does not affect the rate, the reaction is said to be **zero order** with respect to that reagent.

In the reaction between propanone and iodine, catalysed by hydrogen ions, one molecule of propanone reacts with one hydrogen ion in the rate-determining step. Therefore, the reaction is first order with respect to both propanone and hydrogen ions; it is zero order with respect to iodine. Doubling the concentration of propanone or hydrogen ions doubles the rate of the reaction; doubling the concentration of iodine has no effect.

There are some reactions in which two particles of a reagent are involved in the rate-determining step. In this case, the reaction is **second order** with respect to that reagent. Doubling the concentration of this reagent causes the rate to increase four-fold. An example is the breakdown of nitrogen(IV) oxide into nitrogen(II) oxide and oxygen:

$$2NO_2 \rightarrow 2NO + O_2$$

The reaction is second order with respect to nitrogen(IV) oxide. However, it is worth emphasising that it is not possible to judge the order of a reaction from the equation.

The breakdown of nitrogen(V) oxide into nitrogen(IV) oxide and oxygen

$$2N_2O_5 \rightarrow 4NO_2 + O_2$$

is first order with respect to the nitrogen(V) oxide.

The rate equation and the rate constant

The **rate equation** for a reaction between three reagents, A, B and C takes the form:

$$\text{rate} = k[A]^a[B]^b[C]^c$$

where the square brackets represent the concentrations of the reagents (in molar units such as $mol\,dm^{-3}$), and a, b and c are the orders of the reaction with respect to the reagents.

The **rate constant**, k, is a constant of proportionality. Its value varies from reaction to reaction and reflects the ease with which the reaction takes place. It is temperature dependent — for a fast reaction, the value of k is large, and for a slow reaction, k is small. In most circumstances, the rate of reaction is controlled more by the value of the rate constant than by the concentrations of the reagents. As the temperature rises, the value of k increases and the reaction proceeds more quickly. There is a rule of thumb which states that a temperature rise of 10°C doubles the rate of reaction. If the reaction is extremely slow this might not be apparent but in most instances the effect of heating is obvious. More collisions between the reagents involved in the rate-determining step have sufficient energy to overcome the activation energy barrier.

The rate equation for the reaction of iodine and propanone in the presence of hydrogen ions is:

$$\text{rate} = [(CH_3)_2CO]^1[H^+]^1[I_2]^0$$

This is the same as:

$$\text{rate} = [CH_3COCH_3]^1[H^+]^1$$

The overall order of a reaction is sometimes mentioned. This is the sum of the individual orders with respect to the reagents. In a general case, this is $a + b + c + \ldots$ The reaction of iodine, hydrogen ions and propanone has, therefore, an overall order of 2.

The rate equation for the breakdown of nitrogen(IV) oxide into nitrogen(II) oxide and oxygen is:

$$\text{rate} = k[NO_2]^2$$

The rate equation for the breakdown of nitrogen(V) oxide into nitrogen(IV) oxide and oxygen is:

$$\text{rate} = k[N_2O_5]^1$$

The orders of reaction are nearly always whole numbers (integers) and are usually 0, 1 or 2. This is because the orders reflect the number of particles involved in the rate-determining step. It is possible to have non-integer numbers, but this only occurs in the rare case of a reaction having more than one rate-determining step.

If the reaction is first order with respect to a particular reactant it is not necessary to include the power '1'. The rate equation for the reaction of iodine and propanone in the presence of hydrogen ions could be written as:

$$\text{rate} = [CH_3COCH_3][H^+]$$

e In general, the rate equation:
$\text{rate} = k[A]^x[B]^y[C]^z$
indicates that the rate-determining step involves

- x molecules of A
- y molecules of B
- z molecules of C

The units of the rate constant

The rate of a reaction is usually followed by observing either a reduction in one of the reactants or the formation of one of the products. In practice, this may be quite difficult, but there are experiments that allow suitable measurements to be made. For example, if a gas is released, the rate at which it is produced could be measured. Similarly, if there is a change in a property such as colour, acidity or density, it could be monitored.

The rate of a reaction is usually expressed as a change in concentration over a period of time. The units are $\text{mol dm}^{-3}\text{s}^{-1}$ or $\text{mmol dm}^{-3}\text{min}^{-1}$ (mmol stands for millimole, i.e. one-thousandth of a mole) and so on.

The units of the rate constant, k, depend on the orders of reaction.

If a reaction is first order overall, the rate equation is:

$$\text{rate} = k[A]^1$$

If the rate is measured in $\text{mol dm}^{-3}\text{s}^{-1}$ and [A] is in mol dm^{-3}, the unit of k is s^{-1}.

The important thing to remember is that the units of the rate constant, k, ensure that the units on the left-hand side of the equation are the same as those on the right-hand side.

In the example above, the units are:

$$\frac{\text{mol dm}^{-3}}{\text{s}} = \frac{1}{\text{s}} \times \text{mol dm}^{-3}$$

rate = k[A]

If the overall reaction is second order, as in rate = $k[\text{A}]^1[\text{B}]^1$ the units of k are different in order to keep the units consistent.

$$\frac{\text{mol dm}^{-3}}{\text{s}} = \frac{1}{\text{mol dm}^{-3}\,\text{s}} \times \text{mol dm}^{-3} \times \text{mol dm}^{-3}$$

rate = k[A][B]

In this example, k has the units:

$\text{mol}^{-1}\,\text{dm}^{+3}\,\text{s}^{-1}$

It does not matter which way round you write the units. Some people prefer to put the quantity raised to a positive power first in the sequence (e.g. $\text{dm}^{+3}\,\text{mol}^{-1}\,\text{s}^{-1}$).

It looks strange, but these units are required so that the units on the right-hand side of the equation match those on the left.

In general, the units of the rate constant, k, are expressed as $[\text{conc}]^{1-n}[\text{time}]^{-1}$ where n is the overall order of the reaction.

Determining orders of reaction

There are two distinct methods that can be used to determine the order with respect to a reactant:

- In one method, the concentration of each reactant is altered in turn to see what effect the change has on the rate at the start of the reaction. The start of the reaction is chosen because this is the only point in the reaction at which the concentrations of the reactants are definitely known. This type of experiment is called the 'initial rates' method.
- The second method involves monitoring the reaction throughout its course. In this procedure, the amount of a reactant or product is established at various time intervals throughout the course of the reaction.

e It is a common mistake for students to confuse these two methods. When doing rate experiments, you must be clear which method you are using.

It is important to appreciate that the treatment of the results obtained is different for the two methods.

Many chemical reactions occur rapidly and identifying the reaction intermediates can be extremely difficult. The distinguished chemist George Porter was, for many years, Director of the Royal Institution in London where both Davy and Faraday had made important discoveries in the nineteenth century. Porter invented an ingenious and important method for identifying the short-lived radicals produced in gaseous photochemical reactions. By using pulses of light of shorter duration than the existence of the radicals, he was able to identify the spectra that the latter produced. For this work, he shared the Nobel Prize for Chemistry in 1967.

George Porter (1920–2002)

The initial rates method

A typical reaction can be represented graphically as shown below:

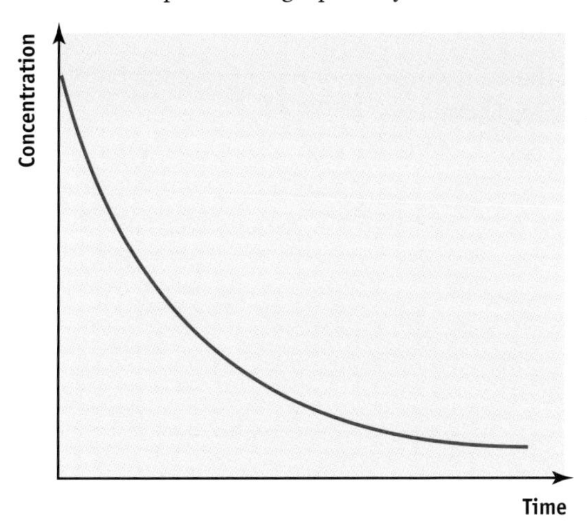

The reaction proceeds quickly at the start and slows down towards the finish. The rate of the reaction is fastest where the curve is steepest and the slope then decreases until the curve becomes horizontal, which indicates that the reaction is complete. To provide a numerical value for the rate, the steepness of the curve is established by finding its gradient. This is be achieved by drawing a tangent to the curve. In Figure 9.3 tangents have been drawn at two points on the graph. The numerical values of their gradients show how much the reaction has slowed down from the initial rate (the gradient at the start of the reaction) to a point about halfway through the reaction.

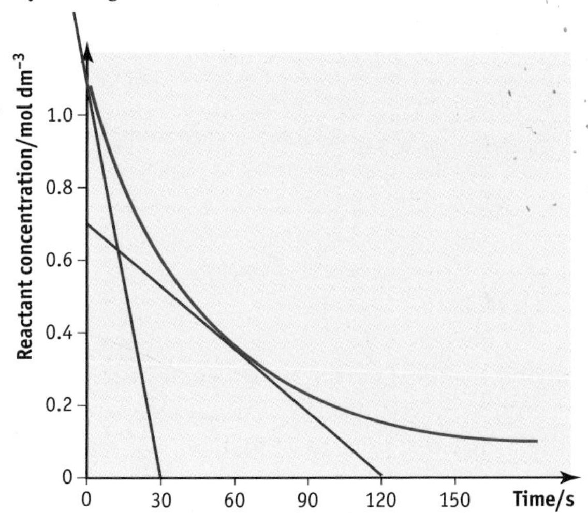

Figure 9.3 *Drawing tangents to calculate reaction rates*

From the graph:

- the initial rate, at time $t = 0$ seconds, is about $1.2/30 = 0.04\,mol\,dm^{-3}\,s^{-1}$
- the rate after 60 seconds is $0.7/120 = 0.006\,mol\,dm^{-3}\,s^{-1}$

These figures show that the rate has slowed considerably as the reaction has progressed.

If the concentrations of the reagents are known at the start of the reaction, a measurement of the initial gradient ($t = 0$) gives a numerical value of the rate at these concentrations.

If the experiment is repeated with the concentration of one of the reactants doubled, a new initial rate is established. Therefore, it is possible to see the effect of this reactant on the overall rate. If it is found that the gradient has doubled, then the reaction is first order with respect to that reactant. Changing the concentrations of each reactant in turn allows the order of reaction with respect to each reagent to be determined.

Worked example

Nitrogen(II) oxide and bromine react together as in the following equation:

$$2NO(g) + Br_2(g) \rightarrow 2NOBr(g)$$

Some data for the initial rates of reaction are shown in the table.

Experiment	$[NO]/mol\,dm^{-3}$	$[Br_2]/mol\,dm^{-3}$	Initial rate/$mol\,dm^{-3}\,s^{-1}$
1	0.01	0.01	0.011
2	0.01	0.02	0.022
3	0.02	0.01	0.044
4	0.03	0.03	0.297

a Use the results of experiments 1–3 to determine the order of reaction with respect to:

(i) bromine

(ii) nitrogen(II) oxide

b Write the rate equation for the reaction.

c Use the results frm experiment 4 to confirm that the rate equation is correct.

Answer

a (i) It can be seen from experiments 1 and 2, that doubling the concentration of bromine doubles the rate. Therefore, the order with respect to bromine is 1.

(ii) Experiments 1 and 3 show that doubling the concentration of nitrogen(II) oxide quadruples the rate. Therefore, the order with respect to nitrogen(II) oxide is 2.

b rate = $k[NO(g)]^2[Br_2(g)]$

c The rate of reaction 4 is 27 times faster than that of reaction 1. The rate has increased 3-fold as a result of the change in the bromine concentration and 9-fold as a result of the change in the nitrogen(II) oxide concentration.

Once all the orders for a reaction have been determined, it is possible to use one set of the results to determine a value for the rate constant, k.

Consider experiment 1 in the worked example above:

$$0.011 = k[0.01]^2[0.01]$$

$$0.011 = k(0.000001)$$

$k = 11\,000 = 1.1 \times 10^4$

The units of k are obtained by comparing the units on either side of the rate equation.

$$\frac{\text{mol dm}^{-3}}{\text{s}} = k \times (\text{mol dm}^{-3})^2 \times \text{mol dm}^{-3}$$

Therefore, the units of k are $\dfrac{1}{\text{s} \times (\text{mol dm}^{-3})^2} = \text{s}^{-1}\,\text{mol}^{-2}\,\text{dm}^6$

An approximation to an initial rate

The experimentation involved in using the method described above to provide data for the initial rate of a reaction is time consuming. For this reason, an approximation to obtain an initial rate is often used. Given that most reactions lead to an integer order, the approximation is usually acceptable. The time taken to reach a specific point in the reaction (which should be soon after the reaction has started) is recorded. The experiment is then repeated with different concentrations to see how the time taken to reach this point changes. The rate to this point is taken to be proportional to 1/time taken for each reaction, as the folllowing example should make clear.

Sodium thiosulphate reacts with dilute hydrochloric acid to form a precipitate of sulphur:

$$Na_2S_2O_3(aq) + 2HCl(aq) \rightarrow 2NaCl(aq) + SO_2(g) + S(s)$$

It is possible to measure the time taken to a point at which a fixed small amount of sulphur has been formed. Provided that this fixed point can be identified each time the reaction is carried out, the experiment can be repeated with different initial concentrations of the reactants and the time taken to reach this point can be measured.

Suppose the 'distance' through the reaction to this point is x and the time taken is t, then the rate can be expressed as:

$$\text{rate} = x/t$$

If the experiment is repeated so that the time is always measured at the point when the same amount of sulphur has been precipitated, then x is the same in each experiment. Therefore, in each case:

$$\text{rate} \propto 1/t$$

Repeating the experiment with different concentrations of the reactants allows the effect of each to be established and the order of reaction with respect to each to be determined. However, to do this we have to compare initial rates for which we know the concentrations of all the reagents. This requires that in each case the reaction has not progressed very far — in this example, the amount of precipitate must be small.

It might be possible to determine the orders of reaction from just a few measurements. However, it is usual to record a range of results and plot a graph of concentration against 1/time. Doing this takes into account the approximations that have been made.

e Remember that the units of k can also be calculated from $[\text{mol dm}^{-3}]^{1-n}\,[\text{time}]^{-1}$ where n is the overall order of the reaction. In this reaction, $1 - n = -2$.

e It might help to think of the rate as the speed of the reaction and recall that speed = distance/time

It is assumed that the reaction proceeds at a constant rate to the point that is measured. This is not strictly true because as soon as the reaction starts the rate begins to slow. However, the approximation is good enough to determine an integer order.

The order of reaction determines the shape of the graph.

If the reaction is **zero order** with respect to a particular reactant then the reactant has no effect on the rate and the graph appears as in Figure 9.4:

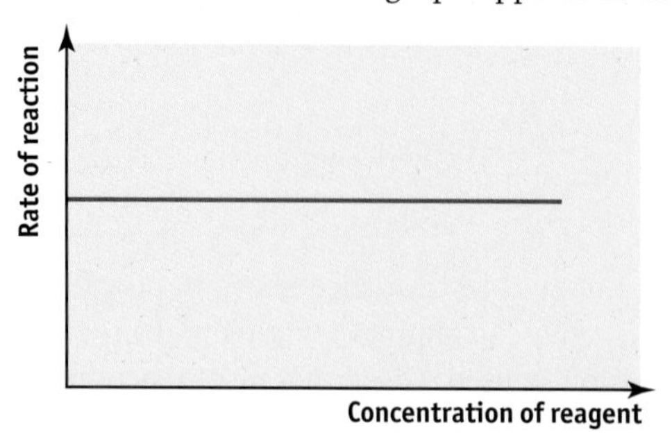

Figure 9.4 *Rate–concentration graph for a zero-order reaction*

If the reaction is **first order** with respect to a particular reactant then the rate is proportional to the concentration and the graph appears as in Figure 9.5. The graph is a straight line that goes through the origin (if there is no reactant, there can be no reaction).

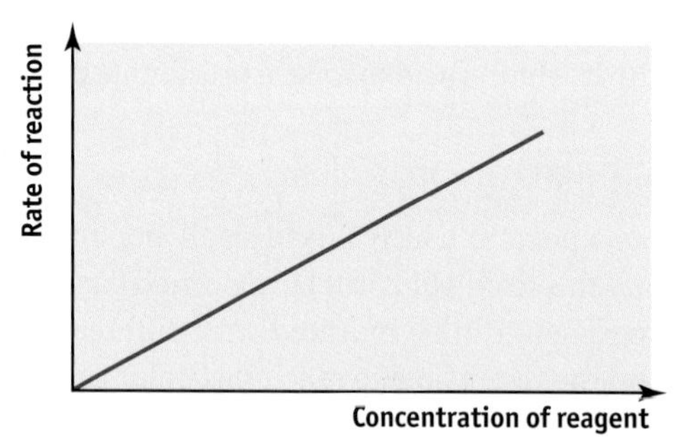

Figure 9.5 *Rate–concentration graph for a first-order reaction*

If the reaction is second order with respect to a particular reactant then the graph is a curve (Figure 9.6). This is because the rate is proportional to the square of the concentration of the reactant.

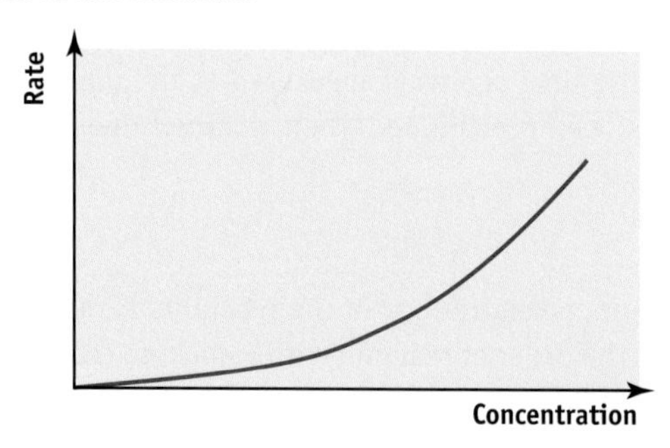

Figure 9.6 *Rate–concentration graph for a second-order reaction*

There are a number of ways of carrying out experiments using this approximation. They are often referred to as 'clock reactions'.

e You will probably have the opportunity to try a 'clock reaction' in class.

Following a reaction through its course

It is sometimes possible to obtain the order of reaction for a particular reagent by recording data throughout the course of the reaction. The reaction must either involve only one reagent or be set up so that the reaction rate depends solely on that reagent. To do the latter, reaction mixtures are made up in which all the other reactants are present in excess. The assumption is that throughout the reaction there will be such an excess of the other reactants that their concentration will be nearly constant and they will have little or no effect on the change in the reaction rate. Therefore, whatever is found is due to the reagent at low concentration.

The interpretation of the results depends on analysing the shape of the graph of concentration of the reagent against time. The theory that explains the likely shapes of these graphs is mathematical and is not required for A2 chemistry. However, the principles can be understood if you have done some calculus in A-level mathematics. An example is included in the additional reading section at the end of this chapter.

For a zero-order reaction, the graph of concentration against time appears as in Figure 9.7. The reactant is consumed in the reaction and therefore its concentration decreases. However, the reactant has no effect on the rate and so the line is not curved.

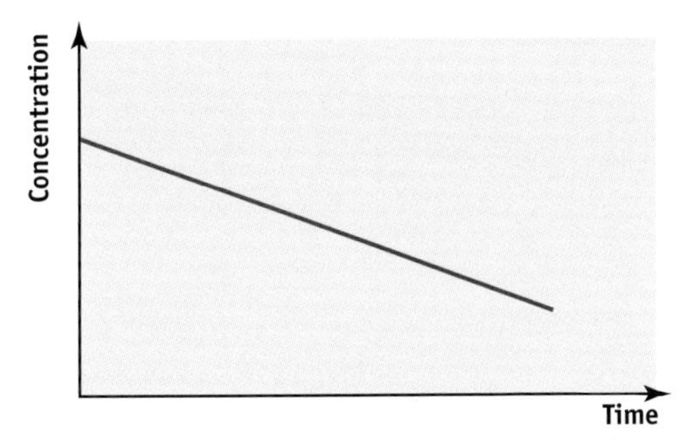

Figure 9.7 *Concentration–time graph for a zero-order reaction*

For a **first-order reaction**, the graph of concentration against time is shown in Figure 9.8.

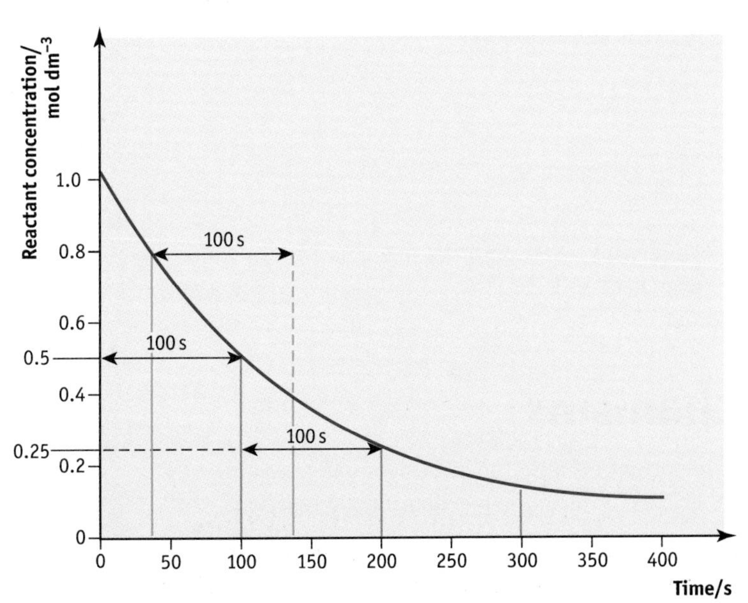

Figure 9.8 *Following the course of a first-order reaction*

The graph is a curve showing how the reagent influences the reaction. It is used up quickly at the start and then its effect diminishes steadily as the reaction proceeds. There is a more subtle feature of the curve that is not immediately apparent but which is a useful way of distinguishing it from other curves, such as that obtained for a second-order reaction. This is that the time taken for the concentration of the reactant to halve is the same whatever the starting concentration. This time is known as the **half-life** of the reaction. You may be familiar with this term, since it is used to indicate the stability of radioactive isotopes; these decay by a first-order mechanism.

Referring to Figure 9.8, it can be seen that the time taken for the concentration to fall to half its original value is 100 s. For example, it takes 100 s for the initial concentration of 1.0 mol dm^{-3} to fall to 0.5 mol dm^{-3}. It also takes 100 s for the concentration to fall from 0.5 mol dm^{-3} to 0.25 mol dm^{-3} and for 0.8 mol dm^{-3} to fall to 0.4 mol dm^{-3}. Whichever starting value is taken, if the reaction is first order, the time taken to halve this concentration is constant.

Although the half-life is the standard figure that is quoted for reactions, the principle can be applied to other measurements of time. Take, for example, the following set of results:

Concentration of the reactant/mol dm^{-3}	0.500	0.450	0.405	0.365	0.328	0.295
Time/s	0	30	60	90	120	150

As is often the case in an experiment, a measurement has been taken at a fixed time interval — 30 seconds in this case. The reaction could be shown to be first order by plotting a graph of concentration against time and extrapolating it to establish a half-life. However, this is unnecessary because the reaction can be shown to be first order by noting that every 30 s the concentration of the reactant has been reduced to 0.9 times its previous value:

$0.9 \times 0.500 = 0.45$

$0.9 \times 0.45 = 0.405$

$0.9 \times 0.405 = 0.3645$ and so on

The reaction has a 'constant 0.9-life'.

There is no need to be able to identify a second-order reaction from its concentration–time graph. If a reaction is neither zero order nor first order then it may be assumed to be second order.

Reaction mechanisms

The order of a reaction is important because the information it gives about the rate-determining step may give a clue as to the mechanism of a reaction.

The hydrolysis of halogenoalkanes by hydroxide ions is a reaction you will have met at AS level. The reaction can be represented as:

$R_3CX + OH^- \rightarrow R_3COH + X^-$

where R is either an alkyl group or hydrogen and X is the halogen.

In some circumstances (for example, when R = CH_3 and X = I), the rate equation is:

rate = $k[R_3CX]$

This shows that the concentration of OH^- is irrelevant to the overall reaction rate. It is the decomposition of the halogenoalkane alone that controls the rate.

The mechanism is:

- Each of the alkyl groups, CH_3, pushes electrons along the σ-bond towards the central carbon atom. This causes the bonded pair in the C–X bond to break, forming a carbonium ion, $(CH_3)_3C^+$. This is the slow step and is the rate-determining step.
- The carbonium ion, although unstable, is stabilised by the inductive effects of the three CH_3 long enough for a nucleophile, OH^- to attack the carbonium ion and form the product, $(CH_3)_3COH$.

The mechanism is illustrated below.

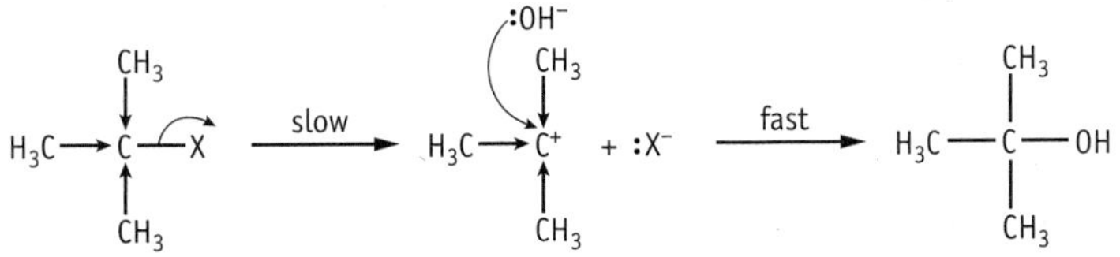

$(CH_3)_3C{-}X \xrightarrow{\text{slow}} (CH_3)_3C^+ + :X^- \xrightarrow{\text{fast}} (CH_3)_3C{-}OH$ (with $:OH^-$ attacking the C^+)

This is known as an S_N1 mechanism (see additional reading, page 193 in the AS book).

In other cases (for example, when R = H and X = Cl) the rate equation is:

rate = $k[R_3CX][OH^-]$

This indicates that both reagents are involved in the rate-determining step, which suggests that it is probably collision between the two reactants that controls the rate.

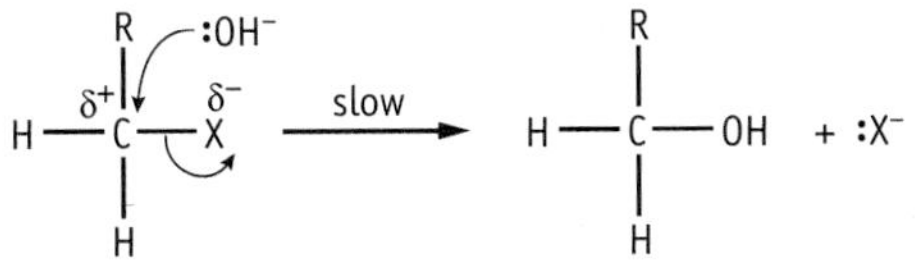

$:OH^- + H{-}\overset{\delta+}{C}(R)(H){-}\overset{\delta-}{X} \xrightarrow{\text{slow}} H{-}C(R)(H){-}OH + :X^-$

This is known as an S_N2 mechanism (see additional reading, page 193 in the AS book).

You may be asked in an examination to suggest a mechanism that is consistent with the orders of reaction that you have calculated. You should supply a sequence that has a rate-determining step consistent with those orders.

As it is a suggested mechanism, it does not matter if the sequence you provide is incorrect.

For example, nitrogen(IV) oxide and carbon monoxide react to form nitrogen(II) oxide and carbon dioxide:

$NO_2 + CO \rightarrow NO + CO_2$

The rate equation for this reaction is:

rate = $k[NO_2]^2$

The rate equation indicates that the slow step involves only two molecules of NO_2.

$rate = k[NO_2]^2$

The order indicates the number of molecules involved in the rate determining step

Indicates that only NO_2 is involved in the rate determining step

The fact that rate = $k[NO_2]^2$ might suggest to you a mechanism with a rate-determining step such as:

$$2NO_2 \rightarrow 2NO + O_2$$

or

$$2NO_2 \rightarrow N_2O_4$$

Both of these are valid suggestions.

You have no way of knowing what is actually involved in the other steps of the mechanism. However, it is essential that the sum of all the steps in the mechanism add up to the overall balanced equation.

For the reaction between nitrogen(IV) oxide and carbon monoxide, the balanced equation is:

$$NO_2 + CO \rightarrow NO + CO_2$$

As the rate equation is rate = $k[NO_2]^2$ we know that the slow step involves two molecules of NO_2 and might be the first step in a two-step mechanism. The slow step and the fast step must combine to give the balanced equation. If we assume:

slow step: $2NO_2 \rightarrow 2NO + O_2$

then by comparing the slow step with the balanced equation, it is apparent that one of the NO molecules and the O_2 must react in the second step.

Fast step: $NO + O_2 + CO \rightarrow CO_2 + NO_2$

Slow step + fast step gives:

Slow step + fast step $\quad 2NO_2 + NO + O_2 + CO \longrightarrow 2NO + O_2 + CO_2 + NO_2$

which cancels out to leave the balanced equation:

$$NO_2 + CO \rightarrow NO + CO_2$$

This mechanism is, in fact, incorrect. However, it is consistent with the information given and would be allowed in an examination just as much as the following version, which is believed to be correct:

- rate-determining step: $2NO \rightarrow NO + NO_3$
- fast step: $NO_3 + CO \rightarrow NO_2 + CO_2$

The important point is to provide a suggestion that is supported by the balanced equation and the rate equation.

Questions

1 If concentrations are measured in $mol\,dm^{-3}$ and rate as $mol\,dm^{-3}\,s^{-1}$, deduce the units of k for a reaction that has the rate equation:

a rate = k[A][B][C]

b rate = $k[A]^2[B]$

2 The table gives data for the reaction:

$2NOCl(g) \rightarrow 2NO(g) + Cl_2(g)$

a Calculate the order with respect to NOCl.

b Calculate the value of the rate constant and give its units.

Initial concentration NOCl/$mol\,dm^{-3}$	Initial rate of reaction/ $mol\,dm^{-3}\,s^{-1}$
0.1	4×10^{-10}
0.2	1.6×10^{-9}
0.4	6.4×10^{-9}

c What is the initial rate of reaction when the initial concentration of NOCl is $0.15\,mol\,dm^{-3}$?

3 Aqueous mercury(II) chloride can be reduced by aqueous ethanedioate ions. A precipitate of mercury(I) chloride is obtained.

The equation is:

$2HgCl_2(aq) + (COO)_2^{2-}(aq) \rightarrow Hg_2Cl_2(s) + CO_2(g) + 2Cl^-(aq)$

The initial rates of reaction at various concentrations of aqueous mercury(II) chloride and aqueous ethanedioate ions were obtained by measuring the amount of mercury(I) chloride precipitated. The initial rate can be expressed in units of $mol\,dm^{-3}\,min^{-1}$ of mercury(II) chloride that has reacted.

The following results were obtained:

Initial concentration of $HgCl_2(aq)$/ $mol\,dm^{-3}$	Initial concentration of $(COO)_2^{2-}(aq)$ /$mol\,dm^{-3}$	Initial rate/ $mol\,dm^{-3}\,min^{-1}$
0.08	0.02	0.48×10^{-4}
0.08	0.04	1.92×10^{-4}
0.04	0.04	0.96×10^{-4}

a Determine the orders of reaction with respect to:

(i) $(COO)_2^{2-}(aq)$

(ii) $HgCl_2(aq)$

b Calculate the value of the rate constant

4 The table below gives data for the reaction:

$BrO_3^-(aq) + 5Br^-(aq) + 6H^+(aq) \rightarrow 3Br_2(aq) + 3H_2O(l)$

$[BrO_3^-(aq)]$/ $mol\,dm^{-3}$	$[Br^-(aq)]$/ $mol\,dm^{-3}$	$[H^+(aq)]$/ $mol\,dm^{-3}$	Initial rate/ $mol\,dm^{-3}\,s^{-1}$
0.1	0.2	0.1	1.64×10^{-3}
0.2	0.1	0.1	1.64×10^{-3}
0.2	0.2	0.1	3.28×10^{-3}
0.2	0.1	0.2	6.56×10^{-3}
0.25	0.25	0.25	x

a Determine the rate equation for this reaction.

b Calculate the value of k, including its units.

c What is the initial rate of reaction, x, for the initial concentrations shown in the last row of the table?

5 Show how the results given in the table indicate that the reaction is first order with respect to the reactant, **B.**

Concentration of B/$mol\,dm^{-3}$	0.800	0.640	0.512	0.410	0.328	0.262
Time/s	0	20	40	60	80	100

6 Two reagents **X** and **Y** react together according to the equation:

$X + 2Y \rightarrow$ products

a The concentration of **X** is measured in the presence of such a large excess of **Y** that it can be assumed that **Y** has no effect on the reaction rate. The results obtained are shown in the table.

Concentration of X/$mol\,dm^{-3}$	0.50	0.43	0.36	0.29	0.22	0.15
Time/s	0	30	60	90	120	150

Deduce the order of reaction with respect to **X**.

b The experiment is repeated, but this time **X** is present in sufficient excess for the rate of reaction to depend only on **Y**. The results obtained are shown in the table.

Concentration of Y/mol dm^{-3}	1.00	0.70	0.49	0.34	0.24	0.17
Time/s	0	30	60	90	120	150

Deduce the order of reaction with respect to Y.

c Using your answers to parts **a** and **b**:

(i) Write the overall rate equation for the reaction.

(ii) Deduce a value for the initial rate of the reaction.

(iii) Deduce a value for the rate constant.

7 Hydrogen peroxide decomposes according to the equation:

$2H_2O_2(aq) \rightarrow 2H_2O(l) + O_2(g)$

The following results were obtained in an experiment to measure the decomposition of hydrogen peroxide in the presence of an acid catalyst.

Concentration of H_2O_2/mol dm^{-3}	0.10	0.09	0.08	0.07	0.06	0.05	0.04	0.03
Time/min	0	792	1678	2682	3841	5210	6889	9052

Determine whether the acid-catalysed decomposition of hydrogen peroxide is zero order, first order or second order.

8 Sulphuryl chloride decomposes in an organic solvent as follows:

$SO_2Cl_2 \rightarrow SO_2 + Cl_2$

The decomposition of sulphuryl chloride is followed by measuring its concentration every 30 min. The results obtained are shown in the table below.

Concentration of SO_2Cl_2/mol dm^{-3}	1.00	0.750	0.563	0.422	0.316	0.237
Time/min	0	30	60	90	120	150

a Determine the order of reaction.

b Calculate the rate constant.

9 Three possible mechanisms for the reaction of nitrogen(II) oxide and oxygen to make nitrogen(IV) oxide ($2NO + O_2 \rightarrow 2NO_2$) are suggested:

Mechanism X:

$2NO + O_2 \rightarrow 2NO_2$ Slow

Mechanism Y:

$NO + O_2 \rightarrow NO_3$ Slow

$NO_3 + O \rightarrow 2NO_2$ Fast

Mechanism Z:

$O_2 \rightarrow 2O$ Slow

$2NO + 2O \rightarrow 2NO_2$ Fast

a Write the rate equation for each mechanism.

b When this reaction is investigated using the initial rates procedure, the initial partial pressures of the gases are used as measures of their concentrations.

The results obtained are shown in the table.

Initial pressure of NO/kPa	Initial pressure of O_2/kPa	Initial rate/kPa hr^{-1}
50	50	0.100
50	20	0.040
100	60	0.480

(i) Calculate the orders of reaction with respect to NO and O_2.

(ii) Which of the above mechanisms **X**, **Y** and **Z** is compatible with your answers to part **(i)**?

Summary

You should now be able to understand what is meant by:

- a reaction profile
- a rate-determining step
- a rate equation and order of reaction
- a rate constant and you should be able to give its units
- the initial rate of a reaction

You should also be able to:

- deduce the value of the initial rate from a concentration–time graph
- use data from measurements of initial rates to determine orders of reaction
- understand how 1/time may be used to approximate to an initial rate
- draw rate–concentration graphs for zero-, first- and second-order reactions
- draw, and know how to interpret, concentration–time graphs for complete reactions and to deduce the order of reaction for zero-order and first-order reactions
- know what is meant by a half-life and use this to identify a first order reaction
- understand that the rate equation provides information about the reagents involved in the rate-determining step
- suggest how the rate-determining step may provide an indication of the mechanism of a reaction

Additional reading

The effect of a change in temperature on the rate of a reaction

Arrhenius was a child prodigy, who reputedly taught himself to read by the age of three. He studied sciences at university and in 1884 submitted a doctoral thesis on the nature of electrolysis to the University of Uppsala. It was almost rejected — despite the fact that it formed the basis of the work for which he was awarded the Nobel Prize in 1903. Arrhenius had a wide range of scientific interests and he formulated the equation that bears his name after working for a time with Boltzmann.

He was one of the first to appreciate the connection between global temperature and the presence of carbon dioxide in the atmosphere. He wrote a number of books to explain the ideas of science to the general public. During the First World War he was instrumental in gaining the release of German scientists who had been made prisoners of war.

Svante Arrhenius (1859–1927)

One of the conditions that greatly affects the rate of a chemical reaction is temperature. All reactions proceed more vigorously as the temperature increases since the reactants have more energy and are more able to exceed the activation energy. However, the relationship between temperature and reaction rate is not simple. A Swedish chemist called Svante Arrhenius studied this in detail and derived the following equation:

$$k = Ae^{-E_a/RT}$$

where:

- k is the rate constant
- E_a is the activation energy of the reaction
- T is the temperature in kelvin, K
- R is the gas constant. Its presence here may seem surprising, but it was required by Arrhenius in his derivation. The value of the gas constant is $0.00831\ \text{kJ K}^{-1}\ \text{mol}^{-1}$.
- 'e' is a number that is also in the derivation. It has the value 2.718 and is sometimes written as 'exp', which is short for exponential.
- 'A' is a constant that is sometimes called the collision or frequency factor but, more often, the pre-exponential factor.

> **e** If you are doing A-level mathematics you may have already met 'exp'. Whether or not you are familiar with 'exp', it is almost certainly present as a button on your calculator.

This may seem complicated, but the equation is important as it is used to predict the effect of a change in temperature on a reaction.

For example, if a reaction has an activation energy of $45\ \text{kJ mol}^{-1}$, what is the effect on the rate constant of raising the temperature from 25°C to 40°C?

Using the Arrhenius equation at 25°C or 298 K gives:

$$k_{298} = Ae^{-45/(0.00831 \times 298)} = A \times 1.28 \times 10^{-8}$$

Using the equation at 40°C or 313 K gives

$$k_{313} = A\,e^{-45/(0.00831 \times 313)} = A \times 3.06 \times 10^{-8}$$

From the example above, you can see that for a 15°C rise in temperature, the rate constant, and hence the rate of the reaction, has more than doubled. It makes it clear how significant temperature is. The rule of thumb mentioned earlier says that 'the rate of reaction doubles for every 10°C temperature rise'. However, this only applies to increases close to room temperature and for reactions with activation energies similar to that selected for the example.

It is of more immediate relevance to consider the effect of a very small temperature change on the rate of a reaction. You might carry out an experiment during which the temperatures of the solutions being used increase by, say 2°C — for example from 19°C to 21°C (292 K to 294 K).

Using the Arrhenius equation and the activation energy of 45 kJ mol^{-1} gives:

$$k_{292} = A\,e^{-45/(0.00831 \times 292)} = A \times 8.83 \times 10^{-9}$$
$$k_{294} = A\,e^{-45/(0.00831 \times 294)} = A \times 1.00 \times 10^{-8}$$

So the rate will have increased by a factor of 1.13. So, for example, a reaction that took 40 s at 19°C would take about 35 s at 21°C — quite a significant difference.

Some detective work using half-lives

The importance of the measurement of half-lives to identify a reaction as first order has been discussed in this chapter. Many people however know the expression 'half-life' in the context of the decay of radioactive isotopes. This is because these isotopes lose their radioactivity according to a first-order pattern. So, for example, uranium with mass number 235 (^{235}U) loses half its radioactivity every 700 million years and will eventually become lead. After 1400 million years a quarter of its radioactivity will remain, and so on.

The steady first-order decay of isotopes has been put to a number of uses in dating ancient material. For example, the decay of ^{14}C has been used to date wooden items. This is based on the fact that the carbon dioxide absorbed by a plant during photosynthesis contains a small fixed amount of $^{14}CO_2$ compared to $^{12}CO_2$. The molecules created within the plant therefore have a constant ratio of ^{14}C atoms to ^{12}C atoms, as long as the plant is alive. Since ^{14}C is radioactive, the plant emits radiation. This has been found to be a constant 15 disintegrations per minute per gram of carbon within the plant.

After the plant has died, the ^{14}C is not replaced. It will decay with a constant half-life of 5750 years. This means that after 5750 years the radioactivity will be only 7.5 disintegrations per minute per gram; it would take 11 500 years to reach 3.75 disintegrations per minute per gram. These fixed points allow a graph to be plotted so that the age of a sample of material can be established.

An entertaining example of using radioactive decay to determine the age of material concerns the identification of an art fraud.

In 1937, a painting called *Christ and the Disciples at Emmaus* was sold as an original Vermeer (1632–75) and fetched a price of around £150 000. In 1945, an artist called Han van Meegeren admitted it was a forgery that he had painted. However, the art critics who had been fooled refused to accept this and still considered the painting to be a genuine Vermeer. Radiochemical evidence was used to establish that van Meegeren was speaking the truth.

Both he and Vermeer had used the same white lead pigment. In nature, the lead ore from which the pigment is made contains amounts of ^{226}Ra and ^{210}Pb at a fixed equilibrium ratio as both are intermediates in the decay of ^{238}U to ^{206}Pb. When the pigment is made, 99% of the ^{226}Ra is removed and the equilibrium balance is upset — the ratio of ^{210}Pb to ^{226}Ra has increased. However, the two isotopes will continue to decay: ^{226}Ra has a half-life of 1600 years; ^{210}Pb has a half-life of 22 days. This is a big difference and the ratio ^{210}Pb:^{226}Ra decreases steadily until the original balance is restored. Calculations show that this takes about 150 years. Therefore, this rebalance would have been achieved in the paint used by Vermeer but not in that used by van Meegeren. In 1968, Bernard Keisch working at the Mellon Institute in Pittsburgh showed beyond any doubt that *Christ and the Disciples at Emmaus* had indeed been painted during the twentieth century.

An extra for those studying mathematics

With a little maths, it is possible to show that first-order reactions should have a constant half-life.

A first-order reaction has the rate equation:

$$\text{rate} = k[\text{Y}]$$

where [Y] is a concentration of the reactant expressed in molar units (i.e. usually mol dm^{-3}).

If this is expressed in calculus notation and in terms of the formation of a product of the reaction X, then the rate equation becomes

$$\frac{d[\text{X}]}{dt} = -k[\text{X}]$$

where t is the time from the start of the reaction

The minus sign is necessary here because as more X is formed the concentration of the reactant Y decreases and the reaction becomes slower.

Therefore:

$$\int \frac{d[\text{X}]}{[\text{X}]} = \int -k dt$$

or $\ln[\text{X}] = -kt + c$ where c is the constant of integration

If, at the start of the reaction when $t = 0$, the concentration of X is $[\text{X}]_0$ then $c = \ln[\text{X}]_0$

and the equation becomes

$$\ln[\text{X}] = -kt + \ln[\text{X}]_0$$

or $kt = \ln[\text{X}]_0 - \ln[\text{X}]$

which is $kt = \ln([\text{X}]_0/[\text{X}])$

This is an important relationship because the half-life, $t_{\frac{1}{2}}$, is the time taken for the concentration X_0 to fall to $\frac{1}{2}[\text{X}]_0$, which when substituted into the equation above shows that

$$kt_{\frac{1}{2}} = \ln 2$$

Therefore, $t_{\frac{1}{2}}$ is independent of the value of $[\text{X}]_0$ and equals $(\ln 2)/t_{\frac{1}{2}}$ or $0.693/t_{\frac{1}{2}}$. This gives a useful method of calculating the value of k when the half-life is known.

You may also appreciate why the time taken for $[X]_0$ to fall to any fixed proportion of its concentration is independent of the value of $[X]_0$. For example, the time taken to fall to 0.8 of the original value, $t_{0.8}$, is given by

$$kt_{0.8} = \ln(1/0.8) = 0.223$$

Stretch and challenge

Alcohol is absorbed through the stomach lining into the bloodstream. It is then slowly removed by oxidation in an enzyme-catalysed reaction.

a A person consumes a pint (568 cm^3) of beer containing 4.1% by volume of ethanol. The density of ethanol is 0.789 $g\,cm^{-3}$.

Show that the amount of ethanol that has been consumed is 0.40 mol

b As soon as the beer has been drunk, ethanol is transferred to the blood from the stomach. The amount of ethanol (in moles) remaining in the stomach over a period of time is shown in the first table.

(i) Show that the transfer of ethanol from the stomach to the blood is first order with respect to ethanol.

(ii) What is the half-life of the transfer?

(iii) Calculate the value of the rate constant.

c Assume that no oxidation takes place during the transfer of the ethanol. Copy and complete the second table to show the concentration of ethanol in the blood.

Assume that there is 40 dm^3 of fluid, including blood, in the body and that the amount of ethanol is distributed uniformly throughout.

d Of course, oxidation begins to take place as soon as ethanol enters the fluid. The oxidation is zero order with respect to ethanol and it has a rate constant equal to 0.2 $g\,dm^{-3}\,h^{-1}$.

(i) Show that the value of the rate constant is 7.25×10^{-5} $mol\,dm^{-3}\,min^{-1}$.

(ii) Using your answers to part **c** and the rate constant, adjust the table to take into account the ethanol removed by oxidation.

e Plot a graph of concentration of ethanol against time. Determine the time when the concentration of ethanol in the body fluid, including the blood, is at its highest value.

f In the UK, the maximum concentration of ethanol allowed in the blood for a person to legally drive a car is 80 mg per 100 cm^3 of blood. Taking your answer from part **e** — with this level of ethanol would the person be fit to drive a car?

e It should be added that this calculation is based on a number of assumptions that differ from person to person and that any amount of alcohol interferes with the capacity to drive safely.

Amount of ethanol/mol	0.40	0.34	0.29	0.24	0.17	0.076	0.014	0.0063
Time/min	0	1	2	3	5	10	20	30

Concentration of ethanol/mol dm^{-3}								
Time/min	0	1	2	3	5	10	20	30

How far?

Chapter 10

During the AS course you will have been introduced to reactions that are in equilibrium and the importance of Le Chatelier's principle will have been discussed. This chapter extends these ideas and focuses on some quantitative aspects of reactions that have reached equilibrium.

Dynamic equilibrium

You will have learned that a reaction in which the reactants and products are in a closed system (i.e. one in which none of the component particles can escape) may reach dynamic equilibrium. The reaction then proceeds in both directions. Le Chatelier's principle can be applied to determine how a change in conditions affects the balance of reactants and products.

Although you will have met Le Chatelier's principle during the AS course it may well be asked as part of an A2 question and it should be revised carefully.

Some students have the impression that reactions that reach equilibrium are encountered only occasionally. It is true that there are relatively few cases in which the equilibrium mixture contains similar amounts of reactants and products. However, equilibrium is a feature of most chemical processes. Consider the energy-profile diagram shown in Figure 10.1.

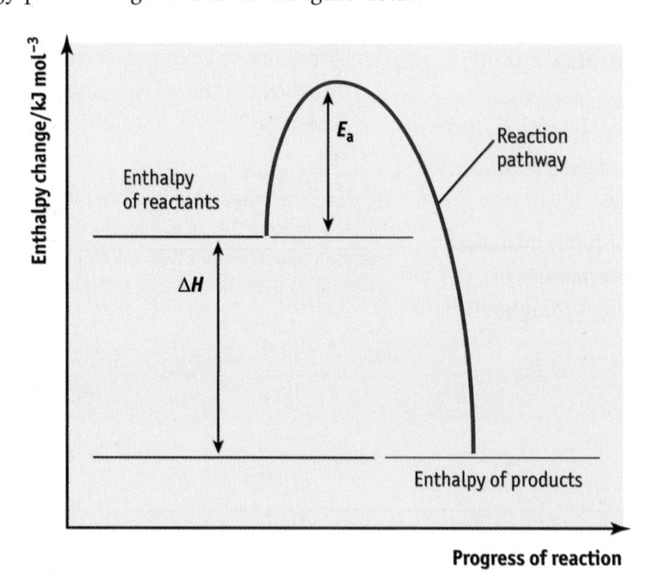

Figure 10.1 *An energy-profile diagram*

Now, remember the Boltzmann energy distribution, which applies both to the reactants and products (Figure 10.2).

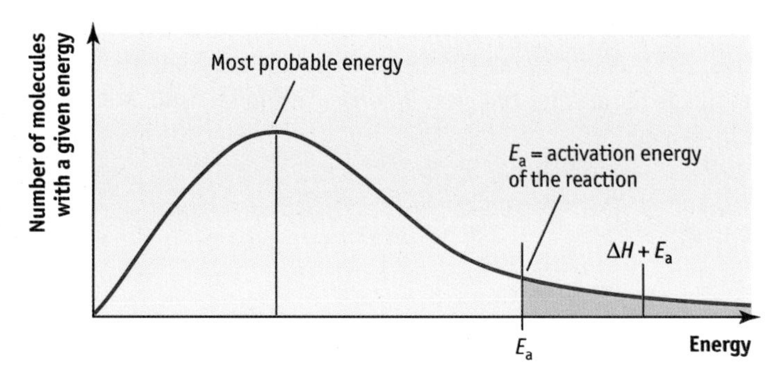

Figure 10.2 *Maxwell–Boltzmann distribution of energies*

The forward reaction requires reactants to have energy greater than the activation energy, E_a. However, the Maxwell–Boltzmann distribution of energies makes it clear that there are always at least a few product molecules that have sufficient energy to enable the reverse reaction to take place to form the reactants, i.e. they possess an energy which on collision with molecules of another product exceeds the energy shown as $\Delta H + E_a$ in Figure 10.2.

Equilibrium will therefore be achieved eventually, as the rate of the forward reaction becomes equal to the rate of the reverse reaction. The attainment of equilibrium is a general feature of reactions that exist in a closed system. The balance of the equilibrium between the reactants and products varies considerably between reactions. This difference is quantified by the equilibrium constant, K.

The equilibrium constant

It can be shown experimentally, or derived theoretically by considering the rates of individual steps in a reaction mechanism, that for the general reaction:

$$aA + bB \rightleftharpoons cC + dD$$

(where a, b, c and d are the number of molecules of the reagents A, B, C and D in a balanced equation) the equilibrium constant, K_c, can be stated as:

$$K_c = \frac{[C]^c[D]^d}{[A]^a[B]^b}$$

For example, the equilibrium constant, K_c, for the reaction:

$$2HI(g) \rightleftharpoons H_2(g) + I_2(g)$$

is 0.02 at 450°C.

Therefore:

$$\frac{[H_2(g)][I_2(g)]}{[HI(g)]^2} = 0.02$$

K_c is a temperature-dependent constant for the equilibrium. The subscript 'c' indicates that the concentrations are expressed in molar units.

Concentrations are used in the definition of an equilibrium constant. If there is a solid reactant or product in the equilibrium then it is not included in the expression for K_c.
For example, in a closed system at high temperature, an equilibrium can be established between calcium carbonate and its decomposition products calcium oxide and carbon dioxide:

$$CaCO_3(s) \rightleftharpoons CaO(s) + CO_2(g)$$

The equilibrium constant is $K_c = [CO_2(g)]$

Units of the equilibrium constant

Notice that the equilibrium constant in the example does not have any units. In this case, this is correct bacause the units of $[H_2]$ and $[I_2]$ on the top of the expression cancel with the units of $[HI]^2$ on the bottom. This always occurs when the total number of molecules of reactants equals the total number of molecules of the products. Of course, this is not always the case. In other instances, units for K_c have to be assigned by inspection.

For the equilibrium,

$$N_2(g) + 3H_2(g) \rightleftharpoons 2NH_3(g)$$

$$K_c = \frac{[NH_3(g)]^2}{[N_2(g)][H_2(g)]^3}$$

If all the reagents have concentrations measured in $mol\,dm^{-3}$:

- the top of the expression has the units $mol^2\,dm^{-6}$ (for $[NH_3]^2$)
- the bottom has the units $mol\,dm^{-3}$ (for $[N_2]$) multiplied by $mol^3\,dm^{-9}$ (for $[H_2]^3$)

Therefore, the units of K_c are $mol^2\,dm^{-6}/mol^4\,dm^{-12}$, which simplifies to $mol^{-2}\,dm^6$.

The units of other equilibrium constants are worked out similarly.

ⓔ Although it is not required for A2, you may come across the equilibrium constant for gaseous reactions defined in terms of the pressure exerted by gases in the equilibrium mixture. This is referred to as K_p. K_p is used widely in industry. If the total number of molecules of gas in the balanced equation is the same on both sides of the equilibrium, then K_p is equal to K_c. Otherwise, the values of K_c and K_p will be different.

The size and measurement of equilibrium constants

Equilibrium constants vary enormously in size. This is to be expected, since the reactants and products may have substantially different energies and one or the other may be the dominant component in an equilibrium mixture. For example, at 25°C, the rate constant for the reaction:

$$N_2(g) + O_2(g) \rightleftharpoons 2NO(g)$$

has a value of 4.7×10^{-31}. This indicates that a mixture of nitrogen and oxygen has virtually no tendency to combine to make nitrogen(II) oxide: a reassuring fact as we live in an atmosphere of these gases.

On the other hand, the stability of ozone with respect to oxygen is minutely small, since the rate constant for the equilibrium:

$$2O_3(g) \rightleftharpoons 3O_2(g)$$

has a value of $10^{55}\,mol\,dm^{-3}$.

There is an important point to be made concerning these values. The equilibrium constant provides an indication of the relative stabilities of the reactants and products but gives no indication of the rate at which this equilibrium will be established. If the activation energy is high, an unstable reactant might exist for a considerable length of time before conversion to a product occurs.

Equilibrium constants can be difficult to measure and few can be obtained easily from a simple laboratory experiment. There are two major difficulties:

- It may take a considerable time for an equilibrium to be established.
- The concentrations of the reactants or products in an equilibrium may be so low that to detect them requires specialised equipment.

Another factor is that if a chemical method is used to measure the concentration of a component in an equilibrium mixture it disturbs the equilibrium, and causes the reagent concentrations within the mixture to change.

One equilibrium constant that can be measured is that for the equilibrium between ethanol, ethanoic acid, ethyl ethanoate and water:

$$C_2H_5OH(l) + CH_3COOH(l) \rightleftharpoons CH_3COOC_2H_5(l) + H_2O(l)$$

A mixture containing known quantities of ethanol, ethanoic acid and an acid catalyst is left to come to equilibrium. It can take a week to do so, but once the equilibrium has been established the amount of ethanoic acid at equilibrium can be measured by titration against a standard alkali. Allowance has, of course, to be made for the acid catalyst. Once the amount of ethanoic acid is known the amounts of the other components in the equilibrium mixture can be calculated and the value of K_c determined.

e The assumption is made that the position of equilibrium will not change as the acid is neutralised during the course of the titration. In this particular case, the equilibrium adjusts only very slowly, so the assumption is acceptable and the value of the equilibrium constant can be considered to be reliable.

Calculations involving K_c

You will be expected to be able to understand calculations involving equilibrium constants. The following examples show what is expected.

Worked example 1

At equilibrium, the concentrations of sulphur dioxide and oxygen in the mixture

$$2SO_2(g) + O_2(g) \rightleftharpoons 2SO_3(g)$$

are found to be $[SO_2] = 0.020\,\text{mol}\,\text{dm}^{-3}$ and $[O_2] = 0.010\,\text{mol}\,\text{dm}^{-3}$.
If the value of K_c is $1.28 \times 10^4\,\text{mol}^{-1}\,\text{dm}^3$, what is the equilibrium concentration of sulphur trioxide?

Answer

$$K_c = \frac{[SO_3(g)]^2}{[SO_2(g)]^2[O_2(g)]}$$

So, $[SO_3(g)]^2 = K_c[SO_2(g)]^2[O_2(g)]$
$[SO_3(g)]^2 = 1.28 \times 10^4 \times (0.020)^2 \times 0.010 = 0.0512\,\text{mol}^2\,\text{dm}^{-6}$
equilibrium concentration of sulphur trioxide $= \sqrt{0.0512} = 0.23\,\text{mol}\,\text{dm}^{-3}$

Worked example 2

0.50 mol of carbon dioxide and 0.80 mol of hydrogen were mixed together and allowed to reach equilibrium as shown below.

$$CO_2(g) + H_2(g) \rightleftharpoons CO(g) + H_2O(g)$$

When analysed, the equilibrium mixture was found to contain 0.04 mol of carbon monoxide.
Calculate the value of K_c for this reaction.

Answer

$$CO_2(g) + H_2(g) \rightleftharpoons CO(g) + H_2O(g)$$

	Carbon dioxide	Hydrogen	Carbon monoxide	Water
Initial amount (mol at time $t = 0$)	a	b	c	d
Equilibrium amount (mol)	$(a-x)$	$(b-x)$	$(c+x)$	$(d+x)$
Initial amount	0.50	0.80	0.00	0.00
Equilibrium amount			0.04	

initial amount of CO = 0.00 mol
equilibrium amount of CO = 0.04 mol
therefore, $x = 0.04$ mol
The equilibrium amount of each reactant and product can now be calculated.
amount (in moles) of CO_2 at equilibrium 0.50 − 0.04 = 0.46 mol
amount (in moles) of H_2 = 0.80 − 0.04 = 0.76 mol
Therefore:

$$K_c = \frac{0.04 \times 0.04}{0.46 \times 0.76} = 4.6 \times 10^{-3}$$

In worked example 2, you should notice that the calculation can be completed using the amounts in moles, rather than concentrations in mol dm^{-3}. This is because there is the same number of molecules on each side of the equation, so the volume term in the expression for K_c would cancel out. If the number of molecules on each side of the equation were different, then the volume of the container would have to be known.

The effect of a change in conditions on the value of K_c

The equilibrium constant is dependent on temperature. It is *not* changed by alterating any other condition. K_c is unaffected by:

- a change in concentration
- the presence of a catalyst
- a change in pressure

The fact that K_c is unaffected by a change in pressure may seem rather surprising and this is considered further in the additional reading at the end of this chapter.

e This is a common question in the exam and is frequently answered incorrectly, so make sure that you learn these points.

It needs to be stressed that K_c gives only information on the concentrations achieved at equilibrium; it tells us nothing about how quickly equilibrium is achieved. The presence of a catalyst and/or an increase in pressure both cause the equilibrium to be established faster because both increase the rate of reaction.

To determine the effect of a change in temperature requires more careful consideration. If the temperature is raised, equilibrium is established faster but, unlike the other changes in condition, the *value* of K_c also changes. Whether it increases or decreases depends on whether the reaction is endothermic or exothermic in a particular direction.

The Haber process for the manufacture of ammonia is exothermic in the forward direction:

$$N_2(g) + 3H_2(g) \rightleftharpoons 2NH_3(g) \quad \Delta H = -93\text{ kJ mol}^{-1}$$

The equilibrium constant for this reaction is:

$$K_c = \frac{[NH_3(g)]^2}{[N_2(g)][H_2(g)]^3}$$

Applying Le Chatelier's principle to the equilibrium, an increase in temperature favours the breakdown of ammonia and the formation of nitrogen and hydrogen. This means that the concentration of ammonia decreases and the concentrations of nitrogen and hydrogen both increase. Therefore, the value of the equilibrium constant will decrease.

Worked example

Consider the reaction:

$$H_2(g) + I_2(g) \rightleftharpoons 2HI(g)$$

At 220°C, the value of the equilibrium constant, K_c, is 160; at 450°C it is 49. Decide whether the breakdown of hydrogen iodide into hydrogen and iodine is exothermic or endothermic. Explain your answer.

Answer

As the temperature rises, the value of the equilibrium constant decreases. This means that less HI is present.

The equilibrium is therefore being moved to the left-hand side of the equation as the temperature increases.

From Le Chatelier's principle, we know that this occurs when the forward reaction is exothermic.

Therefore, the breakdown of hydrogen iodide is endothermic.

Questions

1 Give the expression for the equilibrium constant, K_c for each of the following reactions.

a $N_2O_4(g) \rightleftharpoons 2NO_2(g)$

b $2N_2O(g) \rightleftharpoons 2N_2(g) + O_2(g)$

c $2C(s) + O_2(g) \rightleftharpoons 2CO(g)$

d $2CO(g) + O_2(g) \rightleftharpoons 2CO_2(g)$

2 Consider the reaction:

$$PCl_5(g) \rightleftharpoons PCl_3(g) + Cl_2(g)$$

The numerical value of K_c at a temperature T°C is 1.9.

a Write an expression for K_c for this reaction.

b Give the units of K_c.

c A sample of PCl_5 is heated in a 1 dm^3 container until equilibrium is established. When the equilibrium mixture is analysed, it is found to contain 2.085 g of PCl_5. Calculate the concentration, in g dm^{-3}, of Cl_2 that is present in the equilibrium mixture.

3 An equilibrium is established between ethanol in the blood and ethanol vapour:

$$C_2H_5OH(blood) \rightleftharpoons C_2H_5OH(g)$$

$$K_c = 4.5 \times 10^{-4}$$

Calculate the concentration of ethanol vapour in equilibrium with ethanol in the blood in a person who has a concentration of 80 mg of ethanol in 100 cm^3 of blood (the maximum concentration at which it is legal to drive in the UK).

4 Nitrogen(I) oxide decomposes to nitrogen and oxygen according to the equation:

$$2N_2O(g) \rightleftharpoons 2N_2(g) + O_2(g)$$

In an experiment, 1 mol of nitrogen(I) oxide was heated in a 1 dm^3 container until equilibrium was established. The mixture was then analysed and found to contain 0.1 mol of nitrogen(I) oxide.

a Calculate the concentrations of nitrogen and oxygen present in the equilibrium mixture.

b Calculate the equilibrium constant, K_c.

c If the experiment were repeated using 1 mol of nitrogen(I) oxide in a 2 dm^3 container, how would the value of K_c change?

5 The dissociation of chlorine into its atoms (radicals) can be accomplished in the laboratory by heating chlorine gas to 1200°C.

$$Cl_2(g) \rightleftharpoons 2Cl(g)$$

The equilibrium constant for this reaction increases as the temperature is raised.

a State whether the dissociation is exothermic or endothermic. Explain your answer.

b The dissociation occurs more readily high in the upper atmosphere in the region called the stratosphere. When produced at this height, the presence of chlorine radicals is damaging.

(i) Explain why the breakdown of chlorine occurs more readily in the stratosphere.

(ii) Explain why the production of chlorine radicals is damaging.

Summary

You should now be able to:

- understand the nature of dynamic equilibrium
- define the equilibrium constant, K_c, for a system at equilibrium and give its units
- calculate an equilibrium constant from the amounts of substance present at equilibrium
- deduce the amounts of substance present at equilibrium using appropriate data
- state that changes in concentration or pressure, or the presence of a catalyst, do not change the value of an equilibrium constant
- use Le Chatelier's principle to explain how a change in temperature affects the value of the equilibrium constant

Additional reading

10

Equilibrium constants and pressure

A change in pressure does not change the value of K_c for a reaction in equilibrium. Why then is the pressure increased in industrial processes, such as the Haber process, in which the volume of the products is less than that of the reactants? The first answer is that the increase in pressure results in an increase in the rate of combination of the reactants and the product is formed more quickly, which is an important factor for an industrial process. The product is then extracted long before equilibrium is achieved.

However, the second answer is that even if the reaction is allowed to come to equilibrium the yield of the product will be increased. This seems at first sight to contradict the statement about the constant value of K_c, but, as we shall see, this is not the case.

Consider the reaction:

$$N_2(g) + 3H_2(g) \rightleftharpoons 2NH_3(g)$$

$$K_c = \frac{[NH_3(g)]^2}{[N_2(g)][H_2(g)]^3}$$

The square brackets indicate concentrations. Therefore, if the reaction takes place in a reacting volume, $V\,dm^3$, the equilibrium expression can be rewritten in terms of the amount n (in moles) of the three gases present. In each case, the concentration is the amount (in moles) divided by the volume.

$$K_c = \frac{(n_{NH_3(g)}/V)^2}{(n_{N_2(g)}/V)(n_{H_2(g)}/V)^3}$$

This simplifies to

$$K_c = \frac{(n_{NH_3(g)})^2 \times V^2}{(n_{N_2(g)})(n_{H_2(g)})^3}$$

When the pressure is increased, the volume, V, decreases. The value of K_c is kept constant by the number of moles of ammonia ($n_{NH_3(g)}$) increasing. Therefore, more ammonia is obtained despite the fact that K_c remains constant.

Equilibrium constants and rate constants

Equilibrium is achieved when the rate of the reverse reaction becomes equal to the rate of the forward reaction. It is then easy to see how K_c can be defined. For example, consider the following established equilibrium:

$$A(g) + B(g) \rightleftharpoons C(g) + D(g)$$

Suppose that in both directions the orders with respect to all the reagents is first order, then:

rate of the forward reaction = k_1[A][B]
rate of the reverse reaction = k_2[C][D]
at equilibrium, k_1[A][B] = k_2[C][D]

$$K_c = \frac{k_1}{k_2} = \frac{[C][D]}{[A][B]}$$

However, it would appear that this relationship will only be valid if the mechanism is as simple as the one described above. What, for example, would be the situation for the reaction:

$$2O_3(g) \rightleftharpoons 3O_2(g)$$

where, under appropriate conditions (UV is required), the forward reaction is first order with respect to O_3? Despite this, the equilibrium expression can be written as:

$$K_c = \frac{[O_2(g)]^3}{[O_3(g)]^2}$$

To explain how this is possible, the mechanism of this reaction must be considered. There are two steps:

Step 1: $O_3(g) \rightleftharpoons O_2(g) + O(g)$

Step 2: $O_3(g) + O(g) \rightleftharpoons 2O_2(g)$

- In step 1, the forward reaction is slow.
- In step 2, the forward reaction is fast.

This explains why the reaction is first order with respect to ozone.

However, once an overall equilibrium has been achieved, both individual steps will also be in equilibrium.

Hence in step 1, $K_1 = \frac{[O_2(g)][O(g)]}{[O_3(g)]}$

Hence in step 2, $K_2 = \frac{[O_2(g)]^2}{[O_3(g)][O(g)]}$

If K_1 is multiplied by K_2, a new constant K_c is obtained. It can be seen that:

$$K_c = \frac{[O_2(g)]^3}{[O_3(g)]^2}$$

Therefore, the expression for the equilibrium constant can be stated regardless of the number of steps in the reaction mechanism and which step is rate-determining.

The units of the equilibrium constant

In this chapter, there is a section showing how the units of an equilibrium constant are established. Strictly, however, as is explained below, equilibrium constants should not have units.

e Be aware that this section is added for completeness. You should *always* supply units for equilibrium constants in answers to exam questions.

This is because the equilibrium constant is only quantitatively correct under ideal conditions. In this respect, it is similar to the behaviour of gases. The gas laws apply only if the assumption is made that the gas particles do not have any attraction or repulsion between them or interfere with each other in any way. In the case of a gas, this is only true under particular conditions, such as low pressure.

Within a reaction mixture at equilibrium, there are attractive and repulsive forces between the reactants and products due to the charge separations that occur within the molecules or ions involved. If the conditions are such that this interaction is minimal then the equilibrium constant gets close to the theoretical value. It can be anticipated that this will occur in very dilute solutions or for gases at low pressure. Under all other conditions, the concentrations of the reagents are effectively diminished by a certain amount.

In order to address this problem, equilibrium constants are defined more accurately in terms of 'the activity of the material', i.e. the amount that is actually contributing to the

establishment of the equilibrium. The definition of the equilibrium constant is otherwise the same. So for the Haber process:

$$K_c = \frac{(a_{NH_3(g)})^2}{(a_{N_2(g)}) \times (a_{H_2(g)})^3}$$

where a is the activity of the gas.

All that is different is that concentration has been replaced by 'activity'.

Activity is defined as a comparison of the behaviour of the substance at the concentration being considered against the behaviour under a standard condition. So, in general terms:

$$a = \gamma \frac{[\text{concentration considered}]}{[\text{concentration under standard conditions}]}$$

where γ is a multiplication factor known as the activity coefficient.

The activity cannot have units since it is a comparison between two concentrations. It follows, therefore, that the equilibrium constant is also without units.

Stretch and challenge

The equilibrium constant for the acid-catalysed reaction between ethanol and ethanoic acid to form ethyl ethanoate and water can be determined in the laboratory.

$$C_2H_5OH(l) + CH_3COOH(l) \rightleftharpoons CH_3COOC_2H_5(l) + H_2O(l)$$

In an experiment, 20.0 cm^3 of ethanoic acid, 20.0 cm^3 ethanol and 30.0 cm^3 of ethyl ethanoate are mixed together.

10.0 cm^3 of 0.5 $mol\,dm^{-3}$ sulphuric acid is then added. This provides the necessary acid catalyst and 10.0 cm^3 of water.

The mixture is left for 1 week to come to equilibrium.

1.00 cm^3 of the mixture is then pipetted into a small conical flask. 15.0 cm^3 of water is added to increase the volume and the mixture is then titrated against a 0.250 $mol\,dm^{-3}$ solution of sodium hydroxide. It is found that 14.50 cm^3 of sodium hydroxide solution is required to neutralise the acid in the mixture.

a Given the following densities, calculate the amount (in moles) of ethanol, ethanoic acid, ethyl ethanoate and water that were present at the start of the experiment:

$C_2H_5OH(l) = 0.79\,g\,cm^{-3}$
$CH_3COOH(l) = 1.05\,g\,cm^{-3}$
$CH_3COOC_2H_5(l) = 0.92\,g\,cm^{-3}$
$H_2O(l) = 1.00\,g\,cm^{-3}$

b Use the titration result to determine the amount (in moles) of ethanoic acid that is present in the 1.00 cm^3 sample taken after equilibrium had been established.

c Use your answer from part **b** to determine the amount, in moles, of ethanoic acid present in the full 80 cm^3 of the equilibrium mixture.

d From your answer to part **c** calculate the amount, in moles, of ethanol, ethyl ethanoate and water present in the equilibrium mixture.

e Calculate the value of the equilibrium constant for this reaction.

f Explain why the calculation can be carried out using the amounts in moles, rather than the concentrations in $mol\,dm^{-3}$.

g Calculate the enthalpy change for the forward reaction, given the following values for the enthalpies of formation.

$\Delta H_f(C_2H_5OH) = -277.7\ kJ\ mol^{-1}$
$\Delta H_f(CH_3COOH) = -484.5\ kJ\ mol^{-1}$
$\Delta H_f(CH_3COOC_2H_5) = -485.8\ kJ\ mol^{-1}$
$\Delta H_f(H_2O) = -285.9\ kJ\ mol^{-1}$

h Use your answer to part **g** to comment on the value of the equilibrium constant.

i The answer obtained in part **e** is smaller than the equilibrium constant at 25°C that is usually quoted in data books. It is suggested that a reason for the difference was that the experiment described above was carried out at 20°C. Could this account, at least in part, for the difference in the equilibrium constant obtained? Explain your answer.

Acids, bases and buffers

Chapter 11

You are already familiar with the reactions of acids with oxides, hydroxides and carbonates. You may also have considered the redox reactions of some metals with acids. This chapter looks at the behaviour of acids in more detail and distinguishes between strong acids and weak acids. A definition of pH is given and ways of controlling acidity through the use of buffer solutions are discussed.

In some cases, it is possible to determine acidity quite precisely. The methods required to do this are considered in some detail.

Ionic equations

You will have written some ionic equations during the AS course, However, before proceeding with the A2 work on acids, bases and buffers, it is necessary to extend your knowledge.

e You may be familiar with the content of this section. If so, move on to the explanation of the Brønsted–Lowry theory that follows.

Ionic equations draw attention to the particles that react and ignore those that, although present, take no part in the reaction.

You should recall that ionic bonds are formed by the electrical attraction between positive and negative ions although the ions remain separate from each other. When an ionic compound is melted or made into a solution the ions become detached and the positive and negative ions move around freely. This gives them the capacity to react independently. The bonds in a covalent substance involve the sharing of electrons and the substance stays as a complete entity even when melted or dissolved.

The distinction between ionic and covalent bonding is important. The reactions of ionic substances in solution are usually due to just one of the ions present in the substance whereas covalent compounds react as molecules.

An example of an ionic reaction

When aqueous silver nitrate reacts with aqueous sodium chloride to produce insoluble silver chloride, it is the silver ions from the aqueous silver nitrate that react with the chloride ions from the aqueous sodium chloride to produce a precipitate of silver chloride. The equation for the reaction is:

$$AgNO_3(aq) + NaCl(aq) \rightarrow AgCl(s) + NaNO_3(aq)$$

Writing the equation as:

$$Ag^+(aq) + Cl^-(aq) \rightarrow AgCl(s)$$

makes it clear that it is the aqueous silver ion and aqueous chloride ion that are reacting. The nitrate and sodium ions are present but do not react. Such ions are sometimes referred to as **spectator ions**.

Ionic equations are used extensively in the chapters that follow and are discussed in more detail in Chapter 14.

The Brønsted–Lowry definition of an acid

At a simple level, an acid is defined as a substance that supplies hydrogen ions in aqueous solution and a base is defined as the species that accepts hydrogen ions. The base is often an oxide or a hydroxide. The Brønsted–Lowry theory of acids and bases extends these ideas.

A hydrogen ion, H^+, is formed when a hydrogen atom loses its only electron. It is, therefore, a proton. Brønsted and Lowry proposed the following definitions:

An acid is a species that donates a proton to another species in a reaction.

A base is a species that accepts a proton from an acid in a reaction.

In 1923, Brønsted and Lowry, working independently, both suggested the same theory of acids and bases which is therefore associated with both their names.

www.geocities.com/bioelectrochemistry/bronsted.htm

www.geocities.com/bioelectrochemistry/lowry.htm

Johannes Nicolaus Brønsted (1879–1947) Professor of Chemistry, University of Copenhagen, Denmark

Thomas Martin Lowry (1874–1936) Professor of Physical Chemistry, University of Cambridge, England

At first glance, this may seem to be a repetition of the previous definition. However, some examples should enable you to see that it extends the scope of reactions that are considered to be acid–base.

The reaction between aqueous sodium hydroxide and aqueous hydrochloric acid is:

$$NaOH(aq) + HCl(aq) \rightarrow NaCl(aq) + H_2O(l)$$

The HCl reacts by releasing its hydrogen ion (i.e. its proton), which is received by the OH^- from the NaOH to form water.

This is made explicit by the ionic equation:

$$H^+(aq) + OH^-(aq) \rightarrow H_2O(l)$$

A further example is a reaction that is not so obviously an acid–base reaction. Aqueous sodium carbonate reacts with aqueous hydrochloric acid to form aqueous sodium chloride, carbon dioxide and water:

$$Na_2CO_3(aq) + 2HCl(aq) \rightarrow 2NaCl(aq) + CO_2(g) + H_2O(l)$$

The ionic equation is:

$$CO_3^{2-}(aq) + 2H^+(aq) \rightarrow CO_2(g) + H_2O(l)$$

The acid donates 2 protons to the carbonate which then splits into carbon dioxide and water. The carbonate is, therefore, a base in the reaction.

As explained in Chapter 10, reactions are more properly considered as equilibria between the reactants and products. An example is the dissolving of ammonia in water. Ammonium ions and hydroxide ions are produced:

$$NH_3(g) + H_2O(l) \rightleftharpoons NH_4^+(aq) + OH^-(aq)$$

In the forward reaction, the water donates a proton to the ammonia. Therefore, water is the acid and ammonia is the base. For the reverse reaction, the ammonium ion is the acid and the hydroxide ion is the base.

In summary:

$$\underset{\text{Base 2}}{NH_3(g)} + \underset{\text{Acid 1}}{H_2O(l)} \rightleftharpoons \underset{\text{Acid 2}}{NH_4^+(aq)} + \underset{\text{Base 1}}{OH^-(aq)}$$

If acid–base reactions are considered as equilibria, then each side of the equation contains an acid and a base. The related species are called **conjugate acid–base pairs**.

In the example above, the NH_4^+ and NH_3 are a conjugate acid–base pair; NH_4^+ is the acid and NH_3 is its conjugate base. The H_2O is an acid and OH^- is its conjugate base. The species with the extra proton is *always* the acid and the species without the proton is *always* the base.

In some circumstances, NH_3 is able to lose a proton to form the ion NH_2^-. In this case, the NH_3 and NH_2^- are an acid–base pair, with NH_3 as the conjugate acid and NH_2^- as its conjugate base.

Another example is the formation of the nitration mixture for a reaction with benzene (see pages 10–11). A mixture of concentrated sulphuric and nitric acids is used, which forms the equilibrium:

$$H_2SO_4 + HNO_3 \rightleftharpoons HSO_4^- + H_2NO_3^+$$

Acid 1 Base 2 Base 1 Acid 2

A given species may be an acid in one circumstance and a base in another.

In this reaction:

- H_2SO_4 is an acid; its conjugate base is HSO_4^-.
- HNO_3 is a base; its conjugate acid is $H_2NO_3^+$.

It may seem strange to refer to nitric acid as a base but, in this circumstance, that is what it is.

The Brønsted–Lowry definition extends the idea of acids into a more comprehensive theory of reactions involving proton transfer. This is analogous to the way that the definitions of oxidation and reduction are extended beyond the addition or removal of oxygen to the theory of electron transfer.

Strong and weak acids

Acids vary considerably in the ease with which they are able to release hydrogen ions. When the mineral acids (sulphuric, nitric and hydrochloric) are dissolved in water they break up almost completely into their constituent ions. Although equilibria exist theoretically, the acids are considered to be 100% ionised. They are referred to as **strong acids**. By contrast, when dissolved, organic acids (e.g. ethanoic acid) are only partially ionised and a significant concentration of molecular acid still exists. These acids are referred to as **weak acids**.

- A strong acid is highly ionised in aqueous solution.
- A concentrated acid is made by dissolving a large amount of an acid in a small volume of water.
- A weak acid is partially ionised in aqueous solution.
- A dilute acid is made by dissolving a small amount of an acid in a large volume of water.

e It is important that you do not muddle the use of the word 'strong' with 'concentrated' or 'weak' with 'dilute'.

A similar distinction is made for bases, although in the A2 course you will encounter very few weak bases. Amines and ammonia are weak bases because they are protonated to only a small extent; all metal hydroxides are strong bases.

e Do not make the mistake of using the word 'weak' when you mean 'dilute'. For example, limewater (aqueous calcium hydroxide) is a strong base but, as it has low solubility, a solution of limewater is always dilute.

The usual way of indicating the strength of an acid is to use the equilibrium constant for its ionisation in water. For example, ethanoic acid ionises as:

$$CH_3COOH(aq) \rightleftharpoons CH_3COO^-(aq) + H^+(aq)$$

e In some books you will see $H^+(aq)$ written as H_3O^+. This indicates correctly that the 'aq' attachment represents one H_2O. This course does not require you to represent aqueous hydrogen ions as H_3O^+; it makes no difference to your understanding of the concepts under discussion.

As with the other equilibria (Chapter 10), an equilibrium constant can be defined for this reaction. It is:

$$K_a = \frac{[CH_3COO^-(aq)][H^+(aq)]}{[CH_3COOH(aq)]}$$

To indicate that the reaction involves an acid it is conventional to provide the equilibrium constant K with the subscript 'a'. K_a is called the **acid dissociation constant**.

For ethanoic acid, K_a has the value 1.7×10^{-5} mol dm^{-3}. This makes it clear that a solution of ethanoic acid consists largely of ethanoic acid molecules with relatively few ethanoate ions and hydrogen ions.

The K_a value of methanoic acid is 1.6×10^{-4} mol dm^{-3}, which is almost ten times larger than that for ethanoic acid. This tells us that methanoic acid, although it is a weak acid, is stronger than ethanoic acid.

The mineral acids have much larger K_a values. For example, the K_a for nitric acid is approximately 40 mol dm^{-3} and that for sulphuric acid is often listed simply as 'very large'.

Calculating hydrogen ion concentration

Strong acids are assumed to be ionised completely when they are dissolved in water. This means that the hydrogen ion concentration is related directly to the amount of acid dissolved.

$$HCl(aq) \rightarrow H^+(aq) + Cl^-(aq)$$

For example, 0.0500 mol dm^{-3} HCl will produce 0.0500 mol dm^{-3} of $H^+(aq)$.

To determine the hydrogen ion concentration of a weak acid is more complicated because there is a significant amount of the undissociated (un-ionised) acid present in the solution. This means that it is necessary to refer to the value of K_a for the acid. Using ethanoic acid as an example, the hydrogen ion concentration present in a 0.05 mol dm^{-3} solution is calculated as follows:

$$K_a = 1.7 \times 10^{-5} = \frac{[CH_3COO^-(aq)][H^+(aq)]}{[CH_3COOH(aq)]}$$

Every time a molecule of the acid dissociates, an H^+ ion is formed together with a CH_3COO^- ion. This means that the concentration of CH_3COO^- is always the same as that of H^+.

So, if $[H^+(aq)]$ equals x mol dm^{-3}, then $[CH_3COO^-(aq)]$ also equals x mol dm^{-3}, and the equation can be written as:

$$K_a = 1.7 \times 10^{-5} = \frac{x^2}{[CH_3COOH(aq)]}$$

The concentration of ethanoic acid in this equation is the equilibrium concentration.

However, if the value of K_a is small (as it is in this case) then the concentration of undissociated acid at equilibrium will not be much less than the initial concentration. It is therefore reasonable to take the concentration at equilibrium as being the same as the initial concentration, i.e. 0.05 mol dm^{-3}.

Therefore,

$$K_a = 1.7 \times 10^{-5} = \frac{x^2}{0.05}$$

e If you doubt that this is a fair approximation, then it is possible to solve the precise equation:

$$K_a = 1.7 \times 10^{-5} = \frac{x^2}{0.05 - x}$$

If you carry out this calculation, you will find that the value obtained is $9.1(3) \times 10^{-4}$.
Only for very accurate work would this difference be significant.

$x^2 = 0.05 \times 1.7 \times 10^{-5}$

$x = 9.2(2) \times 10^{-4}\,mol\,dm^{-3}$

You can see that the hydrogen ion concentration in 0.05 mol dm^{-3} ethanoic acid is appreciably less than that present in 0.05 mol dm^{-3} hydrochloric acid.

For a weak acid, the calculation can be summarised thus: $[H^+] = \sqrt{K_a \times c}$

where K_a is the acid dissociation constant and c is the concentration of the solution.

pH and pK_a

The acidity of substances (see Figure 11.1) covers a range of values from concentrations as high as 10 mol dm^{-3} (and occasionally higher) to concentrations as low as 10^{-14} mol dm^{-3}. Although these can be expressed using [H$^+$] values, the pH scale is more convenient.

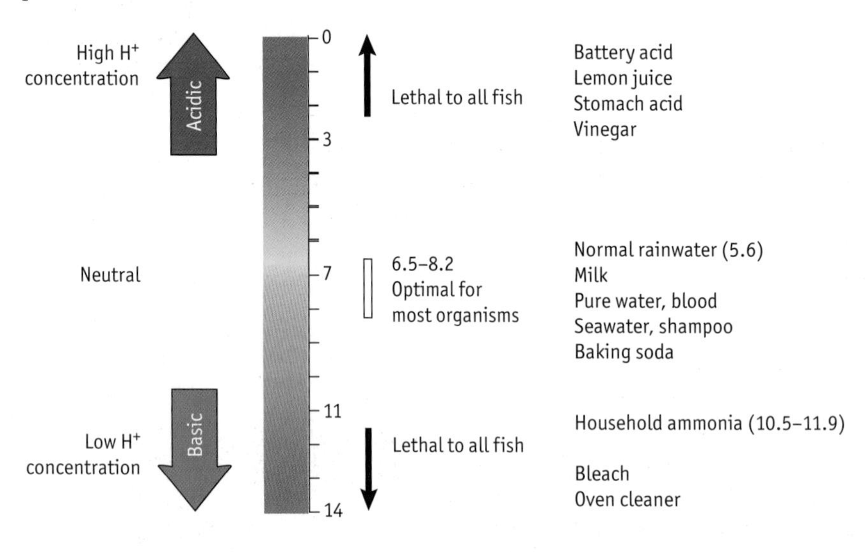

Figure 11.1 *pH scale and pH of some common substances*

$pH = -\log[H^+]$

The pH of 0.1 mol dm^{-3} HCl is therefore equal to 1 (that is $-\log 0.1$)

Do not worry if you are not familiar with the use of logarithms. They are used to convert a scale of numbers in powers of 10 to a linear scale — for example

$\log 100 = 2$ (as 100 is 10^2)

$\log 10 = 1$ (as 10 is 10^1)

$\log 1 = 0$ (as 1 is 10^0)

$\log 0.01 = -2$ (as 0.01 is 10^{-2}) and so on.

Intermediate values, for example $\log 0.05 = -1.3$, can be obtained by using the 'log' button on a calculator.

The pH of 0.05 mol dm^{-3} HCl is 1.3.

The pH of 0.05 mol dm^{-3} ethanoic acid is $-\log \sqrt{K_a c}$

$= -\log \sqrt{(1.7 \times 10^{-5}) \times 0.05}$

$= -\log (9.2 \times 10^{-4}) = 3.04$

The pH scale is useful because it allows a wide range of hydrogen ion concentrations to be expressed as simple positive values.

The conversion of pH back to hydrogen ion concentration is obtained from

$[H^+] = 10^{-pH}$

For example, if a solution has a pH of 2.8 then $[H^+] = 0.0016$ or 1.6×10^{-3} mol dm^{-3}.

e This type of conversion is often carried out incorrectly in exams. Make sure that you understand how to use your calculator to carry out this process.

A similar scale can be used to give the values of the acid dissociation constant.

$\mathbf{pK_a = -\log[K_a]}$

$\mathbf{K_a = 10^{-pK_a}}$

The role of water

So far the role of water in the determination of pH has not been considered. Yet it has a crucial role in controlling the pH of aqueous solutions. First, it is required to allow acids (which are covalent in their anhydrous state) to ionise. Second, it makes a contribution in its own right. Water is substantially covalent but does ionise to a small extent:

$H_2O \rightleftharpoons H^+ + OH^-$

Although it dissociates to produce hydrogen ions, it is neutral because the hydrogen ions are balanced by an equal number of hydroxide ions. It is this

Sørensen was a Danish chemist. He suggested the use of the pH scale to avoid the use of the very small numbers that may result from using hydrogen ion concentrations. As a boy he suffered from epilepsy and he had a pronounced stammer. However, this did not stop him from being appointed director of the Carlsberg Laboratory in Copenhagen in 1901. This institution was financed by the founder of the Carlsberg brewery and its stated purpose was to increase the knowledge of brewing and fermenting. However, it has since widened its scope to cover a range of chemical research topics. Sørensen remained as director for 37 years.

Søren Peter Sørensen (1868–1939)

TOPFOTO

dissociation that defines what is meant by neutral on the pH scale. At room temperature, the concentration of hydrogen ions is 10^{-7} mol dm^{-3}, which means that the pH is 7.

The equilibrium constant for this reaction is:

$$K = \frac{[H^+][OH^-]}{[H_2O]}$$

However, the water is present in such a large excess that its concentration scarcely changes when a substance dissolves in it. So, its concentration is regarded as being a constant and the expresson for the equilibrium constant is simplified to:

$$K_w = [H^+][OH^-]$$

At 298 K, the equilibrium constant has the value 10^{-14} because $[H^+] = [OH^-] = 10^{-7}$ mol dm^{-3}.

K_w is called the **ionic product of water**; the subscript 'w' indicates that the equilibrium constant refers to water.

The pH of strong bases

It is important to realise that, in aqueous solutions, the ionic product of water is often the controlling influence on pH.

When an acid is dissolved in water, it releases so many hydrogen ions that the small contribution of hydrogen ions from the water is completely insignificant. This is particularly so because the addition of hydrogen ions would drive the equilibrium $H_2O \rightleftharpoons H^+ + OH^-$ to the left. On the other hand, water is the reason why even the most alkaline solutions contain some H^+ ions.

A 0.1 mol dm^{-3} solution of sodium hydroxide contains 0.1 mol dm^{-3} of OH^- ions. (It is a strong base and therefore dissociates fully.)

$[H^+][OH^-]$ is always 10^{-14} mol^2 dm^{-6}

So, $[H^+] \times 0.1 = 10^{-14}$ mol dm^{-3}

Therefore, this solution has a pH of 13.

Worked example

What is the pH of a 0.02 mol dm^{-3} solution of potassium hydroxide?

Answer

$[H^+][0.02] = 10^{-14}$

Therefore $[H^+] = 10^{-14}/0.02 = 5 \times 10^{-13}$ mol dm^{-3}

$pH = -\log(5 \times 10^{-13}) = 12.3$

If you tend to make mistakes when dealing with logarithms of small numbers you should note that $[H^+][OH^-] = 10^{-14}$ can be put in a logarithmic form as:

pH + pOH = 14 where pOH is $-\log[OH^-]$

So, in the worked example above:

$$pH = 14 + \log 0.02 = 12.3$$

Buffer solutions

A change in pH has a significant effect on a range of reactions. Examples of processes that depend on the maintenance of a particular pH include:

- the effectiveness of blood in transporting oxygen
- the stability of many molecules in living systems
- the effective use of chromatography to separate some components
- the safety of many products, for example shampoos

There are natural processes that regulate pH in plants and animals, such as the hydrogen carbonate ion system discussed on page 161. It is also possible to construct mixtures in the laboratory that have the effect of resisting changes in pH. A solution that is able to do this is called a **buffer solution.**

A buffer solution is a solution that resists a change in pH when a small quantity of acid or alkali is added.

There are many solutions that are capable of buffering. A simple example is the mixture of a weak acid and the salt of the weak acid. Consider the case of a mixture of ethanoic acid and sodium ethanoate. There are two points to note about this mixture:

- Sodium ethanoate is ionic and will be fully ionised, thereby contributing a large number of ethanoate ions to the mixture.
- The weak ionisation of the ethanoic acid is effectively reduced when the ethanoate ions from the salt are added because the equilibrium

  $$CH_3COOH(aq) \rightleftharpoons CH_3COO^-(aq) + H^+(aq)$$

  is forced to the left (Le Chatelier's principle). This in turn reduces the number of hydrogen, H^+, ions in solution.

Therefore, the mixture will contain a large number of undissociated ethanoic acid molecules, $CH_3COOH(aq)$, and a large number of ethanoate ions, $CH_3COO^-(aq)$.

The second bullet point above means that when sodium ethanoate is added, the pH of the ethanoic acid will be raised because the concentration of hydrogen ions will be reduced.

The mixture acts as a buffer solution. When hydrogen ions are added, ethanoate ions in the mixture combine with the hydrogen ions to form more undissociated ethanoic acid. Therefore, there is only a small effect on the overall pH.

When hydroxide ions are added, they react with the existing hydrogen ions to form water molecules. However, because there is a reserve of ethanoic acid molecules in the mixture, these ionise to replace the lost hydrogen ions (i.e. by Le Chatelier's principle, the equilibrium position shifts to the right). The pH is changed very little.

There is a limit to the range in which the buffer solution will be effective because its action is controlled by the size of the reserve of ethanoate ions and ethanoic acid molecules. However, buffer solutions are effective in providing control of the small changes that occur through the limited addition of hydrogen ions or hydroxide ions.

Lakes are sensitive to acidity. They need to be buffered if aquatic organisms are to be preserved. The chart below shows the ability of various organisms to survive a reduction of pH. Snails are sensitive to pH; frogs can exist in quite acidic water.

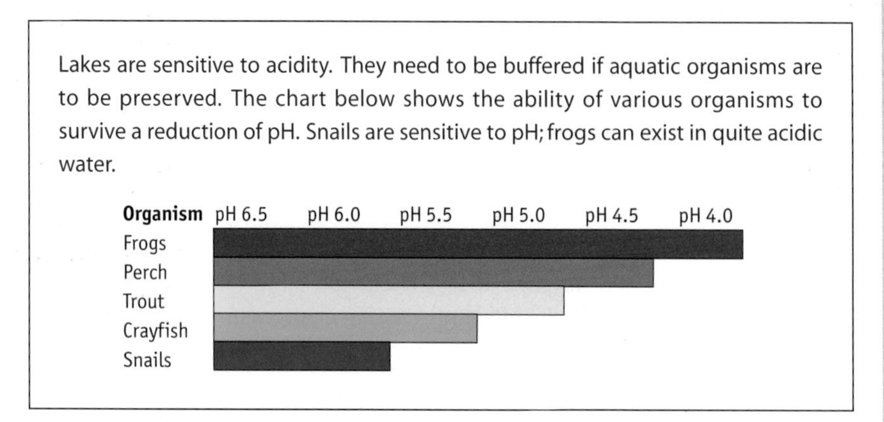

The calculation of the pH of a buffer solution

The ethanoic acid–ethanoate ion buffer system is controlled by the behaviour of the equilibrium:

$$CH_3COOH(aq) \rightleftharpoons CH_3COO^-(aq) + H^-(aq)$$

The equilibrium constant can be written as:

$$K_a = 1.7 \times 10^{-5} = \frac{[CH_3COO^-(aq)][H^+(aq)]}{[CH_3COOH(aq)]}$$

The ethanoic acid in the buffer mixture is present substantially as undissociated molecules and the ethanoate ions come almost wholly from the added salt. To calculate the pH of a buffer solution, the assumption is made that $[CH_3COOH(aq)]$ can be taken to be the original concentration of acid and $[CH_3COO^-(aq)]$ to be the same as the original concentration of the salt.

This is a good approximation that simplifies greatly the calculation of the pH of buffer solutions.

The general equation for a buffer comprising a weak acid and a salt of that weak acid is:

$$K_a = \frac{[\text{salt}][H^+]}{[\text{acid}]}$$

Therefore,

$$\frac{K_a[\text{acid}]}{[\text{salt}]} = [H^+]$$

$$pH = -\log [H^+]$$

Therefore, $pH = -\log\left(\frac{K_a[\text{acid}]}{[\text{salt}]}\right)$

> ***Worked example***
> Calculate the pH of a buffer solution containing 0.50 mol dm^{-3} ethanoic acid and 0.10 mol dm^{-3} sodium ethanoate.
>
> **Answer**
> Substituting into the expression for K_a:
>
> $$K_a = 1.7 \times 10^{-5} = \frac{[0.10][H^+]}{[0.50]}$$
>
> Therefore $[H^+] = 5.0 \times 1.7 \times 10^{-5}$
> $= 8.5 \times 10^{-5}$ mol dm^{-3}
>
> pH $= -\log(8.5 \times 10^{-5})$
> $= 4.07$

An alternative method of carrying out the calculation in the worked example above is to take logs of the expression for the equilibrium constant, which gives:

$$pH = pK_a + \log\left(\frac{[CH_3COO^-(aq)]}{[CH_3COOH(aq)]}\right)$$

So in the worked example above:

$$pH = -\log(1.7 \times 10^{-5}) + \log(0.10/0.50)$$

$$= 4.77 - 0.69$$

$$= 4.07$$

The hydrogen carbonate ion buffer

An example of a natural buffering ion is the hydrogen carbonate ion, HCO_3^-. It is responsible for the maintenance of blood pH and of the pH in lakes and streams. It works as a buffer because in water it forms the following equilibrium system:

$$CO_2 + H_2O \rightleftharpoons H_2CO_3 \rightleftharpoons H^+ + HCO_3^-$$

Addition of acid pushes the positions of both equilibria to the left. Carbonic acid (a weak acid) is formed which can decompose to form carbon dioxide. If more acidity is required, the reverse process takes place and more HCO_3^- and H^+ ions are formed.

In blood plasma, the pH is held at a value about 7.4 by these equilibria.

e A possible trap in working out the pH of buffer solutions is not remembering that when two solutions are mixed there is a dilution. For example, if 25 cm^3 of 0.10 mol dm^{-3} sodium ethanoate were mixed with 75 cm^3 of 0.5 mol dm^{-3} ethanoic acid, the total volume would be 100 cm^3 and the concentrations of the components would be:

$[CH_3COO^-] = (25/100) \times 0.10 = 0.025$ mol dm^{-3}

and

$[CH_3COOH] = (75/100) \times 0.50 = 0.375$ mol dm^{-3}

pH titration curves

During an acid–base titration, the pH of the solution being titrated changes. The way in which it changes depends on the strength of the acid and alkali being used. The most straightforward case is that of a strong acid being titrated against a strong base.

Suppose that 25 cm^3 of 0.1 mol dm^{-3} hydrochloric acid are titrated with 0.1 mol dm^{-3} sodium hydroxide solution. A plot of pH against the volume of sodium hydroxide solution added is shown in Figure 11.2.

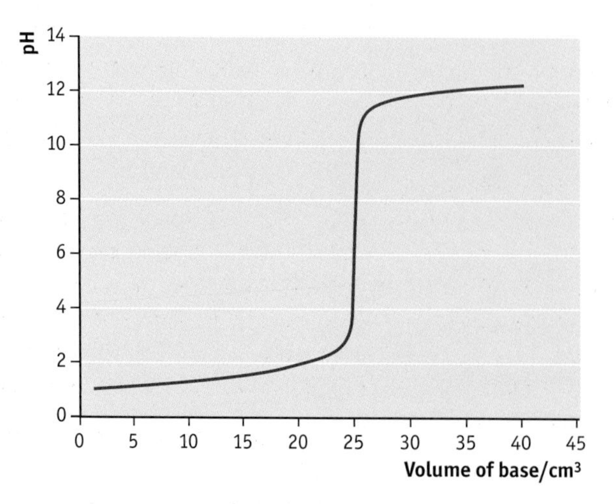

Figure 11.2
Titration of a strong acid with a strong base

The initial pH of 0.1 mol dm^{-3} HCl is 1 (–log 0.1). This begins to rise once the sodium hydroxide is added. What is perhaps surprising is that the increase in pH is quite gradual until about 22 cm^3 of the alkali has been added. There is then a slightly more pronounced increase until, at 25 cm^3, the line of the graph becomes almost vertical as the pH rises from about 3 to 11, indicating that the end point of the titration has been reached. The rise in pH then begins to tail off and reaches an almost steady figure of between 12 and 13.

The reaction is complete when the 25 cm^3 of hydrochloric acid is neutralised by 25 cm^3 of sodium hydroxide solution, so it would be expected that a pH of 7 would be recorded. However, an important feature of the graph is that the addition of a small amount of alkali after the end point results in a sharp change in pH.

A graph illustrating the pH change during the titration of a weak acid with a strong base has a differently shaped curve.

Suppose that 25 cm^3 of 0.1 mol dm^{-3} ethanoic acid are titrated with 0.1 mol dm^{-3} sodium hydroxide, A plot of pH against volume of sodium hydroxide solution added is shown in Figure 11.3.

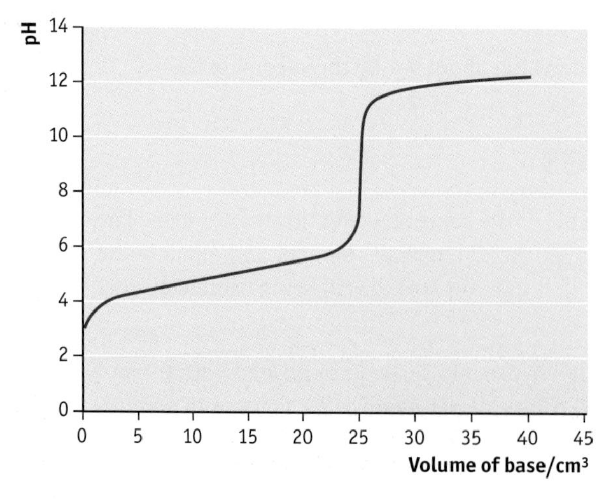

Figure 11.3
Titration of a weak acid with a strong base

It is possible to show by calculation that the pH curve for the titration of a strong acid with a strong base has the shape shown in Figure 11.2.

For example, after 15 cm^3 of 0.10 mol dm^{-3} sodium hydroxide have been added to 25 cm^3 of 0.10 mol dm^{-3} of hydrochloric acid, 10 cm^3 of the acid remains unneutralised. This 10 cm^3 will have been diluted to the total volume of the solution — 25 + 15 = 40 cm^3. Its concentration is therefore (10/40) × 0.10 = 0.025 mol dm^{-3}. It will therefore have a pH of –log 0.025, which is 1.60.

Similar calculations would enable you to calculate pH values up to the end point. At the end point the pH is 7; after the end point it is the hydroxide ions from the sodium hydroxide that control the pH.

After 27 cm^3 of sodium hydroxide have been added, there will be an excess of 2 cm^3 of the sodium hydroxide that remains unreacted. Remembering the dilution, the hydroxide concentration will be (2/52) × 0.10 = 0.0039 mol dm^{-3}. This has a pH of 14 + log 0.0039 = 11.59.

Other points on the pH curve can be calculated similarly.

There is a sharp change of pH at the end point, but its range (from about 6.5 to 11) is not as great as when a strong acid is used. The shape after the acid is neutralised is identical to that in the strong acid–strong base example. However, before neutralisation, the pH rises quite steadily. It is worth looking at this in a little detail.

As soon as some sodium hydroxide is added, neutralisation begins and sodium ethanoate is formed. Therefore, a mixture of ethanoic acid and sodium ethanoate is present. The ethanoate ions from the sodium ethanoate cause the pH to rise as the ethanoic acid equilibrium moves to produce more ethanoic acid molecules. The quite sharp rise in pH as the first drops of sodium hydroxide are added reflects this. Sunsequently, the mixture acts as a buffer solution. Further rises in pH are contained until, close to the end point, the absence of remaining ethanoic acid means that the buffering effect is limited.

The titration curves of a strong acid with a weak base and of a weak acid with a weak base can be inferred from Figures 11.2 and 11.3.

A graph for the titration of a strong acid with a weak base is shown in Figure 11.4.

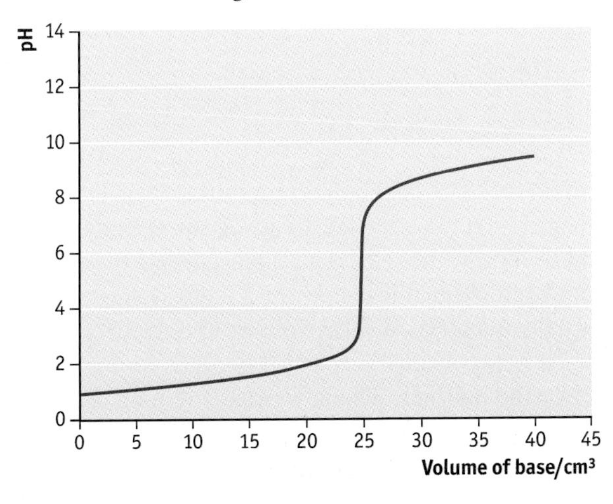

Figure 11.4 *Titration of a strong acid with a weak base*

The acid section has the shape associated with a strong acid, but the weak alkali (such as 0.1 mol dm^{-3} ammonia) part curves more gently after the end point and reaches a maximum of around 10. (The actual value depends on how weak the alkali is.) The vertical section at the end point runs from about pH 3 to pH 7.5.

When both the acid and the base are weak, the change in pH is as shown in Figure 11.5. There is an indistinct change in pH at the end point.

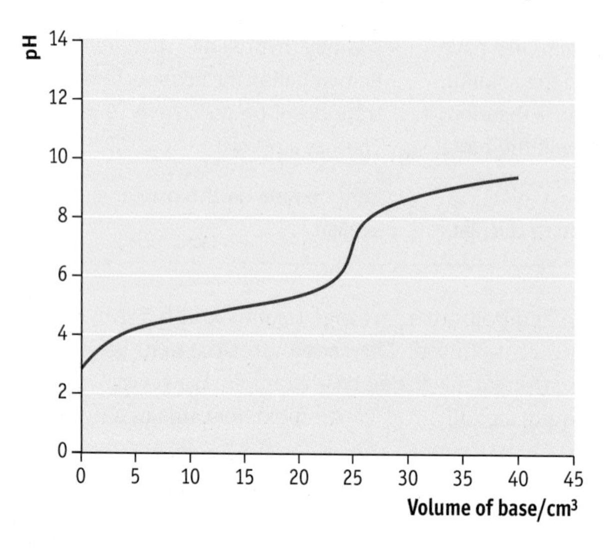

Figure 11.5 *Titration of a weak acid with a weak base*

Indicators

If a pH meter is not used, an acid–base titration requires an indicator to determine the end point. You may not have considered previously how indicators work, but it is now necessary to obtain an understanding of what actually happens when they change colour.

Most indicators are weak acids. An indicator has the particular property that the molecular form is a different colour from that of the anion to which it dissociates — for example:

- The molecular form of methyl orange is red; its anion is yellow.
- The molecular form of phenolphthalein is colourless; its anion is pink.

The equilibrium for methyl orange (HIn) is:

$$\underset{\text{red}}{HIn} \rightleftharpoons H^+ + \underset{\text{yellow}}{In^-}$$

HIn and In^- are a conjugate acid–base pair.

If methyl orange is added to an acidic solution, the equilibrium shifts to favour the molecular form. Therefore, the solution appears red. In alkali, the H^+ ions are removed from the equilibrium to form water. HIn then dissociates further and the indicator shows the colour of In^-, which is yellow. At a halfway stage between red and yellow it shows a combination of those colours and appears orange. The exact point at which the halfway stage is reached depends on the value of the equilibrium constant for the indicator, which is usually written as K_{In}.

For methyl orange, $K_{In} \approx 10^{-4}\,mol\,dm^{-3}$ so

$$K_{In} = \frac{[H^+][In^-]}{[HIn]} = 10^{-4}$$

The halfway point occurs when $[HIn] = [In^-]$. Substituting into the expression for K_{In} shows that this is when $[H^+] = 10^{-4}\,mol\,dm^{-3}$, i.e. when the pH is 4.

Therefore, during a titration, methyl orange changes colour at pH 4. This may seem unsatisfactory, but if you return to the titration curves you will see that methyl orange will correctly show the end point if the colour change occurs on the vertical section of the graph where the pH is changing rapidly. Methyl orange is a suitable indicator for strong acid–strong base titrations and strong acid–weak base titrations. It is not suitable for weak acid–strong base titrations.

Phenolphthalein has a value of $K_{In} \approx 10^{-9}\,mol\,dm^{-3}$ so it changes colour at pH 9. It is a suitable indicator for strong acid–strong base titrations and weak acid–strong base titrations.

If you look in data books you will find that there are many indicators, each of which changes colour at a particular pH. This enables a suitable indicator to be chosen for a particular titration. Usually, there are a number of indicators that could be used and the selection may depend simply on a preferred colour change. An indicator is given a pH range over which the colour change occurs because the eye is not capable of determining the moment when the two colours are exactly balanced.

It can be seen from Figure 11.5 that there is no really decisive change of pH at the end point of a weak acid–weak base titration and no indicator can effectively be used. It is a titration that should not be attempted.

Methyl orange changes colour between pH 3 and pH 5. It can be used for the titration of a strong acid with a weak base (Figure 11.6).

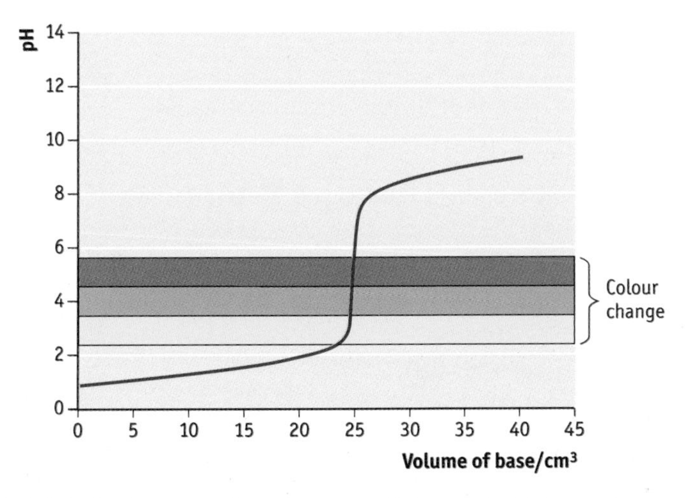

Figure 11.6 *Methyl orange as an indicator*

Phenolphthalein changes colour between pH 8 and pH 10. It can be used for the titration of a weak acid with a strong base (Figure 11.7).

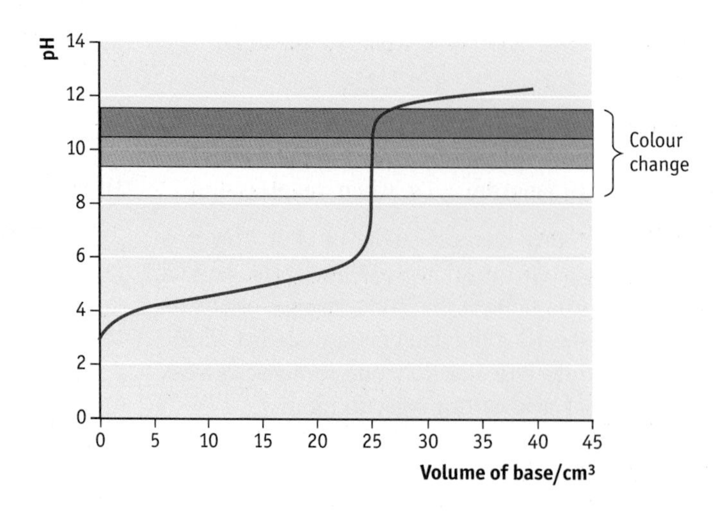

Figure 11.7 *Phenolphthalein as an indicator*

Experiments to determine the strengths of acids and bases

Measuring the pH of a known concentration of an acid or a base would indicate whether it is strong or weak, but there are other ways in which this can be determined. The simplest is to measure the ease with which electricity is conducted. In solution, ions conduct electricity. Conduction is most effective when the ion concentration is high. It follows that if solutions of strong and weak acids and bases of the same concentration are prepared, the strong acids and bases will conduct better than those that are weak. Therefore, the type of acid or base can be identified easily.

There is a further more complicated method that serves as a useful reminder of the enthalpy measurements met at AS. By neutralising an acid of known concentration completely with excess alkali, the standard enthalpy change of neutralisation can be calculated.

The standard enthalpy of neutralisation is the enthalpy change that occurs when 1 mol of water is produced in the reaction of an acid with an alkali under standard conditions.

The standard enthalpy of neutralisation is defined in terms of the amount of water formed in order to take into account the basicity of the acid. If the acid is dibasic — for example sulphuric acid — 1 mol of the acid would require twice as much sodium hydroxide to neutralise it compared with 1 mol of monobasic hydrochloric acid:

The basicity is the value of n in the formula of an acid, H_nX.

$$H_2SO_4(aq) + 2NaOH(aq) \rightarrow Na_2SO_4(aq) + 2H_2O(l)$$
$$HCl(aq) + NaOH(aq) \rightarrow NaCl(aq) + H_2O(l)$$

If the enthalpy of neutralisation is quoted for the formation of 1 mol of water, this difference is removed and the figures obtained for the enthalpy changes can be compared directly. In fact, for sulphuric and hydrochloric acids, the enthalpy change per mole of water produced is the same.

This is not really surprising because the reaction taking place is the same:

$$H^+(aq) + OH^-(aq) \rightarrow H_2O(l)$$

If the experiment is repeated using ethanoic acid and sodium hydroxide, the enthalpy of neutralisation obtained is lower. Since the reaction is the same, the difference must lie in the strength of the acid. The dissociation of ethanoic acid:

$$CH_3COOH(aq) \rightleftharpoons CH_3COO^-(aq) + H^+(aq)$$

is not complete. As the H^+ is neutralised by the base the equilibrium moves to the right. This requires a certain amount of energy which would otherwise have been lost as heat.

The neutralisation reaction of aqueous ammonia and hydrochloric acid has an even lower value for the enthalpy of neutralisation. This is because the process:

$$NH_3(g) + H_2O(l) \rightleftharpoons NH_4^+(aq) + OH^-(aq)$$

requires even more energy than the dissociation of ethanoic acid.

The neutralisation of the weak acid–weak base combination of ethanoic acid and ammonia results in an even lower value for the enthalpy of neutralisation.

Experiments of this type can give some indication of the strengths of acids and bases, even though the measurements are not particularly accurate.

Questions

1 For each of the following equilibria, identify the conjugate acid–base pairs:

a $HCO_3^- + H_2O \rightleftharpoons H_2CO_3 + OH^-$

b $HCO_3^- + OH^- \rightleftharpoons H_2O + CO_3^{2-}$

c $HCO_3^- + HCOOH \rightleftharpoons HCOO^- + H_2O + CO_2$

2 Calculate the pH of each of the following aqueous solutions:

a $0.15\,mol\,dm^{-3}$ HNO_3

b $0.15\,mol\,dm^{-3}$ HCN ($K_a = 4.8 \times 10^{-10}\,mol\,dm^{-3}$)

c $0.15\,mol\,dm^{-3}$ NaOH

d $0.15\,mol\,dm^{-3}$ Na_2SO_4

3 Calculate the pH of a mixture of $20.0\,cm^3$ of $1.00\,mol\,dm^{-3}$ HCl and $10.0\,cm^3$ of $1.00\,mol\,dm^{-3}$ NaOH.

4 Calculate the hydrogen ion concentration of each of the following:

a $0.5\,mol\,dm^{-3}$ KOH

b an aqueous solution of pH 4.0

c an aqueous solution of pH 2.7

d an aqueous solution of pH 11.2

5 A sample of milk has a pH of 6.3. Calculate the concentration of hydrogen ions in the milk.

6 A saturated solution of magnesium hydroxide has a pH of 10.5. Calculate the concentration of magnesium ions in the solution.

7 A sample of lemon juice has a pH of 2.45.

a What is the hydrogen ion concentration in $mol\,dm^{-3}$ in the lemon juice?

b The acid in lemon juice is citric acid, which is tribasic (i.e. its formula is H_3X). Assuming the equilibrium:

$$H_3X \rightleftharpoons 3H^+ + X^{3-}$$

what is the concentration of citrate ion in the juice?

8 Apple juice contains a monobasic acid.

a When 25.0 cm^3 of apple juice is titrated with a 0.120 $mol\,dm^{-3}$ solution of sodium hydroxide, 22.90 cm^3 is required to reach the end point. Calculate the concentration of the acid in the apple juice.

b The pH of apple juice is 3.5. Calculate the hydrogen ion concentration of the apple juice.

c Use your results from parts **a** and **b** to calculate the value of K_a for the acid in apple juice.

9 a In aqueous solution, potassium ethanoate is slightly alkaline. Remembering that ethanoic acid is a weak acid, suggest the reason why aqueous potassium ethanoate has a pH greater than 7.

b A 0.1 $mol\,dm^{-3}$ solution of potassium ethanoate has a lower pH than a 0.1 $mol\,dm^{-3}$ solution of potassium cyanide. Explain why this indicates that hydrocyanic acid is a weaker acid than ethanoic acid.

c Suggest a possible pH for an aqueous solution of ammonium chloride. Explain your answer.

10 a Calculate the pH of a buffer solution made by mixing 50 cm^3 of 0.1 $mol\,dm^{-3}$ potassium propanoate with 50 cm^3 of 0.1 $mol\,dm^{-3}$ propanoic acid. (K_a for propanoic acid = 1.3×10^{-5} $mol\,dm^{-3}$.)

b What would happen to the pH of the buffer solution if more potassium propanoate were dissolved into it.

11 a The pH of a solution of ethanoic acid is 2.70. K_a for the acid is 1.7×10^{-5} $mol\,dm^{-3}$. Calculate the concentration of the ethanoic acid solution.

b Calculate the mass of sodium ethanoate that must be added to the acid to create a buffer solution with a pH of 4.0. (Assume that the sodium ethanoate does not cause an increase in volume as it dissolves.)

12 The pK_{In} of the indicator bromophenol blue is 4.1.

a At what pH value will bromophenol blue show its 'neutral' colour?

b Name two types of titration for which bromophenol blue would be a suitable indicator.

13 The K_{In} value of chlorophenol red is 6.31×10^{-7}. Chlorophenol red is yellow in acid solution and red in alkaline solution.

a Determine the pH which is the mid-point for its colour change.

b Describe how an indicator works using chlorophenol red as your example.

c What will be the colour of chlorophenol red when it is added to the following? Explain your answers.

(i) 0.0001 $mol\,dm^{-3}$ hydrochloric acid

(ii) pure water

14 25.0 cm^3 of a 0.020 $mol\,dm^{-3}$ solution of propanoic acid is is titrated with 0.025 $mol\,dm^{-3}$ sodium hydroxide solution. The reaction is followed by measuring the pH as the sodium hydroxide is added.

a Calculate the volume of sodium hydroxide that will be needed to reach the end point.

b Calculate the pH of 0.020 $mol\,dm^{-3}$ propanoic acid. ($K_a = 1.3 \times 10^{-5}$ $mol\,dm^{-3}$)

c Calculate the pH of 0.025 $mol\,dm^{-3}$ sodium hydroxide.

d Sketch the appearance of the titration curve that would be obtained by plotting the pH against the volume of sodium hydroxide added.

e Suggest a suitable indicator for use in this titration.

15 When 50.0 cm^3 of 1.00 $mol\,dm^{-3}$ sulphuric acid is added to 50.0 cm^3 of 1.00 $mol\,dm^{-3}$ sodium hydroxide at 19.0°C the temperature rises to 23.6°C.

a Write an equation for the reaction and determine which one of sulphuric acid or sodium hydroxide is present in excess.

b Assume that the density of both solutions is 1 $g\,cm^{-3}$ and that the specific heat capacity is 4.18 $J\,g^{-1}\,K^{-1}$.

Calculate the enthalpy of neutralisation for this reaction.

Summary

You should now be able to:

- understand the Brønsted–Lowry theory of acids and bases
- explain the meaning of a conjugate acid–base pair
- explain the difference between strong and weak acids and strong and weak bases
- define K_a for a weak acid
- define pH and pK_a
- calculate the pH of strong and weak acids
- define the ionic product of water
- calculate the pH of strong bases
- explain what is meant by a buffer solution
- give an example of a buffer solution as the combination of a weak acid and its soluble salt and explain how it works
- calculate the pH of a buffer solution
- explain the role of HCO_3^-(aq) as a natural buffering ion in blood
- describe the shapes of the pH titration curves for acids and bases
- explain the action of an acid–base indicator
- explain the selection of an appropriate indicator for an acid–base titration
- define enthalpy of neutralisation
- explain how the measurement of the enthalpy of neutralisation gives an indication of the strength of an acid or a base

Additional reading

The role of water

What is the pH of a solution made by dissolving 1.0×10^{-8} mol of hydrogen chloride in 1.00 dm^3 of water? At first, this may seem to be a routine question asking for the pH of a strong acid, with the answer being $-\log(1.0 \times 10^{-8}) = 8.0$. But a moment's thought will tell you that this cannot be correct: it is predicting that the solution will be alkaline. So why is this calculation incorrect? The error is that the role of the water has not been taken into account. As you are aware, the neutral point of water is at pH 7.00 because water exists in an equilibrium in which there are 1.0×10^{-7} mol dm^{-3} each of hydrogen and hydroxide ions.

By dissolving 1.0×10^{-8} mol of HCl, we are increasing the number of hydrogen ions present so that the water now becomes more acidic. Hence its pH will be less than 7. Is the answer then that the pH is $-\log(1.0 \times 10^{-7} + 1.0 \times 10^{-8}) = -\log(1.1 \times 10^{-7}) = 6.96$? This is also incorrect because the fact that water molecules, hydrogen ions and hydroxide ions are in equilibrium ($H_2O \rightleftharpoons H^+ + OH^-$) has not been taken into account. The addition of 1.0×10^{-8} mol dm^{-3} of hydrogen ions will cause a very small movement of the equilibrium position to the left, in order to to form more H_2O molecules. With a little maths, the correct answer can be obtained.

In the split second before the equilibrium adjusts there are 1.1×10^{-7} mol hydrogen ions. Suppose that after the equilibrium has adjusted x mol have been used to form water molecules. There will be $(1.1 \times 10^{-7} - x)$ mol hydrogen ions remaining.

To be converted into water, x mol of hydrogen ions react with x mol of hydroxide ions. Before the HCl was added there were 1.0×10^{-7} mol of hydroxide ions present. As a result of x mol being used to form H_2O, there will now be $(1.0 \times 10^{-7} - x)$ mol.

Remember that $K_w = 1.0 \times 10^{-14}$ mol^2 dm^{-6} and, so long as the temperature does not change, this is always the case for aqueous solutions. Therefore:

$$(1.1 \times 10^{-7} - x)(1.0 \times 10^{-7} - x) = 1.0 \times 10^{-14}$$

or

$$(1.1 \times 10^{-14}) - (1.1 \times 10^{-7} + 1.0 \times 10^{-7})x + x^2 = 1.0 \times 10^{-14}$$

Rearranging gives the quadratic equation:

$$x^2 - (2.1 \times 10^{-7})x + 1.0 \times 10^{-15} = 0$$

This can now be solved as the lower value from:

$$x = \frac{(2.1 \times 10^{-7}) \pm \sqrt{(2.1 \times 10^{-7})^2 - (4 \times (1.0 \times 10^{-15}))}}{2}$$

The value of $x = 4.9 \times 10^{-9}$ mol dm^{-3}, $[H^+] = 1.051 \times 10^{-7}$ mol dm^{-3} and the pH = 6.98 (2 d.p.).

The point to appreciate is that water can be an important factor in the pH of a solution. Water is normally omitted because the contribution it makes to the acidity is negligible. If a solution contains 0.1 mol dm^{-3} of hydrogen ions, it is unnecessary (and pointless as instruments cannot measure the difference) to worry about the contribution of hydrogen ions from the water. However, as the added hydrogen ions approach a concentration of 10^{-7} mol dm^{-3} then water becomes of greater significance.

The effect of a change in temperature

The value of K_w that you will have used in calculations is 10^{-14} mol^2 dm^{-6}. However the value is temperature dependent and is only valid for temperatures around room temperature.

The equilibrium $H_2O \rightleftharpoons H^+ + OH^-$ is endothermic in the forward direction (ΔH = +55.8 kJ mol^{-1}). So, Le Chatelier's principle predicts that as the temperature is increased the equilibrium position will move to the right and more hydrogen ions will be formed. Therefore, the pH of water will decrease as the temperature is increased. The change is relatively insignificant for a small rise in temperature, but by the time water has been heated to a temperature close to the boiling point, the concentration of hydrogen ions will have fallen to 7.08×10^{-7} mol dm^{-3}, which is a pH of 6.15. This does not mean that water has now become acidic because there will still be an equal number of hydrogen ions and hydroxide ions present.

What has happened is that the definition of the neutral point has changed. You may have noticed that pH meters usually have a dial to make an adjustment for the temperature to take account of this effect.

Stretch and challenge

1 During the processing of apples the skins may be loosened using sodium hydroxide at pH 12. The pH of the sodium hydroxide eventually drops to 11.5 and it becomes too dilute to be effective. This is still a very alkaline pH and so before discarding the solution it is reacted with hydrochloric acid to reduce the pH to the safer value of 10.8. To ensure this has been achieved, an indicator called benzaldehyde 3-nitrophenylhydrazone (NPB) is used. NPB is purple for solutions with a pH greater than 12 and yellow in those with a pH less than 11.

There's more than one way to peel an apple

a Calculate the change in the hydroxide ion concentration (in mol dm^{-3}) that occurs as the pH of the solution falls from pH 12 to pH 11.5.

b Estimate a value for the equilibrium constant, K_{In}, for the indicator NPB.

c Estimate the ratio of the concentration of the un-ionised form of the indicator NPB to the concentration of the anion NPB^- at pH 10.8.

2 Aspirin is an effective painkiller although its use has, to some extent, been discouraged because in some circumstances it can cause stomach bleeding. This appears to be triggered by the molecular form of aspirin dissolving in the covalent lipids of the stomach lining.

Aspirin contains a carboxylic acid group and an ester group. It is hydrolysed readily.

$$C_6H_4(OCOCH_3)CO_2H + H_2O \rightleftharpoons C_6H_4(OH)CO_2H + CH_3COOH$$

COOH O C CH$_3$ O + H_2O

COOH OH + HO C CH$_3$ O

Because of the ease of hydrolysis, aspirin has a limited shelf-life.

a Assuming that the pH of stomach acid is approximately 1, explain why stomach bleeding might be a problem. (K_a for aspirin is $3 \times 10^{-4}\ mol\,dm^{-3}$.)

b The blood is buffered at pH 7.4. Calculate whether aspirin in the blood exists largely in its un-ionised molecular form or as an anion.

Aspirin is usually administered as a calcium salt since this is more soluble. However, as aspirin is hydrolysed rapidly above pH 8.5, care has to be taken in its preparation.

c If a solution of calcium hydroxide containing $0.741\ g\,dm^{-3}$ is used to create the calcium salt by a reaction with aspirin, is this likely to cause hydrolysis?

d A 0.900 g sample of aspirin becomes damp and absorbs 0.100 g of water. An equilibrium is established and analysis shows that 0.117 g of ethanoic acid is present in the equilibrium mixture.

(i) Calculate the value of the equilibrium constant for the hydrolysis of aspirin.

(ii) What percentage of aspirin has been hydrolysed?

(The aspirin would, in fact, be unsafe to use.)

Lattice enthalpy

Chapter 12

It would be helpful if all chemical reactions could be interpreted through an understanding of the behaviour of atoms and electrons. However, it will have become apparent that, at an atomic level, our knowledge and understanding is far from complete. We have a model of an atom in terms of a nucleus and electrons but the detailed way in which these interact has yet to be explained fully.

An alternative approach to deciding whether a reaction might be possible is to consider the energy changes that occur as a result of a reaction. Sometimes these can be measured directly, which is the case for most combustion reactions. In other instances Hess's law can be used to establish an energy change that would otherwise be impossible to determine. In AS Unit 2 you met several examples where Hess's law was used in this way.

The principle on which Hess's law is based is that the energy change for a process is the same whichever pathway is taken to convert the reactants to the products. This can be extended to provide information of a more fundamental kind. This is explored in this chapter, which covers the Born–Haber cycle. It applies to ionic substances. First, however, it is necessary to introduce a few new enthalpy terms.

Lattice enthalpy

You should know already that an ionic lattice consists of ions of opposite charge held together by the strong electrical forces that exist between them.

In ions that have high charge, the electrical forces are great and the crystal is held tightly together. This is confirmed by a high melting point. An example is aluminium oxide. Although you will have understood this in qualitative terms, it would be useful if the information could be obtained quantitatively. In order to do this, a definition of the strength of an ionic crystal must be established. This is known as the **lattice enthalpy**.

Lattice enthalpy is the energy released when one mole of a crystal is formed from its constituent ions in their gaseous state. Standard conditions of 298 K and 101 kPa are applied.

e This is the definition that is used in your exams. However, you should be aware that some sources of data define lattice enthalpy as the energy change for the reverse reaction.

This can be summarised for an ionic crystal, A_xB_y, by the equation:

$$xA^{y+}(g) + yB^{x-}(g) \rightarrow A_xB_y(s)$$

This may look complicated, but for sodium chloride this is simply:

$$Na^+(g) + Cl^-(g) \rightarrow NaCl(s)$$

For sodium oxide, it is:

$$2Na^+(g) + O^{2-}(g) \rightarrow Na_2O(s)$$

Lattice enthalpies can sometimes be estimated by computing all the forces that exist between the ions within the crystal. However, a precise measurement would require an experiment. It is not possible to do this directly because gaseous ions cannot be obtained separately from the ions of opposite charge. However, it is possible to obtain a figure indirectly from an energy cycle. Unlike those that you have met before, this cycle has a number of steps.

It must be emphasised that the cycle is simply a model that allows the determination of lattice enthalpy. It does not represent a reaction mechanism. The actual way in which an ionic crystal is formed is not known fully.

e The definition that OCR uses means that lattice enthalpies always have a negative value. Clearly energy must be released as the ions combine together to form the solid.

The Born–Haber cycle

The following theoretical steps are imagined to take place during the formation of a crystal from its elements in their standard state.

e If you are asked to define the *standard* enthalpy change, you must always quote the standard conditions of 298 K and 101 kPa.

Step 1: The metal lattice breaks into free gaseous atoms. The enthalpy change involved is called the **enthalpy of atomisation.**

$$M(s) \rightarrow M(g)$$

The enthalpy of atomisation of an element is the enthalpy change when 1 mol of gaseous atoms is formed from the element in its standard state.

Step 2: The atoms are then ionised to create ions of the required charge. For a '1^+' ion, this involves just the first **ionisation**.

e Refer to pages 102–105 in the AS book.

For other ions, this involves further ionisatons and the energy required is the sum of all these ionisation energies:

$$M(g) \rightarrow M^+(g) + e^-$$
$$M^+(g) \rightarrow M^{2+}(g) + e^-$$

and so on...

If the formula of the ionic substance involves more than one ion of M, then, since the figures are 'per mole', the numbers may be need to be multiplied appropriately.

The non-metal also creates gaseous ions:

- First, free gaseous atoms are formed. The enthalpy change involved is the enthalpy of atomisation, which (as with the metal) is the enthalpy change for the formation of 1 mol of gaseous atoms.
- Second, negative ions are formed from the gaseous atoms. The enthalpy change involved is called the **electron affinity**.

  The electron affinity is the enthalpy change accompanying the gain of 1 mol of electrons by 1 mol of atoms in the gaseous phase.

As with ionisation energies, there may be more than one step depending on the charge of the negative ion.

First electron affinity:

$X(g) + e^- \rightarrow X^-(g)$

Second electron affinity:

$X^-(g) + e^- \rightarrow X^{2-}(g)$
and so on...

For the non-metal also, care must be taken to remember that all quantities are measured 'per mole'.

To complete the cycle, one further enthalpy term is required, which is the **enthalpy of formation**.

The enthalpy of formation is the enthalpy change when 1 mol of a compound is formed from its constituent elements in their standard states.

A Born–Haber cycle is the result of combining all these factors. An example is shown in Figure 12.1.

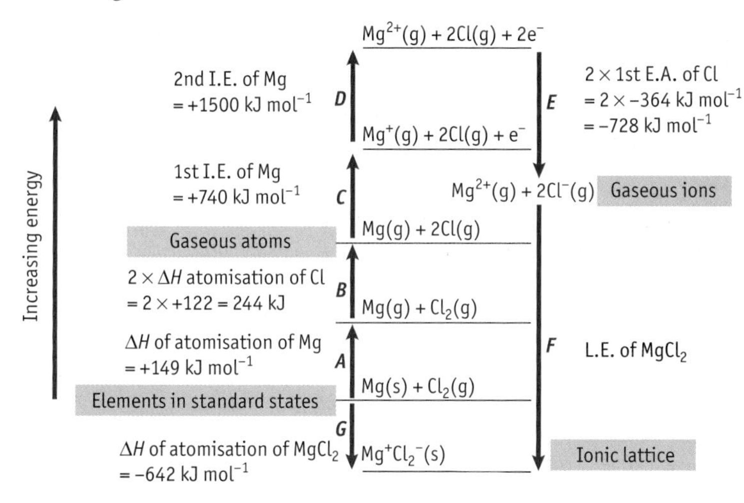

Figure 12.1
The Born–Haber cycle for magnesium chloride

From the cycle in Figure 12.1:

$A + B + C + D + E + F = G$

$149 + 244 + 740 + 1500 + (-728) + F = -642$

So, the lattice enthalpy, F, of magnesium chloride is $-2547\,kJ\,mol^{-1}$.

Starting from the elements in their standard states, you will see that to create the ionic lattice:

- Magnesium is atomised and then ionised to Mg^{2+} (both the first and the second ionisations are needed).
- Chlorine is atomised. Since the atomisation enthalpy is for 1 mol of atoms and 2 mol of atoms are required, the number + 122 (Figure 12.1) must be doubled.
- Likewise, 2 mol of Cl^- are required, so the electron affinity of Cl must be doubled.

The unknown lattice enthalpy can then be calculated by completing the cycle using the enthalpy of formation of magnesium chloride.

Figure 12.1 shows the steps as an 'energy profile'. You should notice that the electron affinity of chlorine is exothermic.

An alternative way of representing the Born–Haber cycle is to construct a cycle as in Figure 12.2. Magnsium chloride is once more used as the example.

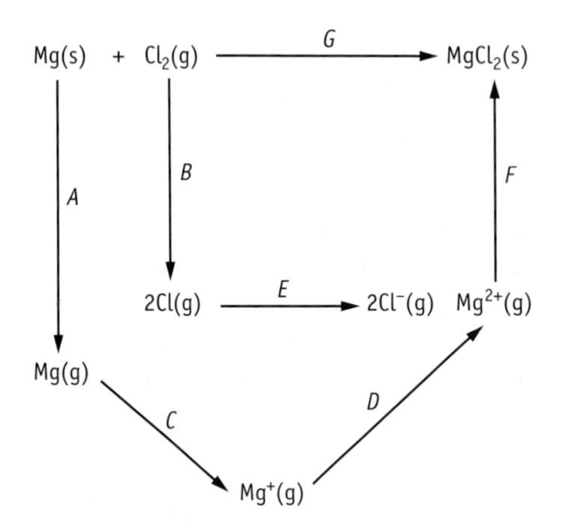

A = enthalpy of neutraslisation of Mg = +149 kJ mol^{-1}
B = 2 × enthalpy of atomisation of Cl = 2 × (+122) = 244 kJ mol^{-1}
C = first ionisation energy of Mg = +740 kJ mol^{-1}
D = second ionisation energy of Mg = +1500 kJ mol^{-1}
E = 2 × electron affinity of Cl = 2 × (−364) = -728 kJ mol^{-1}
F = lattice enthalpy of $MgCl_2$
G = enthalpy of formation = -642 kJ mol^{-1}

Figure 12.2
Alternative representation of the Born–Haber cycle for magnesium chloride

Once you are familiar with them Born–Haber cycles are easy to follow. However, they do require you to look closely at the steps. In exams, most errors are caused by carelessness with the signs and the subsequent arithmetic. Practice at constructing the cycles is essential.

We need not be concerned here with the experimental details of how the energy changes in a Born–Haber cycle are determined. Most have been established quite accurately. It is worth mentioning, however, that a Born–Haber cycle is sometimes used not to determine lattice enthalpies but to provide values for electron affinities, which cannot usually be obtained readily.

Factors affecting the size of lattice enthalpies

The lattice enthalpies of some ionic lattices are given in Table 12.1. The units of all the numbers in the table are kJ mol^{-1}.

The exact values of lattice enthalpies are often in some doubt because of the difficulty in obtaining precise values for the enthalpy changes used to calculate them.

	O^{2-}	Cl^-	Br^-	I^-
Na^+	−2480	−790		
Mg^{2+}	−3790	−2490	−2410	−2310
Al^{3+}	−15 900			
Ca^{2+}	−3500	−2200	−2125	−2040
Sr^{2+}	−3300	−2110	−2050	−1950

Table 12.1
Table of approximate lattice enthalpies

e When referring to lattice enthalpies, you must choose your words carefully. The problem is that if, say, A is numerically greater than B but the numbers are negative then you should really say that –A is less than –B.

There are two major points to note:

(1) As the charge on an ion increases, the lattice enthalpy becomes more exothermic. That is, the numerical value increases and the lattice that is formed is stronger.

If you compare the value of the lattice enthalpy for sodium chloride with that of sodium oxide, you will see that the sodium oxide lattice is stronger (more energy is released when it is formed from its gaseous ions) than that of sodium chloride. This is a result of the stronger forces that exist within the crystal because the charge on the oxide ion (O^{2-}) is higher than the charge on the chloride ion (Cl^-).

Likewise, if you compare the lattice enthalpies of the series Na_2O, MgO and Al_2O_3, you will understand that the lattices become progressively stronger as the charge on the cation increases. There is a further factor involved here which enhances the strength of the aluminium oxide lattice and, to a smaller extent, that of magnesium oxide. This is the polarising effect of a small cation on a large anion.

You will perhaps have met this idea in Unit 1 of the AS course. You may recall that the aluminium ion has 13 protons and 10 electrons and the aluminium atom has 13 protons and 13 electrons. The loss of the three electrons causes a noticeable contraction in radius from 0.14 nm to 0.057 nm. The oxide ion has two more electrons than the oxygen atom, which causes an increase in radius from 0.073 nm to 0.14 nm. As a result, the aluminium ion pulls the electron cloud from the oxide ion towards it and enhances the strength of the link by giving it some covalent character. This applies to an extent to all cases in which the cation is smaller than the anion. However, it is particularly relevant to cations of high charge and small radius (i.e. high charge density) combined with anions with larger radii.

(2) If ions have the same charge, then smaller ions will form a stronger lattice. That is, the numerical value of the lattice enthalpy increases.

If you compare the values of the lattice enthalpies of MgO, CaO and SrO, you will see that the lattices become progressively weaker as the size of the cation increases. It is a less pronounced effect than that caused by a difference in the charge of the ions, but it does make some difference to the stability of the crystal.

You can see the same pattern by comparing the lattice strengths of the group 2 chlorides or the halides of a group 2 metal.

Taken together these factors have their most obvious effect on the melting points of compounds. Some melting points are given in Table 12.2. The units in the table are °C.

Table 12.2 Melting points (°C) of some ionic compounds

	O^{2-}	Cl^-	Br^-	I^-
Na^+	920	808	755	660
Mg^{2+}	2800	714	711	634
Ca^{2+}	2600	782	765	740
Sr^{2+}	2430	875	643	Decomposes

The figures quoted may be slightly different when a solid can exist in more than one crystalline structure. However, there is usually a close relationship between the lattice enthalpy (Table 12.1) and the melting point (Table 12.2).

It should be noted that anomalies do sometimes occur — for example, the melting point of aluminium oxide, which is 2054°C. The reasons, which are related to the exact nature of the crystal structure, need not be considered here.

Solubility and enthalpies of hydration

The concept of a Born–Haber cycle can be extended to provide a partial explanation of what may seem a rather fundamental piece of information — the solubility of substances in water. To understand this, another enthalpy term needs to be introduced. This is the **enthalpy of hydration** of an ion.

The enthalpy of hydration of an ion is the enthalpy change that occurs when 1 mol of a gaseous ion is completely hydrated by water.

It is, therefore, the enthalpy change for the process:

$$X^{n+}(g) \rightarrow X^{n+}(aq)$$

The standard enthalpy of hydration applies to the conditions of 25°C and 101 kPa.

As with lattice enthalpies, it is possible to appreciate trends in the size of enthalpies of hydration. It has already been established that a small highly charged ion exerts a strong force on a neighbouring ions of opposite charge. In the case of hydration, the attraction is either between a cation and the oxygen atom of a water molecule or between an anion and the hydrogen atom of the water molecule. This occurs because of the charge separation across the –OH bond.

e You should be familiar with charge separation from the AS course.

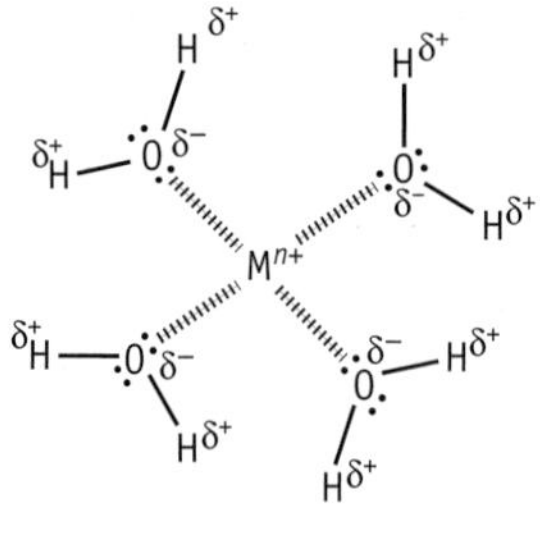

Figure 12.3 The hydration of an M^{n+} ion

Values of lattice enthalpy and enthalpy of hydration relate to the **enthalpy of solution**.

The enthalpy of soluton is the enthalpy change that occurs when 1 mol of an ionic solid dissolves in water.

A typical enthalpy cycle for sodium chloride is as follows.

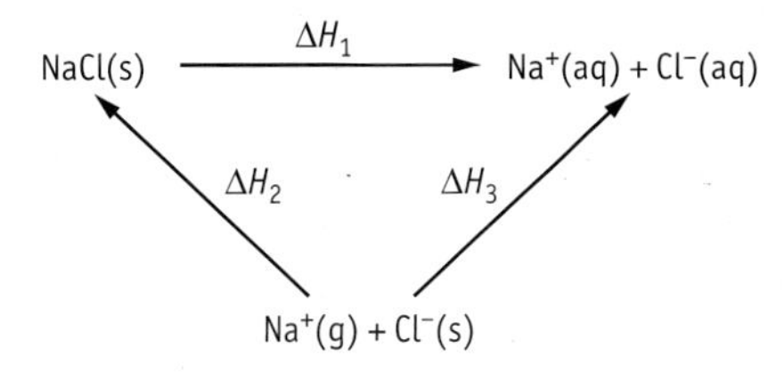

When an ionic solid dissolves it splits into its ions, so it is the ions that are hydrated.

From the diagram above:

- ΔH_1 is the enthalpy of solution.
- ΔH_2 is the lattice enthalpy of sodium chloride ($-781\,kJ\,mol^{-1}$).
- ΔH_3 is the enthalpy of hydration of the sodium ion ($-418\,kJ\,mol^{-1}$) + the enthalpy of hydration of the chloride ion ($-338\,kJ\,mol^{-1}$).

$$\Delta H_2 + \Delta H_1 = \Delta H_3$$

Depending on the information available, this relationship could be used to calculate either ΔH_1 or ΔH_3. In this case, ΔH_1 can be determined as:

$$\Delta H_1 = \Delta H_3 - \Delta H_2 = -418 - 338 + 781 = +25\,kJ\,mol^{-1}$$

Direct measurement of the enthalpy of solution of sodium chloride gives a different figure of about $4\,kJ\,mol^{-1}$. This difference arises because values for ΔH are not always known with complete certainty.

Nevertheless, both figures would suggest that, in enthalpy terms, the dissolving of sodium chloride is not favoured because the reaction is endothermic. Yet this is contrary to our experience, which is that sodium chloride is appreciably soluble in water at 25°C. It suggests that there is some other factor that is encouraging the dissolving to take place. This is an energy-related quantity called **entropy** which is discussed in Chapter 13.

Enthalpy cycles of this type are of most use when comparing the solubilities of substances rather than predicting the likely solubility of an individual substance. Consider the sulphates of group 2 metals — magnesium sulphate is quite soluble in water whereas barium sulphate is virtually insoluble. Enthalpy cycles indicate that there is an appreciable difference in the enthalpy changes predicted when the substances are placed in water.

$MgSO_4(s)$ —ΔH_1→ $Mg^{2+}(aq) + SO_4^{2-}(aq)$

ΔH_2 ΔH_3

$Mg^{2+}(g) + SO_4^{2-}(g)$

$$\Delta H_1 = \Delta H_3 - \Delta H_2$$

$$\Delta H_1 = (-1923 - 1004) + 2833 = -94\,kJ\,mol^{-1}$$

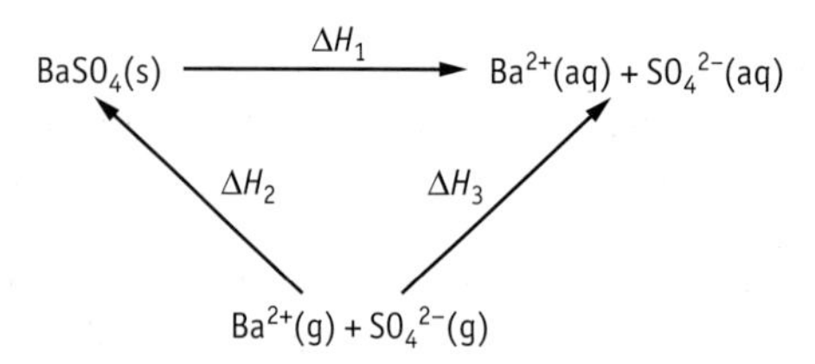

$$\Delta H_1 = \Delta H_3 - \Delta H_2$$

$$\Delta H_1 = (-1364 - 1004) + 2474 = +106\,\text{kJ mol}^{-1}$$

The enthalpy change for magnesium sulphate is quite exothermic; that for barium sulphate is endothermic. This reflects correctly the difference in their solubilities. There is no easy way that this conclusion can be reached, without carrying out the calculation. There are two competing factors. The larger size of the barium ion means that barium sulphate has a weaker lattice than magnesium sulphate. However, it also means that the energy released when the barium ion is hydrated is less than that when the magnesium ion is hydrated. It is the balance between these two factors that determines whether the enthalpy of solution is exothermic or endothermic.

e Remember that it is not possible to predict solubility from enthalpy data with any certainty.

Questions

1 Copy and complete the Born–Haber cycle for the formation of sodium chloride from its elements.

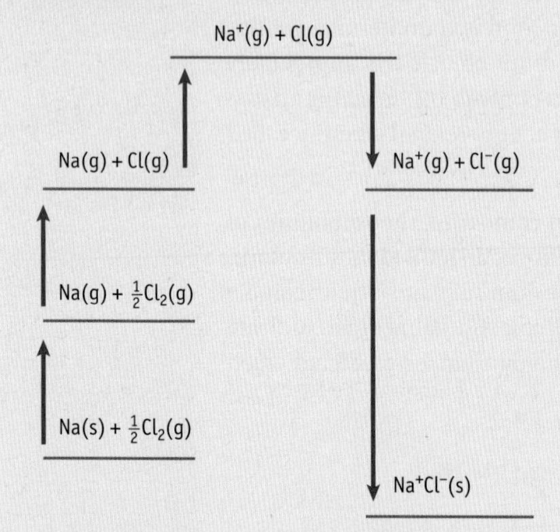

a Label each of the enthalpy changes.

b Use the data in the table to calculate the lattice enthalpy of sodium chloride.

	Enthalpy change/ kJ mol^{-1}
Enthalpy of atomisation of sodium	+108
1st ionisation enthalpy of sodium	+500
Enthalpy of atomisation of chlorine	+122
1st electron affinity of chlorine	−364
Enthalpy of formation of sodium chloride	−411

2 Use the data in the table to:

a construct a Born–Haber cycle for the formation of potassium oxide from its elements

b calculate the lattice enthalpy of potassium oxide

	Enthalpy change/ kJ mol^{-1}
Enthalpy of atomisation of potassium	+89.5
1st ionisation enthalpy of potassium	+420.0
Enthalpy of atomisation of oxygen	+249.4
1st electron affinity of oxygen	−141.4
2nd electron affinity of oxygen	+790.8
Enthalpy of formation of potassium oxide	−361.5

3 Silver fluoride, AgF, is a reasonably stable compound of silver, but gold fluoride, AuF, does not exist under normal conditions. It has been estimated that gold fluoride as AuF would be expected to have a lattice enthalpy of $-772.0 \text{ kJ mol}^{-1}$.

a Use the data in the table to construct Born–Haber cycles and calculate the enthalpies of formation of AgF and AuF.

	Enthalpy change/ kJ mol^{-1}
Enthalpy of atomisation of silver	+286.2
1st ionisation enthalpy of silver	+730.0
Enthalpy of atomisation of fluorine	+79.1
1st electron affinity of fluorine	−332.6
Lattice enthalpy of silver(I) fluoride	−943.0
Enthalpy of atomisation of gold	+369.6
1st ionisation enthalpy of gold	+890.1
Lattice enthalpy of gold(I) fluoride	−772.0

b Comment on how your answers to part **a** may explain why AgF is stable at room temperature but AuF is not.

4 Use the data in the table to calculate the enthalpy of solution of:

a silver chloride

b silver iodide

	Enthalpy change/ kJ mol^{-1}
Enthalpy of hydration of Ag^+	−464.4
Enthalpy of hydration of Cl^-	−384.1
Enthalpy of hydration of I^-	−306.7
Lattice enthalpy of AgCl	−890.0
Lattice enthalpy of AgI	−867.0

Summary

You should now be able to:

- define the terms lattice enthalpy, enthalpy of atomisation and electron affinity
- construct a Born–Haber cycle
- use a Born–Haber cycle to calculate an unknown quantity such as a lattice enthalpy or an electron affinity
- understand how the radius and charge of an ion affect the strength of a crystal lattice
- understand what is meant by the enthalpy of solution and the hydration enthalpy of an ion
- construct enthalpy cycles relating the solubility of an ionic substance to the enthalpies of hydration of its ions
- understand that enthalpies on their own do not provide a complete basis for the prediction of the possibility of a reaction

Additional reading

It is possible to estimate the lattice enthalpy of an ionic crystal if its structure is known. For example, the strength of a sodium chloride crystal which has a simple cubic structure, is due to a combination of the forces of attraction between the oppositely charged ions (Na^+ and Cl^-) and the forces of repulsion between the charges of the same sign. X-rays can be used to establish the positions of these ions, enabling a calculation to be carried out. To simplify the calculation, attempts have been made to provide a general statement that gives a value for the lattice enthalpy where only the sizes of the cation and anion are known. A formula that achieves this was proposed by the Russian scientist Anatoli Federovich Kapustinskii (1906–1960).

The derivation of the formula is complicated. The formula itself might look rather daunting, but its use is fairly straightforward and it can be used in many interesting situations. The formula is:

$$U = -\frac{1202.5n[z^+][z^-]}{(r^+ + r^-)}(1 - 0.345/(r^+ + r^-))$$

where U is the lattice enthalpy in $kJ\,mol^{-1}$

n is the number of ions in the empirical formula of the ionic crystal (e.g. for NaCl, $n = 2$; for Na_2SO_4, $n = 3$)

z^+ is the charge on the cation and z^- is the charge on the anion.

r^+ is the radius of the cation and and r^- is the radius of the anion in angstrom units.

Using the formula to determine the lattice enthalpy of magnesium chloride gives:

1 angstrom = 10 nm; most tables quote radii in nm.

$$U = \frac{-1202.5 \times 3 \times 2 \times 1}{(0.65 + 1.81)} \times (1 - 0.345/(0.65 + 1.81))$$

$$= -2932.9(1 - 0.140)$$

$$= -2521\ kJ\,mol^{-1}$$

This figure is quite close to the value of $-2547\ kJ\,mol^{-1}$ calculated in the example on page 175.

If there are more reliable ways of obtaining lattice enthalpies, the use of the formula is not particularly relevant. Of more interest is the use of the Kapustinskii formula to predict the likelihood of being able to create stable lattices of as yet unknown compounds. Compounds of neon are not known, but might it be possible to create the compound Cs^+Ne^-, for example? Caesium is a promising choice because it is the most readily ionised of the metals, with an ionisation energy of $380\ kJ\,mol^{-1}$.

The radius of Ne^- has been estimated as 0.25 nm (2.5 angstrom), although there is a measure of uncertainty about this figure. Substituting this figure in the Kapustinskii formula gives

$$U = \frac{-1202.5 \times 2 \times 1 \times 1}{(1.69 + 2.5)} \times (1 - 0.345/(1.69 + 2.5)$$

$$= -527\ kJ\,mol^{-1}$$

So the lattice enthalpy suggests that a stable lattice might be possible — it is not much different from that of Cs^+I^-, which $-594\ kJ\,mol^{-1}$.

However, the enthalpy of formation of Cs^+Ne^- should be considered. This has been calculated to be $-45.2\ kJ\,mol^{-1}$. This suggests, therefore, that it might be possible to make crystals of caesium neonide if the correct conditions could be established.

Such conclusions must be regarded as tentative because of the number of assumptions

that have been made. For example, the sum of the ionic radii at 4.19 angstroms is uncertain — other authorities suggest that the value could be as high as 4.9. If this were the case, the enthalpy of formation would be endothermic with a value of +25.2 kJ mol^{-1}. You will learn later in this unit that there is also a substantial loss of entropy as a crystal forms which would contribute significantly to its instability.

Stretch and challenge

Answer the following questions about xenon fluorides.

a In general, the noble gases do not form stable ionic compounds. However, chemists seeking to create a compound containing a noble gas cation thought that one possibility might be a fluoride, M^+F^-.

(i) Why might the fluoride be a sensible initial choice?

(ii) Explain why xenon would be the best noble gas to consider as a possibility.

b Use the data in the table below to calculate the enthalpy change for the process:

$$Xe(g) + \frac{1}{2}F_2(g) \rightarrow Xe^+(g) + F^-(g)$$

	Enthalpy change/ kJ mol^{-1}
1st ionisation enthalpy of xenon	+1170.0
Enthalpy of atomisation of fluorine	+79.1
1st electron affinity of fluorine	−332.6

c Use the internet or a data book to find some values for the lattice energies of fluorides to explain why the formation of Xe^+F^-(s) is unlikely to be possible.

d A solid compound, XeF_2,can be prepared by irradiating xenon and fluorine with UV. This compound boils at 114°C.

What two factors suggest that the bonding in this compound is covalent?

e The covalent molecule, XeF_4, has also been prepared.

Deduce the shape of the XeF_4 molecule.

f XeF_4 reacts with water according to the equation:

$$6XeF_4 + 12H_2O \rightarrow 4Xe + 2XeO_3 + 24HF + 3O_2$$

Discuss whether or not this reaction is:

(i) a redox reaction

(ii) an acid–base reaction

g Explain what is likely to be the shape of the covalent molecule, XeO_3.

h The mass spectrum of XeF_2 has significant peaks for the unfragmented molecule as follows:

Mass peak	166	167	168	169	170	172	174
% occurrence	1.9	26.4	4.1	21.2	26.9	10.4	8.9

(The remaining 0.2% results from extremely small peaks at 162 and 164)

Calculate the relative molecular mass of the molecule.

j Fluorine has only one isotope. What is the relative atomic mass of xenon?

Enthalpy and entropy

Chapter 13

Hess's law and the use of the Born–Haber cycle allow information to be obtained concerning enthalpy changes that occur as a result of chemical reactions. The cycles can often be used successfully to predict whether a reaction should be possible. However, knowledge of enthalpy changes alone is insufficient to provide a definite answer as to whether a reaction is feasible. Further information is needed about another energy-related change that takes place during a reaction; this change is known as **entropy**.

Some aspects of entropy are considered in this chapter. An explanation follows of how enthalpy and entropy changes can be combined to provide what is known as the **free energy change** of the reaction. Knowledge of the free energy change provides a more complete approach to assess the feasibility of a chemical process.

Entropy

Enthalpy changes cannot be used to predict, with certainty, whether a reaction will be feasible. The reason for this is that there is a further energy-related factor — entropy — that must be considered.

When a reaction occurs it is not necessarily the case that all the energy is absorbed or released in a form that can be recognised immediately, such as heat. Some energy is absorbed or released as a result of the re-distribution of the particles when the products are formed. The quantity of energy depends largely on the physical state of the substances and on the temperature. In this course we are concerned only with the reaction itself (known as the reaction system); a more detailed study would include the effect on the surroundings.

The following points should be noted:

- When a solid is formed from a liquid, or precipitated from a solution, then some energy is required to hold the solid in its ordered state. Conversely, when a solid is melted or dissolved then some energy is released. In a liquid or a solution the particles have greater freedom to move than in a solid. This lack of constraint means that energy does not have to be retained as it does within the more rigid structure of a solid.
- When a liquid is formed from a gas then this results in some absorption of energy as the molecules of the liquid are more constrained than those of the gas — the liquid molecules have less freedom of movement than those of a gas. The reverse process releases energy because the particles of a gas are less constrained than those of a liquid and energy is not used because their freedom of movement is not restricted.

e It is worth a reminder that we are referring to whether the products have a lower energy than the reactants. This must not be confused with whether the reaction will occur readily. The rate at which a reaction occurs depends on the activation energy of the reaction.

If both the system and the surroundings are considered, then, in any spontaneous reaction, the total entropy increases. This is known as the second law of thermodynamics.

Entropy is the term used as a measure of this quantity of energy. It is given the symbol, S.

The importance of including the change in entropy when deciding whether a reaction is possible is most easily illustrated, not by a chemical reaction, but by considering the conversion of a solid to a liquid as it melts.

The enthalpy change for the melting of ice to water at 0°C is:

$$H_2O(s) \rightarrow H_2O(l) \quad \Delta H = +6.02\,\text{kJ}\,\text{mol}^{-1}$$

The conversion therefore looks not to be possible — yet ice does melt at 0°C. This is because, as melting occurs, the change in entropy releases sufficient energy to counteract the positive enthalpy. The energy required to hold the rigid structure of the ice in place is released as the less constrained molecules of water are produced.

There are some significant differences between enthalpy and entropy.

We cannot determine what the enthalpy of a substance is — it is only possible to measure an enthalpy change. We must therefore define a 'starting point'. This is done by assigning an arbitrary value of zero to an element in its standard state at 25°C and standard atmospheric pressure (101 kPa). This does not mean that the elements really have zero energy under these conditions; the definition enables values to be given to the enthalpy changes that take place.

Enthalpy is similar to temperature in this respect. We refer to a temperature change, understanding that it is relative to the values given to the freezing and boiling points of water.

However, with entropy it is possible to give a definite value to a substance based on the assumption that at absolute zero entropy is zero. It is understood that at 0 K the particles of a solid are completely unable to move. As the temperature rises from 0 K, entropy steadily increases.

A further difference is that if enthalpy is released when a reaction occurs ΔH is negative, but if entropy is lost then the change in entropy, ΔS, is positive.

The contribution of entropy to a reaction depends on the temperature. As the temperature rises, particles become more able to move in an unrestricted way.

The units of ΔH are $\text{kJ}\,\text{mol}^{-1}$, whereas those of entropy are quoted normally as $\text{J}\,\text{K}^{-1}\,\text{mol}^{-1}$. This indicates the influence that temperature has on entropy. There are two things to note.

- As the 'zero point' for entropy is 0 K, the temperature unit is kelvin.
- The energy unit is usually joules rather than kilojoules. A change in entropy is usually a small effect, although, as it is temperature dependent it becomes increasingly important as the temperature is raised. This is considered further in the next section on 'free energy'.

Entropy may seem to be a rather complicated concept, but it is quite easy to apply to a situation if the following points are remembered. Entropy always increases when there is a greater opportunity for energy to be spread out as a result of a change. Therefore, entropy encourages the change when:

- a solid becomes a liquid
- a liquid becomes a gas
- the temperature rises, even if there is no change in state

- a solid dissolves in a liquid to form a solution
- a reaction produces products with a greater degree of freedom of movement — for example, this could be because a gas is produced when a solid reacts, as in the decomposition of calcium carbonate:

  $CaCO_3(s) \rightarrow CaO(s) + CO_2(g)$

  or when a reaction produces more particles in the same state, as in the combustion of propane:

  $C_3H_8(g) + 5O_2(g) \rightarrow 3CO_2(g) + 4H_2O(g)$

  In the latter case, assuming the temperature is such that the H_2O is a gas, the reaction has produced seven gaseous molecules as products from the six gaseous reactants. (If the H_2O were liquid then the entropy would decrease, as this would be the reverse of a liquid becoming a gas.)

Calculating entropy changes

You will have carried out calculations to determine ΔH for a reaction. Calculations to determine ΔS are similar, although it must be remembered that if ΔS is positive it means that the process releases entropy.

The change in entropy, ΔS, can be calculated using the formula:

$\Delta S = \Sigma$(entropy of products) – Σ(entropy of reactants)

e Do not forget, that in the calculation of an entropy change, the elements in their standard states *do* have values, which must be included.

Worked example 1

Calculate the entropy change when 1 mol of graphite is oxidised to carbon dioxide under standard conditions.

$S^\ominus(C — graphite) = 5.7\,J\,mol^{-1}\,K^{-1}$

$S^\ominus(O_2) = 102.5\,J\,mol^{-1}\,K^{-1}$

$S^\ominus(CO_2) = 213.6\,J\,mol^{-1}\,K^{-1}$

The equation is:

$C(s) + O_2(g) \rightarrow CO_2(g)$

Answer

$\Delta S^\ominus = \Sigma$(entropy of products) – Σ(entropy of reactants)

$\Delta S^\ominus = 213.6 - 5.7 - 102.5 = 105.4\,J\,mol^{-1}\,K^{-1}$

Worked example 2

Calculate the entropy change for the following reaction under standard conditions:

$3O_2(g) \rightarrow 2O_3(g)$

$S^\ominus(O_3) = 237.7\,J\,mol^{-1}\,K^{-1}$

$S^\ominus(O_2) = 204.9\,J\,mol^{-1}\,K^{-1}$

Answer

$\Delta S^\ominus = \Sigma$(entropy of products) – Σ(entropy of reactants)

$\Delta S^\ominus = 2 \times 237.7 - 3 \times 204.9 = -139.3\,J\,mol^{-1}\,K^{-1}$

Free energy

How can enthalpy and entropy be combined to answer the question: 'Is this chemical reaction feasible?'

It turns out that this is quite simple. A new term must be introduced, which is **free energy** (strictly the Gibbs free energy). It is given the symbol G.

The free energy change of a reaction relates to the enthalpy and entropy changes by the equation:

$$\Delta G = \Delta H - T\Delta S$$

ΔG provides a definite answer as to whether a given reaction is feasible.

Notice that ΔS is multiplied by the temperature (which must be in K rather than °C), so $T\Delta S$ has units of $K \times (J\,mol^{-1}\,K^{-1})$, i.e. $J\,mol^{-1}$. This means that ΔG is an energy term with units of $J\,mol^{-1}$ or $kJ\,mol^{-1}$.

- If ΔG is negative, the reaction is feasible.
- If ΔG is positive then, at the particular temperature chosen, the reaction is not feasible.

A reaction will fit one of four possible scenarios.

(1) ΔH is negative and ΔS is positive

In these circumstances, the reaction is always feasible. ΔG is negative since both the terms ΔH and $-T\Delta S$ are negative. It is usual to say that such a reaction will be spontaneous, meaning not that it will necessarily occur quickly (the activation energy will determine that) but that ΔG is certain to be negative. An example of this was given when discussing the solubility of magnesium sulphate in water (page 179). The process has a negative value for ΔH of $-94\,kJ\,mol^{-1}$ and, since a solution is being formed from a solid, ΔS is positive. It is no surprise therefore that it is possible to dissolve magnesium sulphate.

(2) ΔH is positive and ΔS is negative

Here the reaction can never be feasible, since both the terms ΔH and $-T\Delta S$ are positive.

(3) ΔH is negative and ΔS is negative

This situation is more complicated. ΔH favours the reaction but $-T\Delta S$ is positive, thus resisting the change. The sign of ΔG depends on the relative size of the two terms. Although ΔS is likely to have a small value compared with ΔH, when ΔS is multiplied by the temperature in K it could be similar to ΔH. There is no simple answer other than to do the arithmetic. It should be noted that the higher the temperature the greater the value of $-T\Delta S$, so reactions in which both ΔH and ΔS are negative are less likely to be feasible at higher temperatures.

(4) ΔH is positive and ΔS is positive

The reaction will be opposed by ΔH but favoured by ΔS. The outcome can be determined only by establishing the relative sizes of the two terms in the

e Remember that feasible means that the reaction is possible, *not* that it will necessarily take place quickly.

expression for ΔG. The reaction will be favoured by increasing the temperature to make $-T\Delta S$ as large as possible.

Therefore, a reaction that is not feasible under standard conditions of 298 K may be possible at a higher temperature — for example, the decomposition of a solid. Another example is the solubility of sodium chloride (page 179); ΔH has a value of $+25\,kJ\,mol^{-1}$ but the positive entropy value when multiplied by the temperature makes ΔG negative and so the process is spontaneous.

Equilibrium

If ΔG is zero, then the system will be at equilibrium. Even if this does not occur under standard conditions, it is possible that changing the temperature could allow the reaction to achieve equilibrium. If ΔG is equal to zero, then ΔH must be equal to $T\Delta S$.

For the equilibrium between melting ice and water:

$$\Delta S = \Delta H/T$$

where T is 273 K and $\Delta H = 6.02\,kJ\,mol^{-1}$

Therefore, the entropy change = 6020/273 = $22.0\,J\,mol^{-1}\,K^{-1}$

For a chemical reaction, the values of ΔH and ΔS must first be calculated. Then, the value of the temperature, T, for which ΔG is zero can be established. It should be noted that equilibrium can never be achieved for a reaction in which ΔH is negative and ΔS is positive or in which ΔH is positive and ΔS is negative. When ΔH and ΔS have the same sign, it will be possible to find the equilibrium temperature, noting that:

as $\Delta G = 0$, $\Delta H = T\Delta S$

so $T = \Delta H/\Delta S$

The calculation is illustrated in worked example 1 on page 189. The assumption is made that the values of ΔH and ΔS do not change as the temperature is changed. In fact, unless there is a change in state, entropies and enthalpies of formation do not change significantly and so calculations based on this assumption are normally valid.

Calculating free energy change

If tables of information are provided, then calculating the value of ΔG for a reaction is similar to the process of calculating ΔH. This should be clear from by worked examples 2 and 3.

If values for ΔH and S are given for each component in the equation then the calculation is a little more laborious as more arithmetic is required.

The overall ΔG for the reaction can then be determined.

Worked example 1

Use the data below to calculate the temperature at which the reaction:

$2NO(g) + O_2(g) \rightleftharpoons 2NO_2(g)$

reaches equilibrium.

$\Delta H_f^\ominus(NO) = 90.4\,kJ\,mol^{-1}$

$\Delta H_f^\ominus(NO_2) = 33.2\,kJ\,mol^{-1}$

$S^\ominus(NO) = 210.5\,J\,mol^{-1}\,K^{-1}$

$S^\ominus(NO_2) = 240.0\,J\,mol^{-1}\,K^{-1}$

$S^\ominus(O_2) = 204.9\,J\,mol^{-1}\,K^{-1}$

Answer

$\Delta H = \Sigma$(enthalpy of products) $- \Sigma$(enthalpy of reactants)

ΔH for the reaction is $(2 \times 33.2) - (2 \times 90.4) = -114.4\,kJ\,mol^{-1}$

$\Delta S = \Sigma$(entropy of products) $- \Sigma$(entropy of reactants)

$\Delta S = 2 \times 240.0 - (2 \times 210.5 + 204.9) = -145.9\,J\,mol^{-1}\,K^{-1}$

At equilibrium, $\Delta G = 0$

So, $T = \Delta H/\Delta S$

Therefore, $T = -114.4/-0.1459 = 788\,K$ or $515°C$

Remember to convert ΔS from J into kJ,

Worked example 2

a Use the following standard free energies of formation to calculate the free energy change for the following reaction:

$C_2H_4(g) + HCl(g) \rightarrow C_2H_5Cl(g)$

$\Delta G_f^\ominus(C_2H_4) = 68.1\,kJ\,mol^{-1}$

$\Delta G_f^\ominus(HCl) = -95.3\,kJ\,mol^{-1}$

$\Delta G_f^\ominus(C_2H_5Cl) = -59.4\,kJ\,mol^{-1}$

b Calculate the standard entropy of chlorine.

Answer

a $\Delta G = \Sigma$(Gibbs free energy of products) $- \Sigma$(Gibbs free energy of reactants)

$\Delta G = -59.4 - (68.1 - 95.3) = -59.4 + 27.2 = -32.2\,kJ\,mol^{-1}$

b $G_f^\ominus(Cl_2) = -69.4\,kJ\,mol^{-1}$

$G = H - TS$

Under standard conditions of 298K, $\Delta H^\ominus = 0$

Therefore, $-69.4 = -298S^\ominus$

So $S^\ominus = 69.4/298 = 0.233\,kJ\,mol^{-1}$ or $233\,J\,mol^{-1}$

Worked example 3

Use the following data to calculate the standard free energy of combustion of ethane:

$\Delta G_f^\ominus(C_2H_6) = -32.8\,kJ\,mol^{-1}$

$\Delta G_f^\ominus(O_2) = 0\,kJ\,mol^{-1}$

$\Delta G_f^\ominus(CO_2) = -394.6\,kJ\,mol^{-1}$

$\Delta G_f^\ominus(H_2O) = -237.2\,kJ\,mol^{-1}$

The equation for the combustion of ethane is:

$C_2H_6(g) + 3\frac{1}{2}O_2(g) \rightarrow 2CO_2(g) + 3H_2O(l)$

Note that the free energy cycle is the same as that used to calculate the value of ΔH, but values of ΔG are used instead.

Answer

$C_2H_6(g) + 3\frac{1}{2}O_2(g) \xrightarrow{\Delta G} 2CO_2(g) + 3H_2O(l)$

$2 \times \Delta G_f(C_2H_6) + 3\frac{1}{2}\Delta G_f(O_2)$ $2 \times \Delta G_f(CO_2) + 3 \times \Delta G_f(H_2O)$

$2C(s) + 3H_2(g)$

Therefore:

$2 \times (-32.8) + \Delta G = 2 \times (-394.6) + 3 \times (-237.2)$

$-65.6 + \Delta G = -789.2 - 711.6$

$\Delta G = -1435.2\,\text{kJ mol}^{-1}$

It is interesting to compare the result obtained from worked example 3 with the figure for the enthalpy combustion of ethane, which is $-1559\,\text{kJ mol}^{-1}$. The decrease in entropy as water is formed as a liquid is sufficient to reduce the available energy for the reaction. Under most conditions, however, H_2O is obtained as a gas and there is likely to be an increase in entropy as $4\frac{1}{2}$ mol of gaseous reactants are converted to 5 mol of gaseous product, which favours the combustion process.

Questions

1 For each of the following predict whether the reaction will have a positive or negative value for the entropy change.

a $H_2O(g) \rightarrow H_2O(s)$

b $NaOH(s) \rightarrow NaOH(aq)$

c $2Mg(s) + O_2(g) \rightarrow 2MgO(s)$

d $2SO_2(g) + O_2(g) \rightarrow 2SO_3(g)$

2 Sustances **A**, **B** and **C** are iodine, ammonia and methanol (but not necessarily in that order).

Given the following entropies, identify which substance corresponds to which letter. Explain your answers.

A $= 192.5\,\text{J mol}^{-1}\text{K}^{-1}$

B $= 58.4\,\text{J mol}^{-1}\text{K}^{-1}$

C $= 127.2\,\text{J mol}^{-1}\text{K}^{-1}$

3 Calculate the entropy change when sodium reacts with oxygen.

$S^\ominus(Na) = 51.0\,\text{J mol}^{-1}\text{K}^{-1}$

$S^\ominus(O_2) = 204.9\,\text{J mol}^{-1}\text{K}^{-1}$

$S^\ominus(Na_2O) = 72.8\,\text{J mol}^{-1}\text{K}^{-1}$

4 Calculate the entropy change when 1 mol of ethane is burnt in excess oxygen.

$C_2H_6(g) + 3\frac{1}{2}O_2(g) \rightarrow 2CO_2(g) + 3H_2O(l)$

$S^\ominus(C_2H_6) = 229.5\,\text{J mol}^{-1}\text{K}^{-1}$

$S^\ominus(O_2) = 204.9\,\text{J mol}^{-1}\text{K}^{-1}$

$S^\ominus(CO_2) = 213.8\,\text{J mol}^{-1}\text{K}^{-1}$

$S^\ominus(H_2O(l)) = 70.0\,\text{J mol}^{-1}\text{K}^{-1}$

5 Calculate the free energy change for each of the following reactions:

a the addition of hydrogen to ethene

b the addition of steam to ethene

$\Delta G_f^\ominus(C_2H_4) = 68.1\,\text{kJ mol}^{-1}$

$\Delta G_f^\ominus(H_2) = 0\,\text{kJ mol}^{-1}$

$\Delta G_f^\ominus(H_2O(g)) = -228.6\,\text{kJ mol}^{-1}$

$\Delta G_f^\ominus(C_2H_6) = -32.8\,\text{kJ mol}^{-1}$

$\Delta G_f^\ominus(C_2H_5OH) = -174.9\,\text{kJ mol}^{-1}$

6 a Benzene and hydrogen react to form cyclohexane:

$C_6H_6(l) + 3H_2(g) \rightarrow C_6H_{12}(l)$

For the reaction, use the data below to calculate:

(i) the enthalpy change

(ii) the entropy change

$\Delta H_f^\ominus(C_6H_6) = 49.0\,kJ\,mol^{-1}$

$\Delta H_f^\ominus(C_6H_{12}) = -156.2\,kJ\,mol^{-1}$

$S^\ominus(C_6H_6) = 172.8\,J\,mol^{-1}\,K^{-1}$

$S^\ominus(H_2) = 130.6\,J\,mol^{-1}\,K^{-1}$

$S^\ominus(C_6H_{12}) = 204.4\,J\,mol^{-1}\,K^{-1}$

b Use your answers from part **a** to calculate the free energy change for the reaction.

7 At high temperature, it is possible to react carbon with steam to produce a mixture of carbon monoxide and hydrogen known as water gas. This is a useful fuel. The equation for the reaction is:

$$C(s) + H_2O(g) \rightarrow CO(g) + H_2(g)$$

Use the enthalpy and entropy values below to calculate the temperature in °C at which an equilibrium is established between the reactants and products.

$\Delta H_f^\ominus(H_2O(g)) = -241.8\,kJ\,mol^{-1}$

$\Delta H_f^\ominus(CO) = -110.5\,kJ\,mol^{-1}$

$S^\ominus(C) = 5.7\,J\,mol^{-1}\,K^{-1}$

$S^\ominus(H_2O(g)) = 188.7\,J\,mol^{-1}\,K^{-1}$

$S^\ominus(CO) = 197.9\,J\,mol^{-1}\,K^{-1}$

$S^\ominus(H_2) = 130.6\,J\,mol^{-1}\,K^{-1}$

Summary

You should now be able to:

- understand qualitatively what is meant by entropy
- recognise that an increase in the freedom of movement of particles increases their entropy
- explain that the following result in an increase in entropy:
  - solid → liquid → gas
  - solid → solution
  - increase in the number of gaseous molecules
- understand that entropy has an absolute value and is only zero at 0 K
- remember that elements under standard conditions have an entropy value
- calculate entropy changes of reaction given appropriate data
- define ΔG as $\Delta H - T\Delta S$
- explain how the relative values of ΔH and $-T\Delta S$ determine the feasibility of a reaction
- explain why endothermic reactions may yet be spontaneous
- use $\Delta G = \Delta H - T\Delta S$ to calculate free energy changes
- construct free energy cycles to determine an unknown free energy change

Additional reading

More about entropy

The basic knowledge required to answer exam questions does not really give a feel for the nature and importance of entropy. Entropy is a fundamental and significant concept, embedded in the second law of thermodynamics. In simple terms, this states that the entropy of the universe is continuously increasing. The theoretical implication is that if we were able to measure the entropy of the universe at any one moment we would have a means of telling how old it is. Ever since the 'big bang', which scientists believe represented the start of the universe, the entropy within the universe has been increasing as a result of reactions taking place. There is a theory held by some that there may come a moment when entropy goes into reverse and that this will herald the end of the universe.

Without a formal mathematical understanding of the nature of entropy, it is difficult to make sense of the second law of thermodynamics. However, it is possible to get a feel for how it comes about. The following should provide you with a better appreciation of what is, at first sight, an elusive concept.

Any substance can be given a standard entropy value, symbol S. However, the value that is assigned really represents the change in entropy from its zero point at 0 K. Entropy is the accumulation of energy within the substance that is a consequence of its temperature. These are the energies of motion within its structure, including energies of vibration, rotation and, if it is possible, any motion of the particles. This latter energy is called translational energy. Entropy values can be regarded as giving a broad indication of the energies of motion that a substance must have to allow it to exist in a particular state. The figures must not be confused with the (highly significant) energies that are contained as a result of interactions within the bonds that hold the substance together. In fact, you can regard any substance as containing two types of energy;

- stored chemical energy, which relates to enthalpy
- energy due to motion and distribution, which relates to entropy

The size of $S^\ominus$

Entropy values vary depending on the state and structure of a substance. Consider metals — the solid lattice has a firmly held structure with the movement of the atoms being restricted to vibration Their $S^\ominus$ values are relatively low — for example $S^\ominus$ for magnesium is 32.0 J mol^{-1} K^{-1}. Comparing this with other group 2 metals we see that entropy increases down the group; the $S^\ominus$ value for calcium is 41.6 J mol^{-1} K^{-1} and the value for barium is 64.9 J mol^{-1} K^{-1}. This reflects the greater energy required to cause the vibration of heavier atoms. Rigid giant covalent lattices have little entropy, so for graphite, $S^\ominus$ is 5.7 J mol^{-1} K^{-1} and for the even more rigid diamond structure, $S^\ominus$ is 2.4 J mol^{-1} K^{-1}.

Liquids have higher $S^\ominus$ values because of the greater freedom of movement of the particles — for example, the $S^\ominus$ value for water is 70 J mol^{-1} K^{-1}. This is less than that of methanol with $S^\ominus$ = 127.2 J mol^{-1} K^{-1} because methanol has more bonds (three C–H, one C=O and one O–H) that vibrate. Ethanol, as a larger molecule, has an even higher value of 160.7 J mol^{-1} K^{-1}.

Gases have the highest values — for methane, $S^\ominus$ is 186.2 J mol^{-1} K^{-1} and for ethane, which has more bonds than methane, $S^\ominus$ is 229.5 J mol^{-1} K^{-1}.

e These are guiding principles. It should be emphasised that it is not possible to provide rigid rules to allow you to predict the entropy values.

What does an entropy change measure?

In any reaction the entropy change is a measure of how the energy has been spread as a result of the reaction. Entropy always increases if there is the possibility of dispersing the energy more widely. This can be easily appreciated by considering simple examples. The heat of a fire will spread around a room because in doing so more molecules receive the energy and it is, therefore, more widely shared. The melting of ice is another example. Sometimes the energy is not shared, but is dispersed across a wider volume. A compressed gas will expand if it is then allowed to occupy a greater volume.

Chemical reactions and entropy

It is important to remember that in a chemical reaction there are two components, the reacting mixture (the system) and the surroundings. Energy changes occur in both. If a reaction is exothermic, heat is released to the surroundings usually because the bonds of the products are stronger than those of the reactants. The evolved heat is transferred to the surroundings, for example the reaction vessel and the air around it. The surroundings absorb the heat by an increase in the movement of molecules in the form of energy of vibration, rotation or translation (motion). In other words, the entropy of the surroundings increases. The extent of the change depends on the temperature of the surroundings. If it is colder, the effect is greater; if it is hotter, it is less. In fact, the entropy of the surroundings changes by an amount equivalent to $\Delta H/T$ — or, as the surroundings absorb heat from the system, $-\Delta H/T$.

Within the reaction mixture itself, the entropy may either increase or decrease. Which occurs depends on whether there is a greater opportunity to spread the energy through the products or through the reactants.

Overall, the total entropy change as a result of a reaction is equal to the change in entropy of the system plus the entropy change of the surroundings:

$$\Delta S_{total} = \Delta S_{surr} + \Delta S_{system}$$

e This is a justification that the entropy change is $-\Delta H/T$, not a proof. The formal derivation is quite complicated.

Free energy and spontaneity

The quantity ΔG has been defined as:

$$\Delta G = \Delta H - T\Delta S$$

where ΔS refers to a change in the entropy of the system, i.e. it is ΔS_{system}

If we now divide each side of the equation by T, the equation becomes:

$$\Delta G/T = \Delta H/T - \Delta S_{system} \text{ or } -\Delta G/T = -\Delta H/T + \Delta S_{system}$$

However, it has just been established that $-\Delta H/T$ is the enthalpy change of the surroundings, ΔS_{surr}.

$$-\Delta G/T = \Delta S_{surr} + \Delta S_{system}$$

Therefore, $-\Delta G/T$ is in fact an entropy term equal to ΔS_{total}.

So, $\Delta G = -T\Delta S_{total}$

A reaction is spontaneous if ΔG is negative, which can only occur if ΔS_{total} is positive. Therefore, a reaction is spontaneous only if ΔS_{total} increases. This leads to the interesting conclusion that the entropy of the universe must be increasing steadily as a consequence of spontaneous reactions.

Stretch and challenge

Some metals are produced by the reduction of their oxides using carbon (as coke) — for example, iron occurs naturally as an oxide ore. Other metals, for example zinc and copper, occur as sulphides that are roasted initially in air to form the oxide.

The oxide reduction takes place in a furnace. Care must be taken to provide an appropriate temperature to make the process effective. The reactions that occur within the furnace can be quite complex, particularly because carbon can be oxidised to either carbon monoxide or carbon dioxide depending on the air supply and the temperature.

T/K	$\Delta G_f(CO)$/ kJ mol^{-1}	$\Delta G_f(CO_2)$/ kJ mol^{-1}	$\Delta G_f(Al_2O_3)$/ kJ mol^{-1}	$\Delta G_f(Fe_2O_3)$/ kJ mol^{-1}	$\Delta G_f(CuO)$/ kJ mol^{-1}
298	−137.2	−394.5	−1582.4	−740.4	−127.7
500	−155.4	−395.1	−1519.2	−685.0	−109.1
750	−177.8	−395.9	−1440.9	−616.4	−86.0
1000	−200.3	−396.7	−1362.7	−547.9	−63.0
1250	−222.7	−397.5	−1284.4	−479.3	−39.9
1500	−245.1	−398.3	−1206.1	−410.7	−16.8
1750	−267.6	−399.1	−1127.9	−342.1	6.2
2000	−290.0	−399.9	−1049.6	−273.5	29.3
2250	−312.4	−400.7	−971.3	−204.9	52.4
2500	−334.9	−401.5	−893.1	−136.3	75.4
2750	−357.3	−402.3	−814.8	−67.7	98.5
3000	−379.8	−403.1	−736.6	0.8	121.6
3500	−424.6	−404.7	−580.0	138.0	167.7

The table above gives some details of the free energies of formation for some oxides at various temperatures.

a Explain why, as the temperature is increased:

(i) $\Delta G_f(CO)$ becomes more negative

(ii) $\Delta G_f(CO_2)$ changes very little

(iii) ΔG_f for all three metal oxides becomes more positive

b Using the value of $\Delta G_f(CO)$ at 298 K show, by calculation, that the value at 2000 K is −290.0 kJ mol^{-1}. ($\Delta S_f(CO)$ at 298 K = 89.8 J mol^{-1} K^{-1})

c Use appropriate values from the table to calculate the free energy change for the reaction $CO + \frac{1}{2}O_2 \rightarrow CO_2$ at 500 K

d The free energy change for the reaction

$CO + \frac{1}{2}O_2 \rightarrow CO_2$

at 750 K is −218.1 kJ mol^{-1}. Use this result and your result from part **c** to calculate the values of ΔH and ΔS for this reaction. (Assume that ΔH and ΔS do not change with temperature.)

e Estimate the temperature at which the reaction $CO + \frac{1}{2}O_2 \rightleftharpoons CO_2$ reaches equilibrium.

f Copper metal was known to ancient civilisations and many artifacts made of bronze, which contains copper, have been found.

Suggest why copper could be produced so readily while iron was almost unknown. (Some iron was obtained from meteorites but none was extracted from an ore.)

g The blast furnaces used to produce iron operate at temperatures in excess of 1850 K. This is partly because iron melts at 1808 K and so the iron produced is molten and partly because the reduction of the oxide is only effective at temperatures in this region. Explain why the reduction of the oxide is effective only at temperatures in the region of 1850 K.

h Aluminium is extracted by an electrolytic process. Explain why a method using reduction by carbon is not employed. How does this help to explain why aluminium was discovered only relatively recently (in 1808, by Humphry Davy)?

Electrode potentials and fuel cells

Chapter 14

Chapter 11 covered acid–base reactions in detail. This involved a consideration of how hydrogen ions behave in a variety of reactions. Although usually referred to as a hydrogen ion, H^+ is a proton — the processes discussed in Chapter 11 are sometimes called proton-transfer reactions. This chapter looks at reactions that involve the transfer of electrons. These have been met before; they are called redox reactions.

Of all the types of reaction, redox reactions are perhaps the easiest to study. This is because it is possible to assess the energy released in the transfer process by allowing the electrons to move through an external electrical circuit where the energy that they carry can be measured by voltage. The size of the voltage gives an indication of the energy difference between the reactants and products. First, however, it is necessary to become fluent at writing ionic equations which establish the species that are the source and recipient of the electrons.

Fuel cells are an increasingly important development of the use of the electricity that can be generated through chemical reactions. They have been used to power equipment in spacecraft and some road vehicles. The principles behind a hydrogen–oxygen fuel cell, together with the problems that might be associated with their use, are considered.

Ionic equations

You will have written ionic equations for the precipitation reactions of silver nitrate with chlorides, bromides and iodides — for example:

$$Ag^+(aq) + Br^-(aq) \rightarrow AgBr(s)$$

Therefore, you will understand that an ionic equation simply states which ions have reacted and what is produced. The ions present that do not take part in the reaction are excluded from the equation. The ionic equation above makes it clear that *any* source of aqueous silver ions (i.e. any soluble silver salt) will react with *any* source of aqueous bromide ions (i.e. any soluble bromide) to produce a precipitate of silver bromide. It therefore expresses the general case rather than a particular example, such as aqueous silver nitrate reacting with aqueous potassium bromide. You may have written other ionic equations but in case you have not used them before some examples are given below.

To write ionic equations correctly, it is essential to remember the following points:

- Ionic substances that are solid do not have free-moving ions and, therefore, their ions cannot react independently. In an ionic equation their complete formulae must be given.
- Soluble compounds of metals and also strong acids in aqueous solution always split into their ions. These ions react independently.
- Covalent compounds exist as complete molecules and are always shown as complete entities in ionic equations.

The reaction between aqueous sodium hydroxide and hydrochloric acid to produce aqueous sodium chloride and water is a suitable example to illustrate the above points:

$$NaOH(aq) + HCl(aq) \rightarrow NaCl(aq) + H_2O(l)$$

Water is covalent and is present as intact covalent molecules. Sodium hydroxide, hydrochloric acid and sodium chloride are all ionic and, in solution, their ions act independently.

The sodium ions and chloride ions do not change in the reaction; they remain in solution as free and independent ions. The hydroxide ions and the hydrogen ions react to form covalent water molecules. This is summarised by the ionic equation:

$$H^+(aq) + OH^-(aq) \rightarrow H_2O(l)$$

The ionic equation tells us that all hydroxides can react with a source of hydrogen ions (i.e. an acid) and that water is produced as a result.

A further example is the reaction between aqueous sodium carbonate and hydrochloric acid to produce sodium chloride, carbon dioxide and water:

$$Na_2CO_3\,(aq) + 2HCl(aq) \rightarrow 2NaCl(aq) + CO_2(g) + H_2O(l)$$

The sodium ions and the chloride ions are not involved in the reaction; they remain unchanged. The carbonate ions and hydrogen ions react to form the two covalent products, carbon dioxide and water.

This is indicated by the ionic equation:

$$CO_3^{2-}(aq) + 2H^+(aq) \rightarrow CO_2(g) + H_2O(l)$$

Ionic equations must be properly balanced. In this case, the number '2' must be placed before the 'H^+'.

The ionic equation tells us that all carbonates react with acids to form carbon dioxide and water.

Do not forget that ionic substances must be in solution to exist as free independent ions.

With sulphuric acid, the equation is:

$$Na_2CO_3\,(aq) + H_2SO_4(aq) \rightarrow Na_2SO_4(aq) + CO_2(g) + H_2O(l)$$

However, the ionic equation is unchanged:

$$CO_3^{2-}(aq) + 2H^+(aq) \rightarrow CO_2(g) + H_2O(l)$$

The equation for the reaction of solid magnesium oxide with hydrochloric acid to make magnesium chloride and water is as follows:

$$MgO(s) + 2HCl(aq) \rightarrow MgCl_2(aq) + H_2O(l)$$

In this case, the ions of magnesium oxide are held in an ionic lattice and are not free to move. The only ion that is not involved in this reaction is the chloride ion. The ionic equation is:

$$MgO(s) + 2H^+(aq) \rightarrow Mg^{2+}(aq) + H_2O(l)$$

During the reaction the magnesium oxide lattice is broken down and the magnesium ions are released into the solution. The oxide ions combine with hydrogen ions from the acid to form water.

Redox reactions and oxidation numbers

Before proceeding to write more complex ionic equations, it would be wise to make sure that you are able to recognise simple redox reactions by assigning oxidation numbers to the their components. This was covered briefly in Unit 1; the essential points are given below.

In a redox process, electrons are transferred from one reactant to another to create the products. A simple example is the reaction of chlorine with potassium bromide solution to produce potassium chloride and bromine:

$$Cl_2(g) + 2KBr(aq) \rightarrow 2KCl(aq) + Br_2(aq)$$

In the reaction, Cl_2 forms chloride ions, Cl^-, and the bromide ions, Br^-, form bromine, Br_2. Each chlorine atom gains an electron and each bromide ion loses one electron — the electrons are transferred.

Remembering that oxidation is loss of electrons and reduction is gain of electrons, it is clear that the reaction is a redox reaction:

Oxidation

Oxidation numbers $\quad 0 \quad -1 \quad\quad -1 \quad 0$

$$Cl_2 + 2Br^- \longrightarrow 2Cl^- + Br_2$$

Reduction

Remember the mnemonic **oilrig** which refers to electrons: **o**xidation **i**s **l**oss, **r**eduction **i**s **g**ain

Oxidation numbers are a useful way of deciding which component in a reaction has lost or gained electrons.

The basic principles are set out in Table 14.1

The rules supplied do not cover all the metals — a transition metal, in particular, can be present either as a cation or embedded within an anion. This is usually clear within context. For example, in $MnSO_4$ the manganese is a cation with oxidation number +2 whereas in $KMnO_4$ it is within the anion and has the oxidation number +7.

More confusing, if only encountered occasionally, are compounds or ions in which an element is present with two different oxidation numbers. For example, magnetite is a common ore of iron and has the formula Fe_3O_4. There is a problem with applying the oxidation number rules because each iron atom appears to have the oxidation number 8/3. This is not possible. You might be able to guess that the answer is that the compound contains a mixture of Fe^{2+} ions (oxidation number 2) and Fe^{3+} ions (oxidation number 3) in the ratio 1:2. In general, if you are aware that this situation might arise you should be able to guess the likely solution.

Rule number	Rule	Example
1	All elements in their natural state have the oxidation number zero	Hydrogen, H_2; oxidation number = 0
2	Oxidation numbers of the atoms of any molecule add up to zero	Water, H_2O; sum of oxidation numbers = 0
3	Oxidation numbers of the components of any ion add up to the charge on that ion	Sulphate, SO_4^{2-}, sum of oxidation numbers = −2
When calculating the oxidation numbers of elements in either a molecule or an ion you should apply the following order of priority:		
1	The oxidation numbers of elements in groups 1, 2 and 3 are always +1, +2 and +3 respectively.	
2	The oxidation number of fluorine is always −1.	
3	The oxidation number of hydrogen is usually +1.	
4	The oxidation number of oxygen is usually −2.	
5	The oxidation number of chlorine is usually −1.	

Exam questions are not set to trick you. Guidance is always given if the oxidation numbers cannot be deduced using the rules provided.

More redox reactions

The following examples illustrate the range of redox reactions that may be encountered. It most cases it is easier to work with ionic equations.

Example 1 Zinc and aqueous copper sulphate react to form aqueous zinc sulphate and copper metal:

$$Zn(s) + CuSO_4(aq) \rightarrow ZnSO_4(aq) + Cu(s)$$

The ionic equation is:

$$Zn(s) + Cu^{2+}(aq) \rightarrow Zn^{2+}(aq) + Cu(s)$$

The zinc is oxidised; each atom loses two electrons and becomes a zinc ion, Zn^{2+}. The electrons are taken up by a Cu^{2+} ion which is reduced to a copper atom, Cu.

The oxidation number of Zn is 0; it becomes +2 in the zinc ion. The oxidation number of copper in Cu^{2+} is +2; it becomes 0 in the metal.

Example 2 Aqueous iron(II) chloride reacts with chlorine to produce aqueous iron(III) chloride:

$$2FeCl_2(aq) + Cl_2(g) \rightarrow 2FeCl_3(aq)$$

The ionic equation is:

$$2Fe^{2+}(aq) + Cl_2(g) \rightarrow 2Fe^{3+}(aq) + 2Cl^-(aq)$$

The Fe^{2+} ion is oxidised to Fe^{3+} and the Cl_2 is reduced to Cl^-. Two electrons are transferred.

In terms of oxidation number, the oxidation number of iron in the Fe^{2+} ion is +2;

it becomes +3 in Fe^{3+}. the oxidation number of chlorine in Cl_2 is 0; it becomes –1 in Cl^-.

Chlorine gas is the oxidising agent. An oxidising agent is the reactant which causes another component to be oxidised.

Example 3 Iodine and chlorine react together in a two-step reaction:

$$I_2(s) + Cl_2(g) \rightarrow 2ICl(l)$$
$$2ICl + 2Cl_2(g) \rightarrow I_2Cl_6(s)$$

These substances are all covalent, so ionic equations cannot be written.

Using oxidation numbers:

- In the first equation, the oxidation number of iodine starts at 0 and it is oxidised to +1; the oxidation number of chlorine starts at 0 and it is reduced to –1.
- In the second equation, iodine is oxidised further from +1 to +3; chlorine is again reduced from 0 to –1.

These examples are relatively straightforward, but they emphasise that in order to recognise a redox reaction it is sometimes helpful to think about the movement of electrons and sometimes easier to use oxidation numbers. The latter is possibly more systematic, but it can disguise what is really happening in the reaction.

Half-equations

In order to move on to more examples of redox reactions, the use of ionic half-equations is required. These show which reagent is supplying electrons and which is receiving them.

Using example 1 (page 198), the half-equations are:

$$Zn(s) \rightarrow Zn^{2+}(aq) + 2e^-$$
$$Cu^{2+}(aq) + 2e^- \rightarrow Cu(s)$$

For example 2, the half-equations are:

$$Fe^{2+}(aq) \rightarrow Fe^{3+}(aq) + e^-$$
$$2e^- + Cl_2(g) \rightarrow 2Cl^-(aq)$$

This reaction illustrates another use of half-equations, which is to provide a systematic method of constructing a balanced overall equation. In this case, it may seem to be an unnecessarily involved way of arriving at the answer but it is often the best approach when the equations are more complex. The point to appreciate is that the number of electrons released by the reducing agent (Fe^{2+} in this example) must be the same as the number required by the oxidising agent (Cl_2). In the first half-equation Fe^{2+} supplies one electron as it is oxidised to Fe^{3+}, but each Cl_2 molecule requires two electrons to be reduced to two chloride ions. Therefore it is necessary to have two Fe^{2+} ions for every one Cl_2 molecule.

The best way of seeing this clearly is to rebalance the half-equations until the number of electrons is the same. Here, this would be:

$$2Fe^{2+}(aq) \rightarrow 2Fe^{3+}(aq) + \mathbf{2}e^-$$
$$\mathbf{2}e^- + Cl_2(g) \rightarrow 2Cl^-(aq)$$

The overall equation is then obtained by adding the two half-equations together, which eliminates the electrons.

$$2Fe^{2+}(aq) + Cl_2(g) \rightarrow 2Fe^{3+}(aq) + 2Cl^-(aq).$$

Note though that, as in example 3, this method is not usually appropriate when the reaction is wholly between covalent molecules.

Some further examples of more complex redox reactions will help to illustrate the ideas discussed.

Further example 1 Potassium manganate(VII) is an oxidising agent that readily converts $Fe^{2+}(aq)$ to $Fe^{3+}(aq)$. It is used in aqueous solution and must be acidified to provide a source of hydrogen ions, which are required to absorb the oxygen atoms in MnO_4^-. During the oxidation, the manganate(VII) ion, $MnO_4^-(aq)$, is reduced to $Mn^{2+}(aq)$ ions. Potassium manganate(VII) is a powerful oxidising agent that can oxidise a range of inorganic ions as well as organic compounds.

ⓔ This is an important reaction that can be performed as a titration. It is required knowledge in the transition metal module of this unit.

This can be confirmed as a redox reaction by the use of oxidation numbers. The oxidation number of the iron ion is increased from +2 to +3. The oxidation number of Mn in MnO_4^- is +7. During the reaction this is reduced to +2 in the Mn^{2+} ion. The relationship between the amounts of Fe^{2+} and MnO_4^- required can be deduced from the change in oxidation number. The oxidation number of manganese is reduced by 5 (+7 to +2). The reaction requires five times as much Fe^{2+} as MnO_4^- because the oxidation number of iron only increases by 1.

ⓔ This procedure may at first appear daunting but, *with practice*, it can become automatic and will eventually seem quite straightforward.

However, this does not explain fully what is going on in the reaction. To obtain more detail, half-equations are useful.

To obtain the half-equation for the reduction of MnO_4^- to Mn^{2+}, hydrogen ions are added and water is formed. So the basic process is:

$$MnO_4^-(aq) + H^+(aq) \rightarrow Mn^{2+}(aq) + H_2O(l)$$

However, this is not a balanced equation. It must be corrected to:

$$MnO_4^-(aq) + 8H^+(aq) \rightarrow Mn^{2+}(aq) + 4H_2O(l)$$

The equation must also balance for charge. As it stands, the total charge on the left-hand side is made up of 1– (from the MnO_4^-) and 8+ (from the $8H^+$), i.e. it is 7+.

On the right-hand side, the charge is 2+ from the Mn^{2+} ion.

It is clear that $5e^-$ must be supplied, so that the total charge on each side of the half-equation is 2+.

$$5e^- + MnO_4^-(aq) + 8H^+(aq) \rightarrow Mn^{2+}(aq) + 4H_2O(l)$$

5– 1– 8+ → 2+ 0

The half-equation for the oxidation of Fe^{2+} to Fe^{3+} is:

$$Fe^{2+}(aq) \rightarrow Fe^{3+}(aq) + e^-$$

The two half-equations can now be combined. $5Fe^{2+}$ are needed to supply the $5e^-$ required by each MnO_4^-:

$$5Fe^{2+}(aq) \rightarrow 5Fe^{3+}(aq) + 5e^-$$

The overall equation is:

$$5Fe^{2+}(aq) + MnO_4^-(aq) + 8H^+(aq) \rightarrow 5Fe^{3+}(aq) + Mn^{2+}(aq) + 4H_2O(l)$$

You should notice that the total charge on the left-hand side of the overall balanced equation must equal the total charge on the right-hand side. Here it does, as both sides have a net charge of 17+.

Ionic equations are the usual way of writing equations for redox reactions. There is no need to worry about the actual chemicals that would supply the ions. For example, this reaction would work with any soluble iron(II) salt reacting with any soluble metal manganate(VII) in the presence of an acid. The only note of caution would be to ensure that the added chemicals did not have the potential to interfere with the reaction. In this case, sulphuric acid is a better source of hydrogen ions than hydrochloric acid because chloride ions can be oxidised by manganate (VII) ions to produce chlorine gas.

Further example 2 The reaction between potassium manganate(VII) and iodide ions can be summarised by two half-equations:

$$MnO_4^-(aq) + H^+(aq) \rightarrow Mn^{2+}(aq) + H_2O(l)$$
$$I^-(aq) \rightarrow I_2(s)$$

Balancing the half-equations gives:

$$MnO_4^-(aq) + 8H^+(aq) \rightarrow Mn^{2+}(aq) + 4H_2O(l)$$
$$2I^-(aq) \rightarrow I_2(s)$$

Adding electrons to make sure that the charges on either side of each half-equation are the same gives:

$$5e^- + MnO_4^-(aq) + 8H^+(aq) \rightarrow Mn^{2+}(aq) + 4H_2O(l)$$
$$2I^-(aq) \rightarrow I_2(s) + 2e^-$$

To balance the electrons the first half-equation must be multiplied by 2 and the second by 5, so that $10e^-$ are exchanged.

$$10e^- + 2MnO_4^-(aq) + 16H^+(aq) \rightarrow 2Mn^{2+}(aq) + 8H_2O(l)$$
$$10I^-(aq) \rightarrow 5I_2(s) + 10e^-$$

These are added together to give the overall balanced equation:

$$10I^-(aq) + 2MnO_4^-(aq) + 16H^+(aq) \rightarrow 5I_2(s) + 2Mn^{2+}(aq) + 8H_2O(l)$$

The total charge on each side of the equation is 4+.

You should note that the ratio of I^- to MnO_4^- is 10:2 (5:1) which is as predicted by the changes in oxidation number.

Further example 3 The reaction of potassium dichromate(VI) and tin(II) ions can be summarised in two half-equations:

$$Cr_2O_7^{2-}(aq) + H^+(aq) \rightarrow Cr^{3+}(aq) + H_2O(l)$$
$$Sn^{2+}(aq) \rightarrow Sn^{4+}(aq)$$

Balancing the half-equations gives:

$$Cr_2O_7^{2-}(aq) + 14H^+(aq) \rightarrow 2Cr^{3+}(aq) + 7H_2O(l)$$
$$Sn^{2+}(aq) \rightarrow Sn^{4+}(aq)$$

e It is easy to make a mistake somewhere in the procedure and this is a useful check to confirm that everything has been completed correctly.

e Don't forget to check the total charge on both sides of the equation.

Adding electrons to make sure that the charges on either side of each half-equation are the same gives:

$6e^- + Cr_2O_7^{2-}(aq) + 14H^+(aq) \rightarrow 2Cr^{3+}(aq) + 7H_2O(l)$
$Sn^{2+}(aq) \rightarrow Sn^{4+}(aq) + 2e^-$

To balance the electrons the second half-equation must be multiplied by 3 so that $6e^-$ are released:

$6e^- + Cr_2O_7^{2-}(aq) + 14H^+(aq) \rightarrow 2Cr^{3+}(aq) + 7H_2O(l)$
$3Sn^{2+}(aq) \rightarrow 3Sn^{4+}(aq) + 6e^-$

These are added together to give the overall balanced equation:

$3Sn^{2+}(aq) + Cr_2O_7^{2-}(aq) + 14H^+(aq) \rightarrow 3Sn^{4+}(aq) + 2Cr^{3+}(aq) + 7H_2O(l)$

The total charge on each side of the equation is 18+.

This agrees with the change in oxidation numbers. The oxidation number of tin has increased by 2 from +2 to +4; the oxidation number of chromium has decreased from +6 to +3. Since there are two chromium atoms in the dichromate, this is an overall reduction of 6, giving a ratio of Sn^{2+}:$Cr_2O_7^{2-}$ of 3:1.

Further example 4 Iodine reacts with thiosulphate ions ($S_2O_3^{2-}$) to form iodide ions and tetrathionate ($S_4O_6^{2-}$) ions. The reaction may look obscure but, as will be seen in the transition metal module, it forms the basis of a useful titration method (see pages 238–39).

The two balanced half-equations are:

$2S_2O_3^{2-}(aq) \rightarrow S_4O_6^{2-}(aq) + 2e^-$
$I_2(s) + 2e^- \rightarrow 2I^-(aq)$

In this case it can be easily seen that the overall equation is:

$2S_2O_3^{2-}(aq) + I_2(s) \rightarrow S_4O_6^{2-}(aq) + 2I^-(aq)$

The total charge on each side of the equation is 4–.

Electrode potentials

It has already been mentioned that an advantage of redox reactions is that the half-equations of the reactions can be reproduced in electrical cells.

If an electrical circuit is set up as shown in Figure 14.1, an overall chemical reaction takes place, but the half-reactions are kept separate and the transfer of electrons takes place via the external circuit. The energy released by the transfer of electrons can be measured using a voltmeter.

The circuit in Figure 14.1 shows the reaction $Zn(s) + Cu^{2+}(aq) \rightarrow Zn^{2+}(aq) + Cu(s)$ taking place.

At the zinc electrode, a metal atom loses two electrons to become an ion: ($Zn(s) \rightarrow Zn^{2+}(aq) + 2e^-$).

Current and voltage are electrical quantities. The current is a measurement of the flow of electricity (i.e the flow of electrons) within the circuit. In essence, it is how many electrons pass a point in the circuit in 1 second. If 6.24×10^{18} electrons pass in 1 second, then the current is 1 ampere (amp). 6.24×10^{18} electrons is 1 coulomb.

The voltage is a measurement of the energy carried by the electrons. 1 volt indicates that 1 coulomb of electricity is carrying 1 joule of energy.

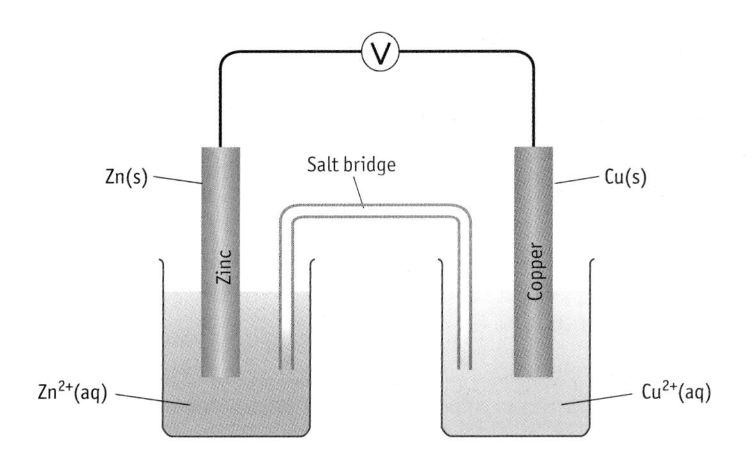

***Figure 14.1** Apparatus to measure an electrode potential*

To measure an electrode potential the voltage should be determined while the flow of current is minimal, otherwise some of the energy which is to be measured is lost. To do this a high resistance voltmeter is used.

The electrons travel around the external circuit through the voltmeter to the copper electrode. Once there, each pair of electrons is taken up by a copper ion and an atom of copper is formed ($Cu^{2+}(aq) + 2e^- \rightarrow Cu(s)$). The voltmeter reading is a measure of the energy of the transferred electrons and, just as the size of a free energy change gives a measure of the energy difference between reactants and products, the voltmeter reading does the same for this reaction in terms of the number of joules per coulomb.

The salt bridge contains a solution or gel of a substance that does not interfere with the reaction. Potassium chloride is commonly used, although this would not be appropriate if the chloride ion formed a precipitate with one of the cations.

Precipitation would occur if, for example, a silver electrode and silver nitrate solution were used. In this case the salt bridge could contain potassium nitrate.

The purpose of the salt bridge needs some explanation. As electrons are released, zinc ions are formed in the solution in the container on the left while copper ions are removed from the solution on the right (see Figure 14.1). The charge of all of the ions (cations and anions combined) in each container must remain zero. The only way for this to happen and to keep the electron flow continuing is to transer excess ions between the containers. This is accomplished by movement of these ions into, and eventually through, the salt bridge.

Any redox reaction can be made to take place in this way and so a wide variety of electrode potentials can be measured. However, to organise and make the information useful it is necessary to measure the voltages against a standard electrode. The electrode chosen for this purpose is the hydrogen electrode, which is illustrated in Figure 14.2. This is used as the standard because it is relatively easy to control its purity and reproducibility. The standard conditions used for the electrode are 298 K with the hydrogen being bubbled over the electrode (which is made of platinum) at the standard pressure of 100 kPa.

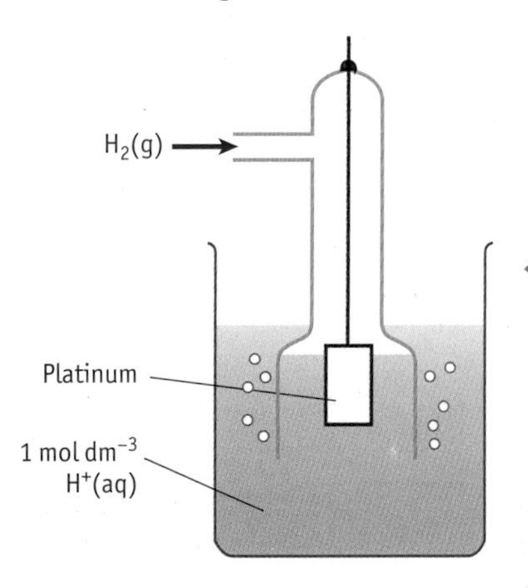

***Figure 14.2** A standard hydrogen electrode*

To provide a better surface for the hydrogen, the platinum electrode is usually coated with very finely divided platinum, known as 'platinum black'.

The standard half-cell is connected via an external circuit and through a salt bridge to the experimental half-cell. The voltage measured is the **electrode potential** of the cell compared with the half-reaction:

$$2H^+(aq) + 2e^- \rightarrow H_2(g)$$

which is given the arbitrary value of zero.

If the concentrations used in the cells are 1 mol dm^{-3}, the temperature is 298 K and all gases are at a pressure of 100 kPa (strictly 101.3 kPa), the voltage measured is the **standard electrode potential**, $E^\ominus$.

For example, if the standard hydrogen electrode is connected to a zinc half-cell, the voltmeter would read 0.76 volts. If it were connected to the copper half-cell it would read 0.34 volts. There is, however, a further issue that must be considered. There is a difference in the way the cells behave. In the case of zinc, electrons flow from the zinc to the hydrogen electrode, equivalent to the reaction:

$$Zn(s) + 2H^+(aq) \rightarrow Zn^{2+}(aq) + H_2(g)$$

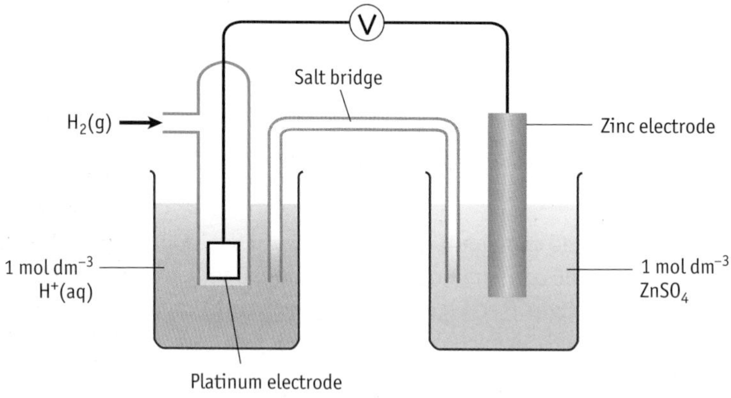

Figure 14.3 *A zinc–hydrogen cell*

When the copper half-cell is connected, the electrons flow in the reverse direction, i.e. from the hydrogen to the copper, equivalent to the reaction:

$$Cu^{2+}(aq) + H_2(g) \rightarrow Cu(s) + 2H^+(aq)$$

This is recognised by the use of a sign convention. A positive sign (+) indicates that the voltage measured was as a result of a flow of electrons to the hydrogen electrode; a negative sign (–) indicates a flow of electrons away from the hydrogen electrode.

The results can then be recorded in a table of 'standard electrode potentials'. The standard electrode potentials for the examples used in this chapter are shown in Table 14.1.

A positive value for the electrode potential indicates that the process as written is energetically favourable compared with the hydrogen half-reaction; a negative sign indicates that it would be unfavourable. This is used in the next section to predict the feasibility of reactions.

Table 14.1 *Some standard electrode potentials*

Reaction	$E^\ominus$/V
$Zn^{2+}(aq) + 2e^- \rightleftharpoons Zn(s)$	–0.76
$2H^+(aq) + 2e^- \rightleftharpoons H_2(g)$	0.00
$Cu^{2+}(aq) + 2e^- \rightleftharpoons Cu(s)$	+0.34

The scale of electrode potentials can be extended to include all processes in which half-reactions occur with the loss or gain of electrons.

***Table 14.2** Standard reduction potential values at 298 K*

Alphabetical list		Alphabetical list	
Electrode reaction	**$E^{\ominus}$/V**	**Electrode reaction**	**$E^{\ominus}$/V**
$Ag^+(aq) + e^- \rightleftharpoons Ag(s)$	+0.80	$H^+(aq) + e^- \rightleftharpoons \frac{1}{2}H_2(g)$	0.00
$Al^{3+}(aq) + 3e^- \rightleftharpoons Al(s)$	−1.66	$\frac{1}{2}I_2(s) + e^- \rightleftharpoons I^-(aq)$	+0.54
$Ba^{2+}(aq) + 2e^- \rightleftharpoons Ba(s)$	−2.90	$Li^+(aq) + e^- \rightleftharpoons Li(s)$	−3.04
$\frac{1}{2}Br_2(l) + e^- \rightleftharpoons Br^-(aq)$	+1.07	$MnO_4^-(aq) + 8H^+(aq) + 5e^- \rightleftharpoons Mn^{2+}(aq) + 4H_2O(l)$	+1.52
$Ca^{2+}(aq) + 2e^- \rightleftharpoons Ca(s)$	−2.87	$MnO_4^-(aq) + e^- \rightleftharpoons MnO_4^{2-}(aq)$	+0.56
$\frac{1}{2}Cl_2(g) + e^- \rightleftharpoons Cl^-(aq)$	+1.36	$MnO_4^{2-}(aq) + 2H_2O(l) + 2e^- \rightleftharpoons MnO_2(s) + 4OH^-(aq)$	+0.59
$HOCl(aq) + H^+(aq) + e^- \rightleftharpoons \frac{1}{2}Cl_2(g) + H_2O(l)$	+1.64	$\frac{1}{2}H_2O_2(aq) + H^+(aq) + e^- \rightleftharpoons H_2O(l)$	+1.77
$Cr^{3+}(aq) + e^- \rightleftharpoons Cr^{2+}(aq)$	−0.41	$\frac{1}{2}O_2(g) + 2H^+(aq) + 2e^- \rightleftharpoons H_2O(l)$	+1.23
$Cr_2O_7^{2-}(aq) + 14H^+(aq) + 6e^- \rightleftharpoons 2Cr^{3+}(aq) + 7H_2O(l)$	+1.33	$O_2(g) + 2H^+(aq) + 2e^- \rightleftharpoons H_2O_2(aq)$	+0.68
$Cu^+(aq) + e^- \rightleftharpoons Cu(s)$	+0.52	$Pb^{4+}(aq) + 2e^- \rightleftharpoons Pb^{2+}(aq)$	+1.69
$Cu^{2+}(aq) + 2e^- \rightleftharpoons Cu(s)$	+0.34	$Sn^{2+}(aq) + 2e^- \rightleftharpoons Sn(s)$	−0.14
$Cu^{2+}(aq) + e^- \rightleftharpoons Cu^+(aq)$	+0.15	$Sn^{4+}(aq) + 2e^- \rightleftharpoons Sn^{2+}(aq)$	+0.15
$\frac{1}{2}F_2(g) + e^- \rightleftharpoons F^-(aq)$	+2.87	$S(s) + 2H^+(aq) + 2e^- \rightleftharpoons H_2S(g)$	+0.14
$Fe^{2+}(aq) + 2e^- \rightleftharpoons Fe(s)$	−0.44	$Zn^{2+}(aq) + 2e^- \rightleftharpoons Zn(s)$	−0.76
$Fe^{3+}(aq) + e^- \rightleftharpoons Fe^{2+}(aq)$	+0.77		

If a second gas is involved then the two half-cells look similar (Figure 14.3).

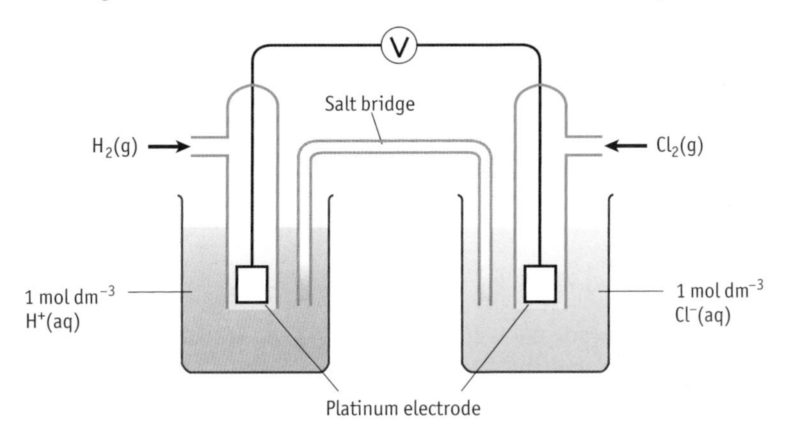

***Figure 14.4** A cell with two gas half-cells*

If the other reaction involves the transfer of electrons between two ions, then the solution in the half-cell must contain both ions and the electrode should be platinum.

For a *standard* electrode potential to be measured the concentration of all the ions involved should, in theory, be 1 mol dm^{-3}. In practice, it is rare that substances are sufficiently soluble to achieve this. However, it has been found that if all the ions are at the same concentration (in mol dm^{-3}) the measured electrode potential has the same value. For example, the standard electrode potential of the redox half-reaction $Fe^{3+}(aq) + e^- \rightarrow Fe^{2+}(aq)$ can be measured with both ions at a concentration of 0.1 mol dm^{-3}.

Cell potentials

Electrode potentials can be used to predict the voltage or cell potential of any combination of half-cells.

A redox reaction is feasible only if the cell potential is positive. Using the example of a zinc and copper electrode, the zinc electrode provides the electrons and the copper ions receives them.

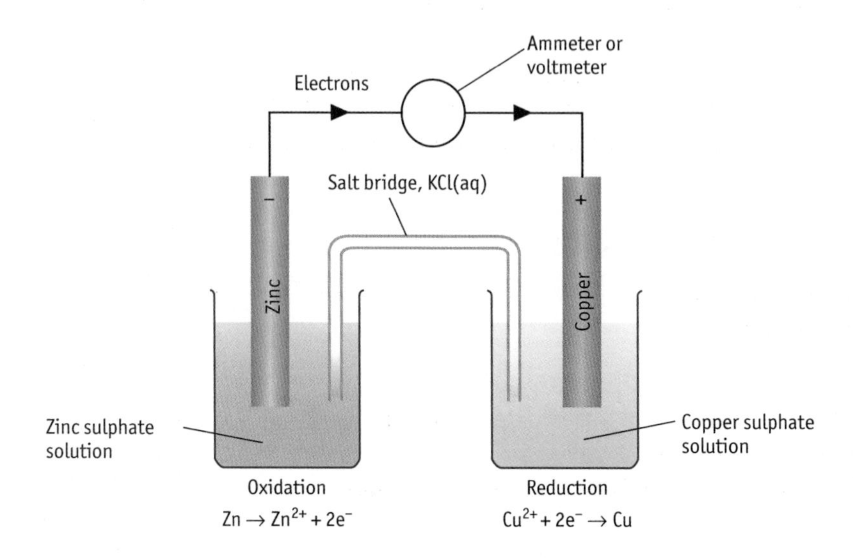

Figure 14.5
An electrochemical cell of Zn/Zn^{2+} and Cu/Cu^{2+}

Ⓔ Care has to be taken to remember that in any overall cell, one component of the cell must supply electrons and the other must receive them. This rather obvious point can be easily forgotten.

The standard electrode potential of $Zn^{2+}(aq) + 2e^- \rightleftharpoons Zn(s)$ is –0.76 V.

So, the standard electrode potential of the reverse reaction $Zn(s) \rightleftharpoons Zn^{2+}(aq) + 2e^-$ is +0.76 V.

The standard electrode potential of $Cu^{2+}(aq) + 2e^- \rightarrow Cu(s)$ is +0.34 V.

Therefore, the overall cell potential is +0.76 + 0.34, which is +1.10 V.

You may wonder how it is possible to decide if it is zinc, rather than copper, that provides the electrons. The answer is that the overall cell potential must be positive. The electrode potential for the half-reaction $Cu(s) \rightarrow Cu^{2+}(aq) + 2e^-$ would be –0.34 V (the reverse of the standard electrode potential) and that for $Zn^{2+}(aq) + 2e^- \rightarrow Zn(s)$ is –0.76 V. Combined, these give an overall figure of –1.10 V, which has the correct numerical value for the cell potential but the wrong sign.

If the equation of a half-cell is reversed, the sign of the electrode potential is reversed.

Summary

Both half-cells can move either to the right or to the left. If they move to the left, the sign is reversed.

$Zn^{2+}(aq) + 2e^- \rightleftharpoons Zn(s)$	$E^\ominus = -0.76\,V$	$Cu^{2+}(aq) + 2e^- \rightleftharpoons Cu(s)$	$E^\ominus = +0.34\,V$
(i) $Zn^{2+}(aq) + 2e^- \longrightarrow Zn(s)$	$E^\ominus = -0.76\,V$	**(a)** $Cu^{2+}(aq) + 2e^- \longrightarrow Cu(s)$	$E^\ominus = +0.34\,V$
$Zn^{2+}(aq) + 2e^- \longleftarrow Zn(s)$	$E^\ominus = +0.76\,V$	$Cu^{2+}(aq) + 2e^- \longleftarrow Cu(s)$	$E^\ominus = -0.34\,V$
which is better written as		which is better written as	
(ii) $Zn(s) \longrightarrow Zn^{2+}(aq) + 2e^-$	$E^\ominus = +0.76\,V$	**(b)** $Cu(s) \longrightarrow Cu^{2+}(aq) + 2e^-$	$E^\ominus = -0.34\,V$

In any reaction, one half-cell loses electrons and the other half-cell gains electrons. The two possible combinations in the diagram above are **(i)** with **(b)** and **(ii)** with **(a)**. To obtain the cell potential, the two half-cell values are added together:

- **(i)** with **(b)** gives $(-0.76) + (-0.34) = -1.10\,V$
- **(ii)** with **(a)** gives $(+0.76) + (+0.34) = +1.10\,V$

A reaction is only feasible if the cell potential is positive. Therefore, the combination of **(ii)** with **(a)** is correct.

Worked example 1

Use the following standard electrode potentials to predict the cell potential for a combination of chromium and cadmium electrodes.
In which direction will the electrons flow in the external circuit?

Reaction	$E^\ominus/V$
$Cr^{3+}(aq) + 3e^- \rightleftharpoons Cr(s)$	−0.74
$Cd^{2+}(aq) + 2e^- \rightleftharpoons Cd(s)$	−0.40

Answer

Electrons must be supplied by one half-reaction and received by the other half-reaction. The overall cell potential must be positive. This can only be achieved if the electrons are supplied by chromium.
The cell potential is obtained by the combination of:

$Cr(s) \rightarrow Cr^{3+}(aq) + 3e^-$ $\quad E^\ominus = +0.74\,V$

$Cd^{2+}(aq) + 2e^- \rightarrow Cd(s)$ $\quad E^\ominus = -0.40\,V$

The cell potential is $+0.74 - 0.40 = +0.34\,V$
The electrons flow from chromium metal to cadmium ions.

Worked example 2

Use the following standard electrode potentials to predict the cell potential for a combination of copper and silver electrodes.
In which direction will the electrons flow in the external circuit?

Reaction	$E^\ominus/V$
$Cu^{2+}(aq) + 2e^- \rightleftharpoons Cu(s)$	+0.34
$Ag^{+}(aq) + e^- \rightleftharpoons Ag(s)$	+0.80

Answer
Electrons must be supplied by one half-reaction and received by the other half-reaction. The overall cell potential must be positive. This can only be achieved if the electrons are supplied by copper.
The cell potential is obtained by the combination of:

$Cu(s) \rightarrow Cu^{2+}(aq) + 2e^-$ $E = -0.34\,V$

$Ag^+(aq) + e^- \rightarrow Ag(s)$ $E = +0.80$

The cell potential is $-0.34 + 0.80 = +0.46\,V$.
The electrons flow from copper metal to silver ions.

ⓔ When you look at the electrode potentials in worked example 2 and notice that they are both positive be careful not to just add them together. Both processes release energy, but within a cell, one process must provide electrons and the other process must receive them. Therefore, one half-reaction is forced to proceed in an energetically unfavourable direction.

In worked example 2 above, do not be tempted to adjust the cell potentials to match the number of electrons transferred. It is true that an atom of copper supplies two electrons and that a silver ion only requires one electron to form a silver atom. However, this simply means that two atoms of silver are deposited on the electrode for each atom that is lost from the copper electrode. Do *not* multiply the silver half-cell potential by 2.

To obtain the overall potential, just add or subtract the electrode potentials of the half-reactions as usual.

ⓔ The reason that electrode potentials can be treated in this simple way is because a volt is a unit of energy (joule) carried per coulomb of electricity. So for each half-cell, the energy is for a fixed number of electrons (the coulomb) compared with the same fixed number of electrons from the other half-cell.

A copper coil in silver nitrate solution

The net reaction indicates that if Cu(s) is placed in a solution of $Ag^+(aq)$, Ag(s) will be deposited and $Cu^{2+}(aq)$ will be formed.

Worked example 3
Use the following standard electrode potentials to predict the cell potential for a combination of a standard half-cell containing a platinum electrode in contact with $Fe^{2+}(aq)$ and $Fe^{3+}(aq)$, with a half-cell consisting of a lead electrode and $Pb^{2+}(aq)$.
In which direction will the electrons flow in the external circuit?

Reaction	$E^{\ominus}/V$
$Fe^{3+}(aq) + e^- \rightleftharpoons Fe^{2+}(aq)$	+0.77
$Pb^{2+}(aq) + 2e^- \rightleftharpoons Pb(s)$	−0.13

Answer

Electrons must be supplied by one half-reaction and consumed in the other half-reaction. The overall cell potential must be positive. This can only be achieved if the electrons are supplied by lead.

The cell potential is obtained by the combination of:

$Pb(s) \rightarrow Pb^{2+}(aq) + 2e^-$ $E^\ominus = +0.13\,V$

$Fe^{3+}(aq) + e^- \rightarrow Fe^{2+}(aq)$ $E^\ominus = +0.77\,V$

The cell potential is $+0.77 + 0.13 = +0.90\,V$.

The electrons flow from lead to Fe^{3+} ions.

If you are given a list of electrode potentials, you will notice that the cell potential is always the difference between the two electrode potentials, taking into account the sign. Although this provides a quick method of obtaining the potential, it is better (for reasons that will become clear when you have considered the next section on the feasibility of reactions) to consider the chemical processes that take place.

Using electrode potentials to predict the feasibility of a reaction

The reactions that take place in electrical cells parallel the redox reactions that can be performed in the laboratory. By measuring the voltage, the energy change for the process can be established. Because of this, it is possible to use electrode potential data to establish whether a chemical reaction is feasible (capable of taking place) or not.

To illustrate this we will predict whether $Fe^{3+}(aq)$ can be reduced to $Fe^{2+}(aq)$ under standard conditions using:

- $I^-(aq)$
- $Br^-(aq)$

The relevant standard electrode potentials are shown in the table below.

Reaction	$E^\ominus/V$
$I_2(aq) + 2e^- \rightleftharpoons 2I^-(aq)$	+0.54
$Fe^{3+}(aq) + e^- \rightleftharpoons Fe^{2+}(aq)$	+0.77
$Br_2(aq) + 2e^- \rightleftharpoons 2Br^-(aq)$	+1.09

The reaction will be feasible if the overall electrode potential is positive.

With $I^-(aq)$: Is the reaction $2Fe^{3+}(aq) + 2I^-(aq) \rightarrow 2Fe^{2+}(aq) + I_2(aq)$ possible? The electrode potentials for the half-reactions are:

$Fe^{3+}(aq) + e^- \rightarrow Fe^{2+}(aq)$ $E = +0.77\,V$

$2I^-(aq) \rightarrow I_2(aq) + 2e^-$ $E = -0.54\,V$ (the reverse of the standard electrode potential)

Adding these together, gives a cell potential of $+0.77 - 0.54 = +0.23\,V$.

e There is a direct relationship between $E^\ominus$ values and ΔG:

$\Delta G = -nFE^\ominus$

where F is a constant (known as the Faraday constant) and n is the difference in oxidation state between the oxidised and reduced forms in the half-equation.

Therefore a positive value of $E^\ominus$ corresponds to a negative value for ΔG, which is the condition that is required for a reaction to be feasible.

The positive answer indicates that the reaction is feasible.

With $Br^-(aq)$: the half-reaction required is $2Br^-(aq) \rightarrow Br_2(aq) + 2e^-$. This is the reverse of the standard potential and therefore has a value of −1.09 V.

Adding this to $E = +0.77\,V$ (the potential for $Fe^{3+}(aq) + e^- \rightarrow Fe^{2+}(aq)$ gives an overall potential of $+0.77 - 1.09 = -0.32\,V$.

The negative answer indicates that the reaction is not feasible.

In fact, you may be able to see that $Br_2(aq)$ will oxidise $Fe^{2+}(aq)$ to $Fe^{3+}(aq)$.

Worked example 1

Under standard conditions, can acidified aqueous potassium manganate(VII) ions oxidise $Br^-(aq)$ to $Br_2(aq)$?

The standard electrode potentials are given in the table.

Reaction	$E^{\ominus}/V$
$Br_2(aq) + 2e^- \rightleftharpoons 2Br^-(aq$	+1.09
$MnO_4^-(aq) + 8H^+(aq) + 5e^- \rightleftharpoons Mn^{2+}(aq) + 4H_2O(l)$	+1.51

Answer

The process $2Br^-(aq) \rightarrow Br_2(aq) + 2e^-$ is the reverse of that given by the $E^{\ominus}$ value, and therefore has a value of −1.09 V. The reduction of the manganate(VII) has an $E^{\ominus}$ of +1.51 V.

Adding these together gives a cell potential of $-1.09 + 1.51 = +0.42\,V$.

The positive answer indicates that the reaction is feasible.

Worked example 2

Under standard conditions, can aqueous Ag^+ ions oxidise $Fe^{2+}(aq)$ ions to $Fe^{3+}(aq)$ ions?

The standard electrode potentials are shown in the table.

Reaction	$E^{\ominus}/V$
$Ag^+(aq) + e^- \rightleftharpoons Ag(s)$	+0.80
$Fe^{3+}(aq) + e^- \rightleftharpoons Fe^{2+}(aq)$	+0.77

Answer

The process $Fe^{2+}(aq) \rightarrow Fe^{3+}(aq) + e^-$ is the reverse of that given by the $E^{\ominus}$ value, and therefore has a value of −0.77 V. The reduction of the $Ag^+(aq)$ to $Ag(s)$ has an $E^{\ominus}$ of +0.80 V.

Adding these together gives an overall value of +0.03 V.

The positive answer indicates that the reaction is feasible.

Worked example 3 above illustrates a potential trap that must be avoided. It is tempting to suggest that the low value of the electrode potential (+0.03 V) will mean that the reaction will be slow. However, it is not possible to deduce this. As with the value of ΔG, the overall potential indicates the relative energies of the reactants and products. It gives *no* information about the rate of the reaction. Whether a reaction occurs quickly or slowly depends on the activation energy of the reaction and activation energy is not related to the energy difference between the reactants and products. A clear distinction must be made between thermodynamic stability (which relates to the relative energies of the reactants and products) and kinetic stability (which refers to how fast the conversion will take place).

The effect of concentration on the feasibility of reactions

The predictions made so far have been based on the use of standard electrode potentials, which specify particular conditions of temperature, pressure and concentration. Normal laboratory conditions of temperature and pressure are usually close to the standard conditions. However, concentration is an important factor — reactions are not often carried out using concentrations of $1\,mol\,dm^{-3}$ and, in any case, if a reaction does occur, the concentrations change as the reaction takes place.

For concentrations usually used in the laboratory, the values of electrode potentials do not change significantly. A reduction in concentration from $1\,mol\,dm^{-3}$ to $0.1\,mol\,dm^{-3}$ only changes the electrode potential of a half-reaction by 0.06 V or less. Therefore, predictions based on standard electrode potentials are nearly always correct for other concentrations used in the laboratory.

Using an argument based around Le Chatelier's principle, it is possible to decide qualitatively whether a change in concentration will increase or decrease the value of the electrode potential. Electrode potentials refer to equilibria and the position of the equilibrium varies in accordance with Le Chatelier's principle. A change in concentration will, therefore, cause a change in the balance of equilibrium and hence the value of $E^{\ominus}$.

For example, if the equilibrium:

$$Fe^{3+}(aq) + e^- \rightleftharpoons Fe^{2+}(aq) \quad \text{for which } E^{\ominus} = +0.77\,V$$

is carried out with a reduced concentration of $Fe^{2+}(aq)$, this will encourage a shift in the equilibrium position from left to right, which will cause the electrode potential to increase. If there is a reduced concentration of $Fe^{3+}(aq)$ then the equilibrium position will move to the left and the value of $E^{\ominus}$ will decrease. In both cases, it would require a large change in concentration for the difference to be apparent.

More detail on the quantitative effect of a change in concentration is given in the 'additional reading' section.

Storage cells and fuel cells

Storage cells

Storage cells are used commonly in appliances to supply electricity. There is a wide range available as manufacturers strive to provide improved reliability and a longer life. The principles discussed above can be used to anticipate the voltage that would be expected from such cells.

For obvious reasons, it is better for a battery that is going to be sold to the general public not to contain liquid. The cells usually contain pastes that surround the

There is no need to remember the details or construction of any particular cell, but you may be asked in an exam to interpret provided data.

electrodes. Some examples will help you appreciate the relative complexity of the reactions taking place.

An 'alkaline battery' has a cathode made from graphite and manganese(IV) oxide and an anode made of either zinc or nickel-plated steel.

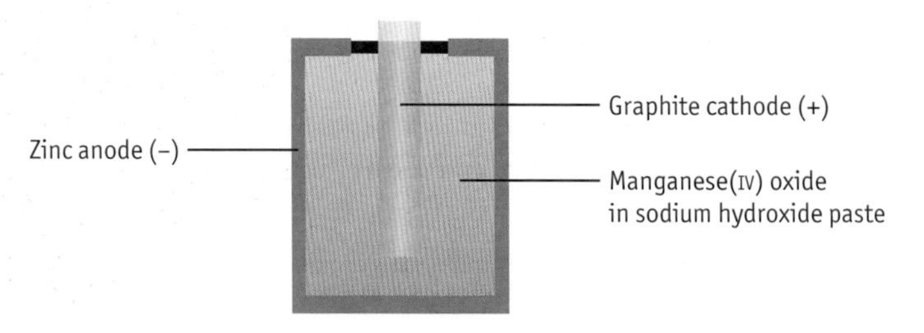

Figure 14.6 *An alkaline battery*

The half-cell that provides the electrons to the external circuit is the anode; the half-cell that receives them is the cathode. The paste or solution is the electrolyte.

The reactions that take place are:

At the anode: $Zn + 2OH^- \rightarrow ZnO + H_2O + 2e^-$

At the cathode: $2MnO_2 + H_2O + 2e^- \rightarrow Mn_2O_3 + 2OH^-$

The electrolyte is potassium hydroxide.

The overall equation taking place is found by combining the two half-reactions:

$$Zn + 2MnO_2 \rightarrow ZnO + Mn_2O_3$$

Both OH^- and H_2O can be eliminated from the equation because their concentrations should remain constant. (In practice, some loss does occur.) The overall voltage is about 1.5 V and is a combination of the electrode potentials of the two half-reactions.

Another battery often encountered is the rechargeable nickel–cadmium (Ni–Cd) cell.

While it is supplying electricity the reactions taking place are:

At the anode: $Cd + 2OH^- \rightarrow Cd(OH)_2 + 2e^-$

At the cathode: $2NiO(OH) + 2H_2O + 2e^- \rightarrow 2Ni(OH)_2 + 2OH^-$

The electrolyte is aqueous potassium hydroxide.

The overall reaction is:

$$Cd + 2NiO(OH) + 2H_2O \rightarrow Cd(OH)_2 + 2Ni(OH)_2$$

The battery can be recharged by applying an external voltage that reverses the reactions shown above. A disadvantage of this type of battery is that cadmium is toxic and care needs to be taken when disposing of the batteries. A voltage of around 1.2 V can be obtained during use.

Many other storage cells are manufactured and the intense research into the development of batteries for use in vehicles has led to a number of different constructions. The lead–acid accumulator is still employed widely.

However, its considerable weight is a disadvantage. Alternatives such as a sodium–sulphur cell are much lighter, but can be expensive and more risky in use.

A car battery

Fuel cells

Perhaps more promising than storage cells as a source of energy, are fuel cells. A fuel cell produces electrical power from the chemical reaction of a fuel (for example hydrogen, hydrocarbons or alcohols) with oxygen. The fuel cell operates like a conventional storage cell, except that the fuels are supplied externally as gases. The cell will therefore operate more or less indefinitely so long as the fuel supply is maintained.

The hydrogen–oxygen fuel cell is used widely and illustrates the principles behind fuel cells in general.

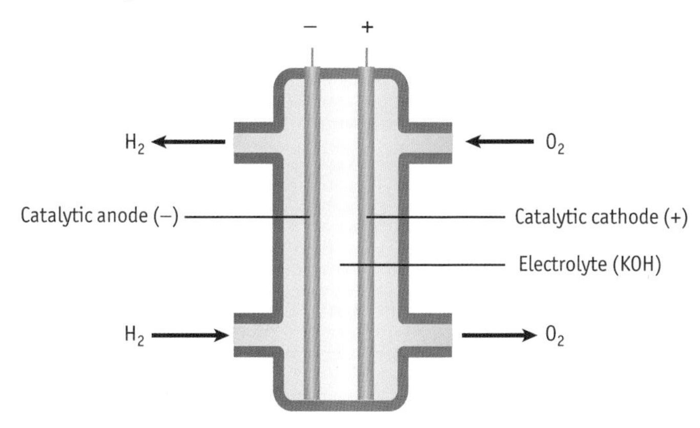

Figure 14.7 *A hydrogen–oxygen fuel cell*

You will not be asked to draw a diagram in an exam. It is the principles of how a fuel cell operates that are required.

The electrodes are made of a material such as a titanium sponge coated in platinum. The electrolyte is an acid or alkaline membrane that allows ions to move from one compartment of the cell to the other. (In other words it acts like a salt bridge (see page 203).

In alkaline solution, hydrogen reacts with hydroxide ions at the anode to form water while, at the cathode, oxygen reacts with water to form hydroxide ions:

$$H_2(g) + 2OH^-(aq) \rightleftharpoons 2H_2O(l) + 2e^- \quad E^\ominus = +0.83\,V$$

$$\tfrac{1}{2}O_2(g) + H_2O(l) + 2e^- \rightleftharpoons 2OH^-(aq) \quad E^\ominus = +0.40\,V$$

In an acidic solution, hydrogen is converted to hydrogen ions at the anode while, at the cathode, oxygen reacts with hydrogen ions to make water:

$$H_2(g) \rightleftharpoons 2H^+(aq) + 2e^- \quad E^\ominus = 0.00\,V$$

$$\tfrac{1}{2}O_2(g) + 2H^+(aq) + 2e^- \rightleftharpoons H_2O(l) \quad E^\ominus = +1.23\,V$$

The overall reaction in both cases is the same:

$$H_2(g) + \tfrac{1}{2}O_2(g) \rightarrow H_2O(l)$$

and the voltage produced is 1.23 V.

As in acidic solution, the overall reaction is:

$$H_2(g) + \frac{1}{2}O_2(g) \rightarrow H_2O(l)$$

and the voltage produced is again 1.23 V.

Vehicles utilising fuel cells are often referred to as FCVs.

There are a number of variations on the basic structure of a fuel cell. Those intended for vehicles may produce the hydrogen when it is required from hydrogen-rich materials such as methanol or natural gas. A reformer is required to strip the hydrogen from the supplied fuel.

A zero-emission London bus

Hydrogen–oxygen fuel cells:

- offer an alternative to the use of fossil fuels (petrol or diesel) that will eventually run out
- avoid the production of products that are pollutants, for example carbon monoxide, carbon dioxide or oxides of nitrogen
- are relatively light and are more efficient than engines that use fossil fuels

If hydrogen is to be used directly it has to be carried on board the vehicle. This requires a great deal of care because hydrogen is explosive. The possibilities are:

- to compress the gas until it liquefies
- to adsorb it onto the surface of a suitable solid material
- to absorb it into a suitable material

Compressing the gas may seem the simplest way, but it has disadvantages in terms of safety. The transportation of hydrogen under pressure is potentially particularly hazardous.

Many metals can adsorb hydrogen, so the choice of storage materials is wide. No firm conclusion has been reached as to which might be the best. A number of transition metal alloys have been tried — for example, an alloy of iron and titanium which absorbs hydrogen quite well, but is too heavy to be practical. Another possibility is to use carbon nanotubes. These are tiny lightweight structures containing arrangements of carbon atoms that have small, cylindrical pores which capture the hydrogen. This material is light enough but needs further investigation before it could be considered to be a commercial proposition.

Many metals can also absorb hydrogen to form metal hydrides, which can then, under the right conditions, release hydrogen. Particular interest has been shown in the use of lighter metals — for example magnesium, which forms the hydride MgH_2. A problem with this is that the hydride needs to be quite hot before the hydrogen is released. In addition, the manufacture of these materials requires energy and this must be included in assessing their viability, as must the expense of disposing of them after use. In fact, hydrides tend to have a rather limited lifetime.

As will be appreciated, safety is a major issue concerning the use of hydrogen (and hydrocarbons) as fuels. Research is continuing into safe and effective methods of production, transportation and storage of hazardous materials. These issues, along with finding methods of production of hydrogen that are economically feasible and environmentally acceptable, will shape the possible future use of hydrogen as a fuel.

Questions

1 Write ionic equations for each of the following reactions:

a aqueous potassium carbonate with nitric acid

b solid calcium carbonate with hydrochloric acid

c the precipitation of calcium carbonate from aqueous calcium chloride and aqueous sodium carbonate

d the neutralisation of aqueous calcium hydroxide by hydrochloric acid

e the precipitation of copper (II) hydroxide from aqueous copper nitrate and aqueous potassium hydroxide

f zinc oxide with nitric acid

2 Give the oxidation number of the element in **bold** in each of the following.

a $\mathbf{H}_2$

b $K\mathbf{Cl}O_3$

c $Na_3\mathbf{P}O_4$

d $Mg\mathbf{S}O_4$

e $Ca(\mathbf{Cl}O)_2$

f $K_2\mathbf{Cr}_2O_7$

g $\mathbf{I}Cl_3$

h $Al(\mathbf{N}O_3)_3$

i $\mathbf{S}O_3^{2-}$

j $\mathbf{Cl}O_4^-$

k $\mathbf{N}O_2^-$

l $\mathbf{V}O_3^-$

m $\mathbf{V}O^{2+}$

n $\mathbf{Al}(OH)_4^-$

o $\mathbf{Al}F_6^{3-}$

p $\mathbf{S}_2O_3^{2-}$

3 Complete each of the following pairs of half-equations and construct an overall equation for the reaction that occurs.

a $Fe^{3+}(aq) \rightarrow Fe^{2+}(aq)$

$I^-(aq) \rightarrow I_2(s)$

b $MnO_4^-(aq) + H^+(aq) \rightarrow Mn^{2+}(aq) + H_2O(l)$

$V^{2+}(aq) \rightarrow V^{3+}(aq)$ (V is vanadium)

c $MnO_4^-(aq) + H^+(aq) \rightarrow Mn^{2+}(aq) + H_2O(l)$

$Sn^{2+}(aq) \rightarrow Sn^{4+}(aq)$

d $MnO_4^-(aq) + H^+(aq) \rightarrow Mn^{2+}(aq) + H_2O(l)$

$V^{2+}(aq) + H_2O \rightarrow VO_3^-(aq) + H^+(aq)$

e $Cr_2O_7^{2-}(aq) + H^+(aq) \rightarrow Cr^{3+}(aq) + H_2O(l)$

$Br^-(aq) \rightarrow Br_2(aq)$

f $Cr_2O_7^{2-}(aq) + H^+(aq) \rightarrow Cr^{3+}(aq) + H_2O(l)$

$SO_2(aq) + H_2O(l) \rightarrow SO_4^{2-}(aq) + H^+(aq)$

g $NO_3^-(aq) \rightarrow NO(g)$

$Cu(s) \rightarrow Cu^{2+}(aq)$

4 Use the standard electrode potentials in the table below to calculate the cell potential for each of the following pair of half-cells:

a $Mg(s)/Mg^{2+}(aq)$ and $Zn(s)/Zn^{2+}(aq)$

b $Sn^{4+}(aq)/Sn^{2+}(aq)$ and $Fe^{3+}(aq)/Fe^{2+}(aq)$

c $I_2(aq)/2I^-(aq)$ and $Br_2(aq/2Br^-(aq)$

d $Zn(s)/ Zn^{2+}(aq)$ and $I_2(aq)/2I^-(aq)$

e $Sn^{4+}(aq)/Sn^{2+}(aq)$ and $Br_2(aq)/2Br^-(aq)$

Reaction	$E^\ominus/V$
$Mg^{2+}(aq) + 2e^- \rightleftharpoons Mg(s)$	−2.37
$Zn^{2+}(aq) + 2e^- \rightleftharpoons Zn(s)$	−0.76
$Sn^{4+}(aq) + 2e^- \rightleftharpoons Sn^{2+}(aq)$	+0.15
$I_2(aq) + 2e^- \rightleftharpoons 2I^-(aq)$	+0.54
$Fe^{3+}(aq) + e^- \rightleftharpoons Fe^{2+}(aq)$	+0.77
$Br_2(aq) + 2e^- \rightleftharpoons 2Br^-(aq)$	+1.09

5 Dental amalgam contains about 40% mercury (Hg) combined with an alloy that is made largely of silver and tin. (Small amounts of copper and zinc are also present.)

A number of redox reactions are possible with the amalgam as one electrode and saliva in the mouth as the electrolyte.

Two examples are:

$Ag^+ + e^- \rightleftharpoons Ag/Hg$ (amalgam) $E^\ominus = +0.85\,V$

$Sn^{2+} + 2e^- \rightleftharpoons Sn/Hg$ (amalgam)

$E^\ominus = -0.13\,V$

If a piece of aluminium foil is bitten by teeth containing an amalgam, an unpleasant sharp pain is experienced. This results from a temporary cell being set up between the amalgam and the aluminium.

Describe what happens in the cell and why it results in a pain being felt.

6 Use the standard electrode potentials in the table below to decide whether the following mixtures are likely to react.

a $Co(s) + Pb^{2+}(aq)$

b $Pb(s) + Ni^{2+}(aq)$

c $I_2(aq) + Ni^{2+}(aq)$

d $Cl^-(aq) + MnO_4^-(aq) + H^+(aq)$

e $I^-(aq) + Cr_2O_7^{2-}(aq) + H^+(aq)$

f $Cl^-(aq) + Cr_2O_7^{2-}(aq) + H^+(aq)$

g $Cr^{3+}(aq) + H_2O(l) + MnO_4^-(aq) + H^+(aq)$

Reaction	$E^\ominus$/V
$Co^{2+}(aq) + 2e^- \rightleftharpoons Co(s)$	−0.28
NOi$^{2+}(aq) + 2e^- \rightleftharpoons Ni(s)$	−0.25
$Pb^{2+}(aq) + 2e^- \rightleftharpoons Pb(s)$	−0.13
$I_2(aq) + 2e^- \rightleftharpoons 2I^-(aq)$	+0.54
$Fe^{3+}(aq) + e^- \rightleftharpoons Fe^{2+}(aq)$	+0.77
$Cr_2O_7^{2-}(aq) + 14H^+(aq) + 6e^- \rightleftharpoons 2Cr^{3+}(aq) + 7H_2O(l)$	+1.33
$Cl_2(aq) + 2e^- \rightleftharpoons 2Cl^-(aq)$	+1.36
$MnO_4^-(aq) + 8H^+(aq) + 5e^- \rightleftharpoons Mn^{2+}(aq) + 4H_2O(l)$	+1.51

7 Iron(III) chloride is a stable compound but iron(III) iodide cannot be made. Consider the relevant electrode potentials and explain why iron(III) iodide cannot be formed.

8 A cell that can be used for heart pacemakers contains a zinc anode and a platinum cathode in which zinc ions are discharged and oxygen in the blood reacts with water to form hydroxide ions.

a Write half-equations for the reactions taking place at the two electrodes.

b Write an overall equation for the reaction in the cell.

c The standard electrode potential for Zn^{2+}/Zn is −0.76 V and that for the reaction of oxygen is +0.40 V.

What does this suggest the overall voltage of the cell would be if it were operated under standard conditions?

d Why is the actual operating cell voltage likely to be different from that in your answer to part **c**?

9 Lead–sulphuric acid accumulators are used widely as electrical power supplies in vehicles. They do however have the disadvantages of being heavy and the soft lead metal being damaged easily. More expensive cells have been developed that are lighter, stronger and, in some cases, more reliable. An example is a zinc–silver cell in which concentrated potassium hydroxide is the electrolyte.

a When supplying current, the chemical change at the anode of a lead–sulphuric acid accumulator is:

$Pb + HSO_4^- \rightarrow PbSO_4 + H^+$.

Balance this half-equation by adding electrons.

b The overall equation in the accumulator is:

$Pb + PbO_2 + 2H^+ + 2HSO_4^- \rightarrow 2PbSO_4 + 2H_2O$

Give the half-equation for the reaction at the cathode.

c The overall reaction taking place in the zinc–silver cell with a potassium hydroxide electrolyte is:

$Ag_2O + Zn + H_2O \rightarrow 2Ag + Zn(OH)_2$

(Note that Ag_2O is formed because AgOH is unstable and cannot exist in the cell.)

Suggest half-equations for the reactions at the anode and cathode within the cell.

10 A fuel cell developed by NASA in 1980 used an arrangement similar to the hydrogen–oxygen fuel cell described in this chapter. However, instead of pumping gases into the anode and cathode compartments, solutions were run from external storage tanks over inert electrodes within each compartment.

One tank supplied Cr^{2+}(aq) and the other tank supplied Fe^{3+}(aq).

A redox reaction occurs and by continually supplying fresh solutions the cell is able to run for an indefinite length of time.

Electrode potentials that may be relevant are given in the table.

Reaction	$E^{\ominus}/V$
$Cr^{3+}(aq) + e^- \rightleftharpoons Cr^{2+}(aq)$	−0.41
$Cr^{2+}(aq) + 2e^- \rightleftharpoons Cr(s)$	−0.74
$Fe^{3+}(aq) + e^- \rightleftharpoons Fe^{2+}(aq)$	+0.77
$Fe^{3+}(aq) + 3e^- \rightleftharpoons Fe(s)$	−0.037

a Use the electrode potentials in the table to suggest the overall reaction. State the voltage produced by the cell.

b The solutions are supplied as chlorides and the membrane of the cell allows the passage of chloride ions through it.

Explain why it is essential for the membrane to allow the passage of chloride ions.

Summary

You should now be able to:

- write ionic equations for precipitation reactions and for reactions of acids with metals, oxides, hydroxides and carbonates
- assign oxidation numbers to elements, compounds and ions
- use oxidation numbers to determine whether a reaction is redox
- explain and use the terms oxidising agent and reducing agent
- write ionic half-equations for redox reactions involving:
  - metals and their cations
  - non-metals and their anions
  - the oxidation or reduction of an element from one oxidation state to another
- combine half-equations to produce an overall balanced equation
- explain what is meant by an electrode potential
- state the conditions necessary for the measurement of a standard electrode potential
- draw a labelled diagram of a standard hydrogen electrode
- draw a labelled diagram to illustrate the measurement of the standard electrode potential of:
  - a metal or non-metal in contact with its ion
  - a pair of ions of the same element in different oxidation states
- determine an overall cell potential by a combination of redox half-cells
- use electrode potentials to predict the feasibility of a reaction
- understand qualitatively the effect of a change in concentration on an electrode potential

- understand that a prediction of feasibility does not provide information about the rate of a reaction
- explain how the principles applied to cells are utilised in the production of storage cells and fuel cells
- describe the importance of a fuel cell based on hydrogen and oxygen
- list the advantages and problems in the use of fuel cells based on hydrogen and oxygen

Additional reading

14

It has been mentioned earlier in the chapter that the value of an electrode potential does not depend significantly on the concentration of the solutions. However, following extensive work in the field of electrochemistry, Walther Nernst formulated an equation that allows this change to be calculated. The electrode potential, E, obtained if the concentrations of the reagents in a half-equation are changed is:

$$E = E^{\ominus} + \frac{c}{n} \log \frac{[\text{concentration of oxidised species}]}{[\text{concentration of reduced species}]}$$

where $E^{\ominus}$ is the standard electrode potential
n is the number of electrons involved in the half-equation

At the standard temperature of 298 K, the value of c is 0.06.

Walther Nernst (1864–1941)

Example 1 The easiest case is that of a metal being oxidised to its ion. In this case, because it is a solid, the reduced form does not have a concentration. It is assigned a value of 1. Copper has an $E^{\ominus}$ value of +0.34 V and since the half-equation is $Cu^{2+}(aq) + 2e^- \rightarrow Cu(s)$, it can be seen that $n = 2$.

Therefore, the electrode potential, E, for this half-cell with a solution of copper ions of concentration 0.1 mol dm^{-3} is:

$$E = 0.34 + (0.06/2) \times \log(0.1/1) = 0.34 + 0.03 \log(0.1) = 0.31\text{ V}$$

Clearly the electrode potential changes by only a small amount. Even if the concentration were as low as10^{-4} mol dm^{-3}, the value of the electrode potential would only fall to 0.22 V.

Example 2 This example considers the electrode potential of the redox system:

$$Fe^{3+}(aq) + e^- \rightarrow Fe^{2+}(aq) \quad E^{\ominus} = +0.77\text{ V}$$

The Nernst equation is

$$E = 0.77 + \frac{0.06}{1} \log \frac{[Fe^{3+}(aq)]}{[Fe^{2+}(aq)]}$$

(n =1 in this case)

You may recall that for the measurement of redox potentials the solutions do not have to be 1 mol dm^{-3}, provided that the oxidised and reduced forms have the same concentration. The equation makes it clear that this will be so because the electrode potential will be the standard value when $\log([Fe^{3+}(aq)]/[Fe^{2+}(aq)]) = 0$. This occurs when $([Fe^{3+}(aq)]/[Fe^{2+}(aq)]) = 1$.

If, however, we had a situation in which, for example, $[Fe^{3+}(aq)] = 1$ mol dm^{-3} but $[Fe^{2+}(aq)] = 0.05$ mol dm^{-3} then the value of electrode potential would be:

$$E = 0.77 + 0.06 \log(1/0.05) = 0.77 + 0.06 \log(20) = 0.77 + 0.08 = 0.85\text{ V}$$

Example 3 To see another important consequence of the Nernst equation, it is worth looking at an overall reaction:

$$Fe^{3+}(aq) + V^{2+}(aq) \rightarrow Fe^{2+}(aq) + V^{3+}(aq)$$

The relevant electrode potentials for the half-reactions are:

$Fe^{3+}(aq) + e^- \rightarrow Fe^{2+}(aq)$ $\quad E^\ominus = +0.77\,V$

$V^{3+}(aq) + e^- \rightarrow V^{2+}(aq)$ $\quad E^\ominus = -0.26\,V$

Combining the electrode potentials gives a value of +1.03 V, which means that under standard conditions the reaction is feasible.

However, what is of interest is what happens as the reaction proceeds.

The E value of each half-reaction is given by the Nernst equation.

For iron:

$$E_{Fe} = 0.77 + 0.06 \log([Fe^{3+}(aq)]/[Fe^{2+}(aq)])$$

For vanadium:

$$E_V = -0.26 + 0.06 \log([V^{3+}(aq)]/[V^{2+}(aq)])$$

During the course of the reaction, the concentration of Fe^{2+} will steadily increase. This means that $\log([Fe^{3+}(aq)]/[Fe^{2+}(aq)])$ will be negative because $([Fe^{3+}(aq)]/[Fe^{2+}(aq)])$ decreases.Therefore the value of E_{Fe} will also decrease.

At the same time, the concentration of V^{3+} increases. This will make $\log([V^{3+}(aq)]/[V^{2+}(aq)])$ larger, so the value of E_v will increase.

These two effects mean that the reaction will become increasingly less favourable. So at what point will the reaction cease? It must be when E_{Fe} and E_v become equal, i.e. when:

$$0.77 + 0.06 \log([Fe^{3+}(aq)]/[Fe^{2+}(aq)]) = -0.26 + 0.06 \log([V^{3+}(aq)]/[V^{2+}(aq)])$$

or

$$1.03 = 0.06 \log([V^{3+}(aq)]/[V^{2+}(aq)]) - 0.06 \log([Fe^{3+}(aq)]/[Fe^{2+}(aq)])$$

Therefore:

$$1.03/0.06 = \log \frac{([V^{3+}(aq)]/[V^{2+}(aq)])}{([Fe^{3+}(aq)]/[Fe^{2+}(aq)])}$$

or

$$17.2 = \log \frac{[V^{3+}(aq)][Fe^{2+}(aq)]}{[V^{2+}(aq)][Fe^{3+}(aq)]}$$

$\dfrac{[V^{3+}(aq)][Fe^{2+}(aq)]}{[V^{2+}(aq)][Fe^{3+}(aq)]}$ is the equilibrium constant, K, for the reaction.

So, $17.2 = \log K$ and $K = 10^{17.2}$

This logic can be applied to all redox reactions to determine the value of their equilibrium constants. It is particularly useful because equilibrium constants of any size are difficult to determine by simple experiments.

The close relationship between ΔG and an electrode potential was mentioned in passing in this chapter ($\Delta G = -nFE$). This further calculation illustrates the interlinking of ideas relating to these quantities.

Stretch and challenge

1 The health of most plants requires good circulation of air through the soil in which they grow. This is in part because a good supply of oxygen maintains ions in the soil in the fully oxidised form that may be necessary to support plant growth.

$E^\ominus$ for the half-reaction $O_2(g) + 4H^+(aq) + 4e^- \rightleftharpoons 2H_2O(l)$ is 1.23 V.

However, the conditions in the soil are considerably different from standard conditions and, as a result, the electrode potential falls to about 0.80 V.

a In what ways do the conditions in soil differ from standard conditions?

b Explain, qualitatively, why the different conditions lead to a lower value for the electrode potential.

c Iron is usually present in soils in the form of Fe^{3+} ions. However, if the oxygen circulation is restricted because the soil becomes compacted or waterlogged, the iron is found as Fe^{2+} and the soil takes on a characteristic green–grey colouration.

Explain why this occurs.

d In a landfill site, rubbish is compacted heavily and circulation of air is almost impossible. Under these circumstances, sulphate ions are reduced to sulphide and the foul-smelling, toxic gas hydrogen sulphide, H_2S, can be released. Organic matter in the waste is also reduced with the result that methane is formed. Hydrogen ions are required for these reductions and they are provided by surrounding decomposing materials.

Write balanced half-equation for the reductions:

(i) $H^+ + SO_4^{2-} \rightarrow H_2S + H_2O$

(ii) $H^+ + CH_3COOH \rightarrow CH_4 + H_2O$ (to illustrate the reduction of organic matter)

2

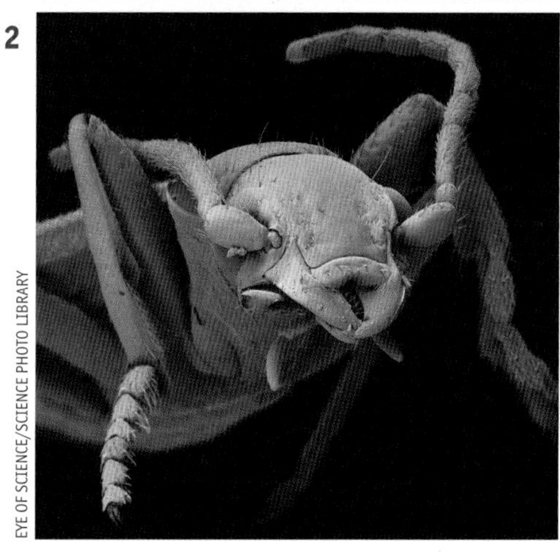

A bombardier beetle (×27)

The remarkable bombardier beetle has a unique method of defending itself against attack. The beetle stores two separated solutions:

- a 25% solution (by mass) of hydrogen peroxide, H_2O_2
- a 10% solution (by mass) of hydroquinone

HO—⟨benzene ring⟩—OH

When the beetle is threatened, the solutions mix, oxygen is generated and an explosion takes place! The surroundings are sprayed with an unpleasant cocktail of chemicals and steam. The explosion is caused by redox reactions that proceed via an intermediate complex called quinhydrone and then result in the formation of quinone as the final product:

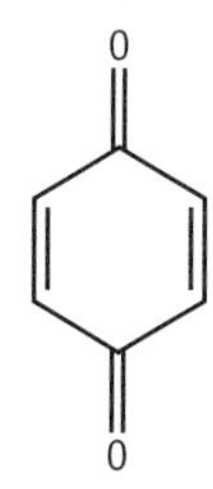

The mixture that is sprayed contains both quinhydrone and quinone.

Oxygen, which comes from the catalytic decomposition of some of the hydrogen peroxide, provides the propulsion for the explosion.

The half-equations and standard electrode potentials are:

quinone + $2H^+$ + $2e^- \rightleftharpoons$ quinhydrone $E^\ominus = +0.75\,V$

quinhydrone + $2H^+$ + $2e^- \rightleftharpoons$ hydroquinone $E^\ominus = +0.63\,V$

$H_2O_2 + 2H^+ + 2e^- \rightleftharpoons 2H_2O$ $E^\ominus = +1.77\,V$

$O_2 + 2H^+ + 2e^- \rightleftharpoons H_2O_2$ $E^\ominus = +0.69\,V$

a It is not entirely clear how the catalytic breakdown of hydrogen peroxide occurs. The likely answer is that it is catalysed by an enzyme, but it is possible that hydrogen peroxide undergoes an autocatalytic process in which one redox reaction of hydrogen peroxide promotes the other.

Use the two hydrogen peroxide half-equations to explain how autocatalysis might be possible.

b (i) A further possibility is that a metal ion such as Fe^{3+} initiates the hydrogen peroxide decomposition. Use the half-equation $Fe^{3+}(aq) + e^- \rightarrow Fe^{2+}(aq)$ ($E^\ominus = +0.77\,V$) to explain how this could be possible.

(ii) Suggest why Fe^{3+} might effectively be a catalyst for this process.

c Calculate the concentration in $mol\,dm^{-3}$ of:

(i) a 10% solution of hydroquinone (assume this is 10 g of hydroquinone in 100 cm^3 of solution)

(ii) a 25% solution of hydrogen peroxide (assume this is 25 g of hydrogen peroxide in 100 cm^3 of solution)

d (i) If just 0.1 cm^3 of hydrogen peroxide were converted to oxygen using the redox reaction with $Fe^{3+}(aq)$ ions, what volume of oxygen would be produced at 100°C (the temperature at which the beetle expels liquids at the point of the explosion)
(Assume that 1 mol of gas occupies a volume of 30.1 dm^3 at 100°C.)

(ii) How does this volume of oxygen differ from the volume of oxygen that would be collected at 100°C by heating 0.1 cm^3 of hydrogen peroxide?

e Use the electrode potentials to explain how quinhydrone and quinone can both be formed as part of the beetle's spray.

3 Using the data below, explain why $E^\ominus$ for the half-reaction $Fe^{3+}(aq) + 3e^- \rightarrow Fe(s)$ is $-0.037\,V$?

$E^\ominus$ for the half-reaction $Fe^{3+}(aq) + e^- \rightarrow Fe^{2+}(aq)$ is $+0.77\,V$.

$E^\ominus$ for the half-reaction $Fe^{2+}(aq) + 2e^- \rightarrow Fe(s)$ is $-0.44\,V$.

4 $Cu^{2+}(aq)$ reacts with $I^-(aq)$ to form $I_2(s)$, which can then be titrated with thiosulphate, $S_2O_3^{2-}(aq)$.

The relevant electrode potentials for the generation of iodine are:

$Cu^{2+}(aq) + e^- \rightleftharpoons Cu^+(aq)$ $E^\ominus = +0.15\,V$

$I_2(s) + 2e^- \rightleftharpoons 2I^-(aq)$ $E^\ominus = +0.54\,V$

a Using these electrode potentials, show that you would not expect this reaction to be feasible.

b However, the reaction does occur quite rapidly. Explain why this is so.

Transition elements

Chapter 15

Some features of the periodic table — for example, the trends in properties of the elements of the second and third periods — were studied in the first unit of the AS course. Coverage of some features of the group 2 and group 7 elements should have illustrated the advantage of studying these elements according to their position in the table. Mention has been made of the transition elements that occupy the central section of the periodic table. You will be familiar with some of these (e.g. iron and copper) but so far there has been no systematic coverage of their properties.

Some of the properties of the transition metals are reviewed in this chapter, which explains how they can be considered as a group with features in common. As with the elements in the main groups, their behaviour stems from similarities in outer electron configuration. This is considered first and then the features of some of the elements in the first row of the transition metal series are introduced.

The electron configuration of the *d*-block elements

The *d*-orbitals first appear in the ground state of the elements in period 4 of the periodic table. These are the elements from scandium (atomic number 21) to zinc (atomic number 30). You should be able to recall the sequence of orbital filling as 1*s*, 2*s*, 2*p*, 3*s*, 3*p*, 4*s*, 3*d*..., so the electron configuration of these elements can be deduced. The ground state electron configurations of these elements are given in Table 15.1. However, there are two places where the configuration is not quite what might have been expected.

The ground state is the electron arrangement that has the lowest energy for the element.

Table 15.1 *The ground state electron configuration of the d-block elements*

		4s	3d				
Sc	[Ar] $4s^2\,3d^1$	↑↓	↑				
Ti	[Ar] $4s^2\,3d^2$	↑↓	↑	↑			
V	[Ar] $4s^2\,3d^3$	↑↓	↑	↑	↑		
Cr	[Ar] $4s^1\,3d^5$	↑	↑	↑	↑	↑	↑
Mn	[Ar] $4s^2\,3d^5$	↑↓	↑	↑	↑	↑	↑
Fe	[Ar] $4s^2\,3d^6$	↑↓	↑↓	↑	↑	↑	↑
Co	[Ar] $4s^2\,3d^7$	↑↓	↑↓	↑↓	↑	↑	↑
Ni	[Ar] $4s^2\,3d^8$	↑↓	↑↓	↑↓	↑↓	↑	↑
Cu	[Ar] $4s^1\,3d^{10}$	↑	↑↓	↑↓	↑↓	↑↓	↑↓
Zn	[Ar] $4s^2\,3d^{10}$	↑↓	↑↓	↑↓	↑↓	↑↓	↑↓

[Ar] represents the electron configuration $1s^2\,2s^2\,2p^6\,3s^2\,3p^6$ as found in an argon atom

The ground state electron configuration of the *d*-block elements

When the *p*-orbitals were introduced it was noted that, when there is more than one orbital of equivalent energy, these are occupied initially by a single electron to avoid the repulsion that occurs when the second electron is added to create a pair. For example, in titanium, two separate *d*-orbitals are used, rather than a single *d*-orbital accommodating two electrons. The configuration for chromium extends this principle. In chromium, the pair of electrons in the 4*s*-orbital of an atom become separated and one of the electrons moves to enter the last of the five *d*-orbitals. The ground state is therefore:

- Cr: $1s^2\ 2s^2\ 2p^6\ 3s^2\ 3p^6\ 4s^1\ 3d_1{}^1\ 3d_2{}^1\ 3d_3{}^1\ 3d_4{}^1\ 3d_5{}^1$

The behaviour of the chromium atom is not unique — similar behaviour occurs for all orbitals of greater energy than 4*s*. It occurs because the difference in energy between the 4*s*-orbital and the 3*d*-orbitals is small and the re-arrangement avoiding the repulsion of the electron pairing is more favourable.

Notice also that the behaviour of the electrons in the ground state of a copper atom, where pairing of the electrons in the *d*-orbitals is favoured over retaining the pair of electrons in the 4*s*-orbital.

e The suffixes 1, 2, 3, 4, 5 can be used in an exam, rather than the correct labels of xy, yz, xz, $x^2 - y^2$, z^2.

e Because the 4*s*-electrons are lost before the 3*d*-electrons, some people prefer to write the ground state electron configuration of the element with the 4*s*-electrons after the 3*d*. So, the electron configuration of Fe is written as [Ar] $3d^6\ 4s^2$. Either way of writing the configuration is accepted in exams.

Knowledge of the shapes of the *d*-orbitals is not required for the A2 examination. However, it is an interesting feature, and the shapes are shown in the diagram.

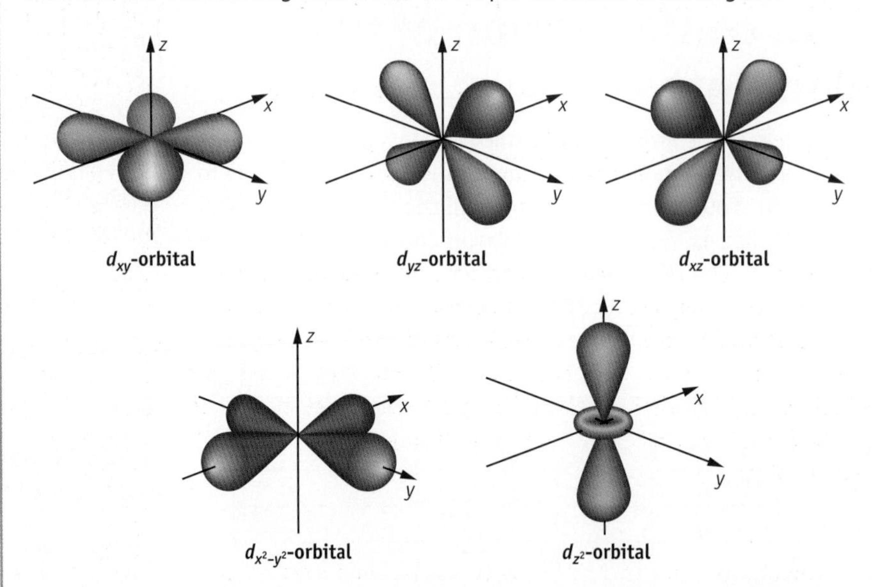

Three are identical — d_{xy}, d_{yz}, d_{xz}. They differ only in the direction in which they lie in space. The $d_{x^2-y^2}$ orbital looks similar, but the fifth orbital, the d_{z^2}, appears to be very different. This is a consequence of the mathematics on which the shapes of the orbitals have been established. The important thing to remember is that all five orbitals have identical energies or to use the correct term, they are degenerate.

The electron configuration of the *d*-block element ions

The close similarity in the energies of the 4*s*- and 3*d*-orbitals becomes even more apparent when the elements form ions, because when a *d*-block element forms an ion, the 4*s*-electrons are lost before the 3*d*-electrons. For example, the electron configurations of the ions of iron are:

- Fe^{2+} is [Ar] $3d^6$, *not* [Ar] $4s^2\ 3d^4$.
- Fe^{3+} is [Ar] $3d^5$, *not* [Ar] $4s^2\ 3d^3$.

Transition elements

It is not unusual for the *d*-block elements to be referred to as the transition elements. However, this is not strictly correct because the elements scandium and zinc do not show many of the properties that are listed below as being characteristic of transition elements. This is because most of these properties depend on the presence of ions that contain *d*-orbitals that are not completely full.

When a scandium atom forms an ion it does so by losing three electrons to make Sc^{3+}. This has the electron configuration $1s^2\ 2s^2\ 2p^6\ 3s^2\ 3p^6$, which has no *d*-electrons. (The configuration is the same as that of argon.)

A zinc ion, Zn^{2+}, does have *d*-electrons. The configuration $1s^2\ 2s^2\ 2p^6\ 3s^2\ 3p^6\ 3d^{10}$ — the *d*-orbitals are completely filled.

e Don't forget that it is the 4*s*-electrons that are lost when a zinc atom forms its ion.

A transition element is a *d*-block element that forms an ion with at least one partly filled *d*-orbital.

The above definition of a transition element excludes scandium and zinc.

Typical properties of transition elements

The transition elements of period 4 have many features in common. As solids, their atoms pack together closely and because of the ease with which electrons can move from orbital to orbital the transition elements are all metals that conduct electricity well. Their atoms have relatively small radii and pack together quite closely, which means that there are strong bonds between the metal ions and the delocalised electrons. This contributes to their high density and high melting and boiling points.

Variable oxidation states

The transition elements all exhibit two or more oxidation states in their compounds. Some of the many examples are:

- iron: Fe(II) and Fe(III)
- chromium: Cr(III) and Cr(VI) (e.g. in $Cr_2O_7^{2-}$)
- manganese: Mn(II), Mn(IV), Mn(VII) (e.g. in MnO_4^-)

The reasons for the existence of multiple oxidation states are quite complicated. As you will have learned when considering the Born–Haber cycle, a number of energy processes are involved in the formation of stable compounds and their aqueous ions. Nonetheless there are features that are worth noticing. Partly because of the similarity of the sizes of their atoms, the ionisation energies of the transition elements are similar, with reasonably low values (see Figure 15.1).

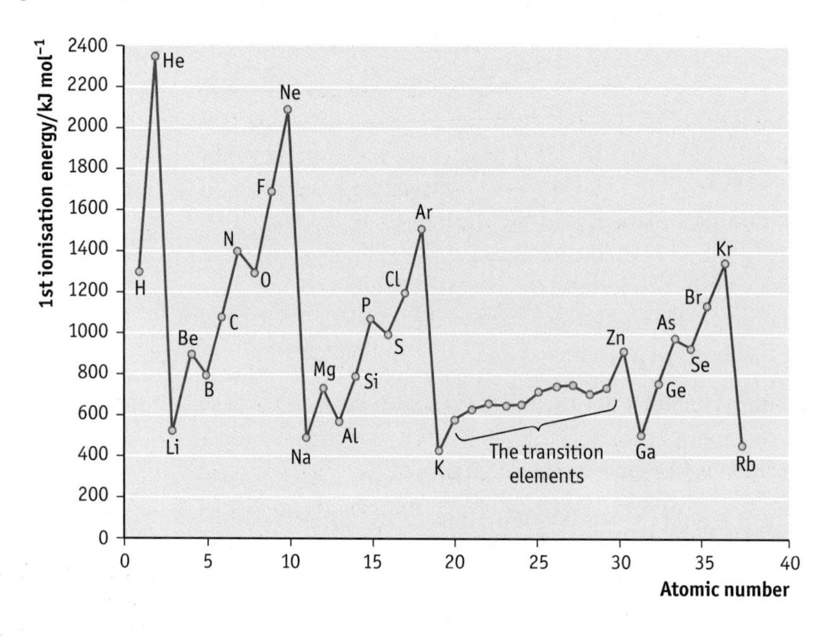

Figure 15.1
First ionisation energies

The values of the subsequent ionisation energies are higher, but the compensation of the extra stability of the crystal lattices formed as a result is often sufficient to allow a compound to be formed.

You do not need to know all the variations in the oxidation states that are possible but two general points should be appreciated:

- All the transition metals can form an ion of oxidation state 2, which represents the loss of the two 4*s*-electrons.
- The maximum oxidation state cannot exceed the total number of 4*s*- and 3*d*-electrons present in the electron configuration. (This point should be obvious, because once these electrons have been lost the stable electron configuration of argon is reached.)

The formation of coloured ions

A particularly noticeable feature of the transition elements is the characteristic colours of their aqueous ions. Aqueous copper ions are blue, aqueous chromium(III) ions are green, aqueous dichromate(VI) ions are orange and aqueous manganate ions are purple: the list could be extended. The reasons why the ions are coloured are considered in the additional reading section at the end of this chapter. Here, it is sufficient to say that it is a consequence of the presence of an ion containing incompletely filled *d*-orbitals that has an ion or molecule attached to it. Water is a common attachment and it is this that makes aqueous copper(II) salts appear blue and iron(II) salts appear light green.

e Be careful not to say that transition metal ions are coloured. This is not true — they must have an attached species to appear coloured. Without such an attachment they are usually white.

Use as catalysts

Transition metals are often used as heterogeneous catalysts. This is because a transition metal can use its *d*-orbitals to bind other molecules or ions to its surface. Examples include the use of iron as a catalyst in the Haber process to produce ammonia, the use of platinum, palladium and rhodium in the catalytic converter of a car and the use of nickel in the hydrogenation of alkenes.

Transition metal ions may also act as homogeneous catalysts. You do not need to know an example but you should appreciate that, in these cases, the ions take part in the reaction and are then re-generated. The transition metal ion is usually converted to a different oxidation state during the course of the reaction and it is the relative ease with which this is possible that makes them so useful.

Precipitation reactions

Precipitation reactions are not confined to transition metal ions. However, many transition metals undergo reactions that produce characteristic coloured precipitates, which can be useful in identification of the ion. There are four examples given here to illustrate the point but many other reactions of this type occur. In all cases, a solution containing an ion of one charge is mixed with a solution containing an ion of opposite charge to create a compound that is insoluble in the solvent used (water in all the examples below).

e You could be asked in an exam for a description of these precipitation reactions, so you should be ready with the necessary detail.

Example 1: aqueous sodium hydroxide and aqueous copper(II) ions

Copper(II) hydroxide is insoluble in water. Aqueous hydroxide ions from the sodium hydroxide react with the aqueous copper(II) ions from the copper salt solution to instantly form a blue precipitate of copper(II) hydroxide.

ANDREW LAMBERT PHOTOGRAPHY/SCIENCE PHOTO LIBRARY

$$Cu^{2+}(aq) + 2OH^-(aq) \rightarrow Cu(OH)_2(s)$$

The blue solution of the copper salt solution fades in colour as the ions are used in the formation of a light blue precipitate of copper(II) hydroxide.

Example 2: aqueous sodium hydroxide and aqueous cobalt(II) ions

Pink aqueous cobalt(II) ions react with aqueous hydroxide ion to produce a blue–green precipitate of cobalt(II) hydroxide. During the reaction, the pink colour of the Co^{2+}(aq) ion fades.

MARTYN F.CHILLMAID/SCIENCE PHOTO LIBRARY

$$Co^{2+}(aq) + 2OH^{-}(aq) \rightarrow Co(OH)_2(s)$$

Example 3: aqueous sodium hydroxide and aqueous iron(II) ions

Pale green aqueous iron(II) ions react with aqueous hydroxide ions to form a dark green precipitate of iron(II) hydroxide.

ANDREW LAMBERT PHOTOGRAPHY/SCIENCE PHOTO LIBRARY

$$Fe^{2\rightarrow}(aq) + 2OH^{-}(aq) \rightarrow Fe(OH)_2(s)$$

Example 4: aqueous sodium hydroxide and aqueous iron(III) ions

Aqueous iron(III) ions are yellow if the solution is dilute and appear orange in more concentrated solutions. The addition of aqueous hydroxide ions creates a rust-coloured precipitate of iron(III) hydroxide. (Indeed rust is largely composed of iron(III) hydroxide which forms readily when iron metal is exposed to wet conditions. It forms easily because oxygen in the air promotes oxidation on the surface of the iron to form iron(III) ions and there are enough hydroxide ions in water to start the precipitation.)

ANDREW LAMBERT PHOTOGRAPHY/SCIENCE PHOTO LIBRARY

$Fe^{3+}(aq) + 3OH^{-}(aq) \rightarrow Fe(OH)_3(s)$

Ligands and the formation of complex ions

In Chapter 12 it was explained how water encourages ionic compounds to dissolve by hydrating their ions. The spare pairs of electrons on the oxygen atom of a molecule of water are attracted strongly to positive ions in the ionic lattice

Alfred Werner founded the principles of coordination chemistry. It is claimed that, at the age of 24, he had a dream in which he imagined the structures of the coordination compounds and that the next day he completed the paper which has provided the basis for all the future work on this subject. In 1913 he was the first Swiss chemist to receive a Nobel Prize. Sadly, almost immediately afterwards his health deteriorated as a result of arteriosclerosis. This could have been the result of overwork, but he also drank heavily. He was a cheerful sociable man who enjoyed billiards, cards and chess.

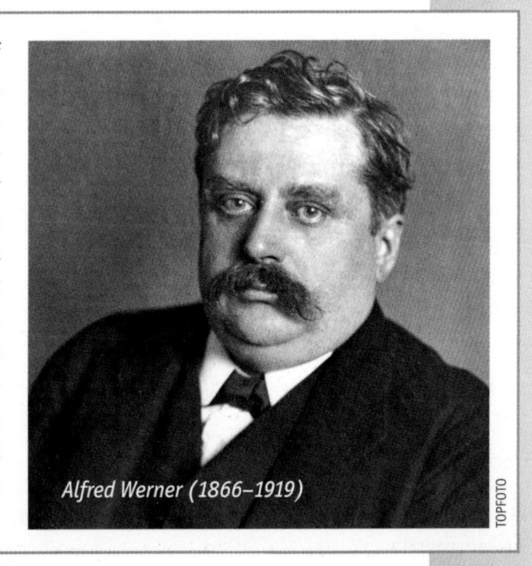

Alfred Werner (1866–1919)

TOPFOTO

and bind with them firmly enough to allow the breakdown of the lattice to occur. This can happen with any ion that possesses spare pairs of electrons, although the effectiveness of the attraction may not be as strong as it is for water.

The ion making the attachment via a spare pair of electrons is called a **ligand**.

When the bond is formed between the cation and the ligand, the ligand supplies both electrons. Therefore, the bond is a **coordinate bond** (or dative covalent bond). The resulting ion is called a complex ion.

An example of a complex ion with water as a ligand is $[Fe(H_2O)_6]^{3+}$:

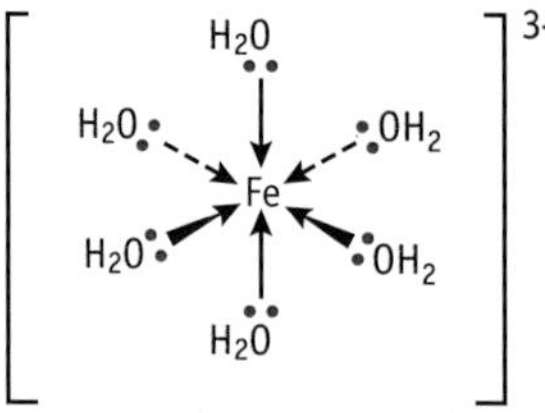

A range of complex ions can be formed and, as with precipitation reactions, this is not just a feature of transition metal ions. However, it is of particular interest for these elements because of their characteristic colours and sometimes for their practical uses.

Ammonia is another effective ligand. It binds to a cation in the way illustrated below for the nickel ammonia complex ion:

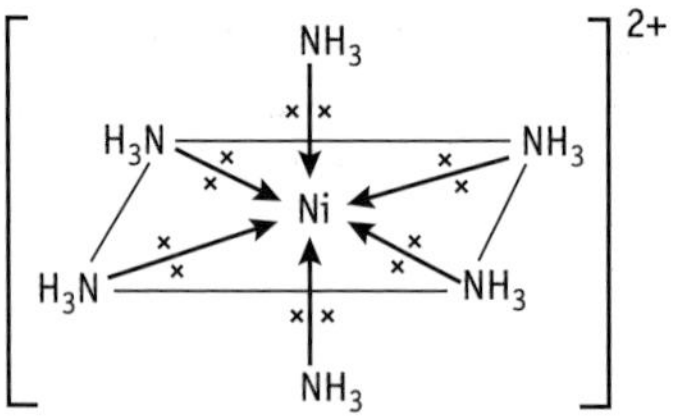

Note that the word used for the ammonia ligands is 'ammine' with two letter 'm's in the name. Do not confuse this with amine, which indicates a $-NH_2$ group.

Since six ammonia molecules are attached, the complex is a nickel hexaammine ion. As the ammonia molecules have no charge, the overall complex ion retains the 2+ charge of the nickel ion.

The overall charge of a complex ion is always the sum of the individual charges of the ions or molecules from which the complex is formed. If, for example, the complex has a cyanide ion as a ligand (the cyanide ion is a powerful ligand) then it must be remembered that the ion, CN^-, has a 1– charge. Therefore, the complex formed between Fe^{3+} and CN^- ions, in which there are six CN^- ions for each Fe^{3+} ion, has the formula, $[Fe(CN)_6]^{3-}$ because there are six 1– charges and a 3+ charge.

Complexes can have various shapes. The commonest shape is octahedral, as illustrated above by $[Ni(NH_3)_6]^{2+}$ and $[Fe(H_2O)_6]^{3+}$.

There are some complex ions that have only four ligands, in which case there are two possible shapes:

- If all four ligands that surround the central cation lie in the same plane, the shape is called square planar. An example is $Ni(NH_3)_2Cl_2$, (which has no overall charge):

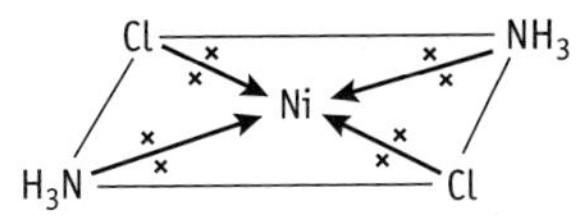

- Alternatively, the shape of the ion might be tetrahedral, as in $[CoCl_4]^{2-}$.

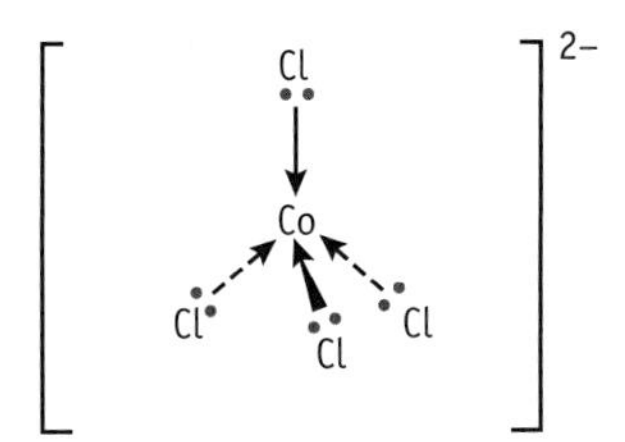

There is generally no need to remember the shapes of complex ions although it is necessary to know a couple of examples of octahedral complexes and to be ready to illustrate an exam answer with one example of both the other shapes.

An octahedral complex is said to show sixfold coordination; square planar and tetrahedral complexes show fourfold coordination.

Most complexes are formed by ions. However, it is also possible to form a complex using the uncharged metal atom. For example, nickel forms a complex with carbon monoxide that has the formula $Ni(CO)_4$.

Monodentate and bidentate ligands

All the examples so far have involved **monodentate** ligands. These ligands have a single point of attachment to the central cation. It is possible however to have ligands that attach at more than one point. For example, the **bidentate** ligand 1,2-diaminoethane (often abbreviated in this context to 'en') has the structure:

$CH_2—CH_2$
H_2N NH_2

The nitrogen atoms at the ends of the molecule each have a spare pair of electrons. This allows the molecule to attach to a central cation at two points:

$CH_2—CH_2$
H_2N NH_2

This is often drawn simply as:

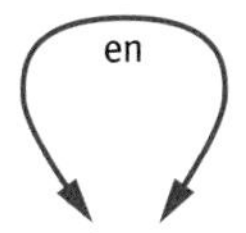

with the two arrows representing the pairs of electrons on the nitrogen atoms. An octahedral complex of nickel ion coordinated with the 1,2-diaminoethane can be represented as:

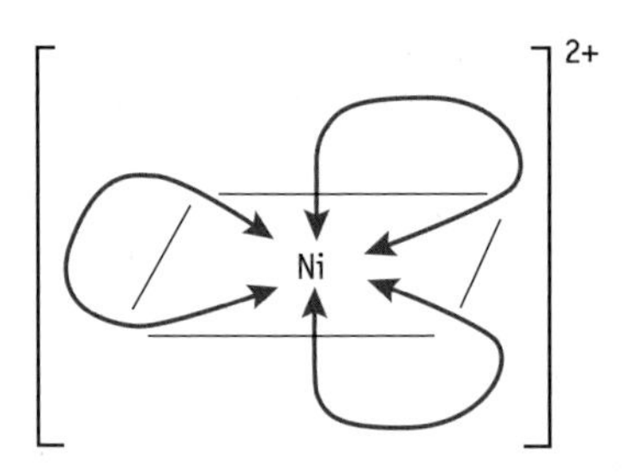

Polydentate ligands (i.e. those with several points of attachment) also exist. One of the most commonly encountered is EDTA, which has six points of attachment — it is hexadentate. The structure of EDTA is:

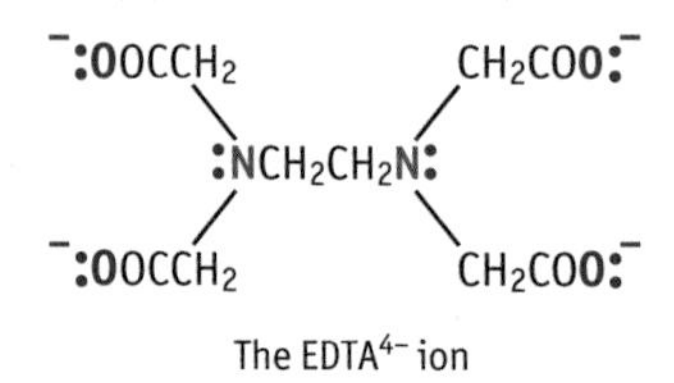

The $EDTA^{4-}$ ion

Each of the lone pairs is able to form a coordinate (dative) bond and is therefore able to attach in six places.

This is not required knowledge for the A2 exam but it gives an indication the variety of complexes that can be formed.

Isomerism of complex ions

Isomerism is commonplace in organic compounds. It also occurs in some inorganic substances.

The square planar structure of $Ni(NH_3)_2Cl_2$ has been illustrated above (page 231). In fact, the compound has two isomeric forms with the ammonia molecules and chloride ions being either on opposite sides of the complex ion (the *trans* form) or alongside each other (the *cis* form).

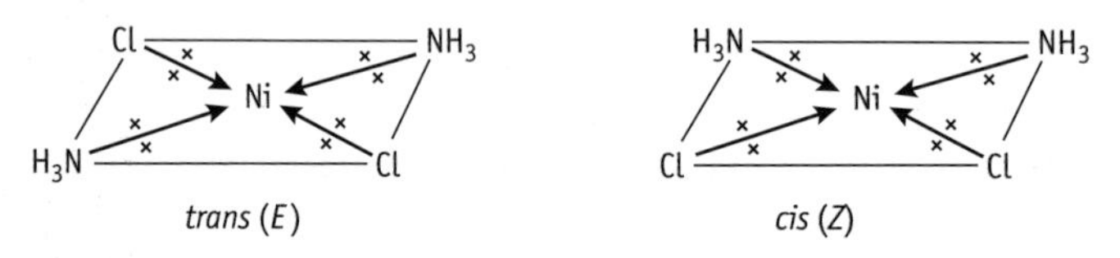

You will have met this type of isomerism in both AS and A2 organic chemistry. In organic chemistry, the *cis* isomer is designated as *Z* and the *trans* isomer as *E*.

Optical isomerism is also possible in complexes coordinated with polydentate ligands. The nickel(II) 1,2-diaminoethane complex ion is an example:

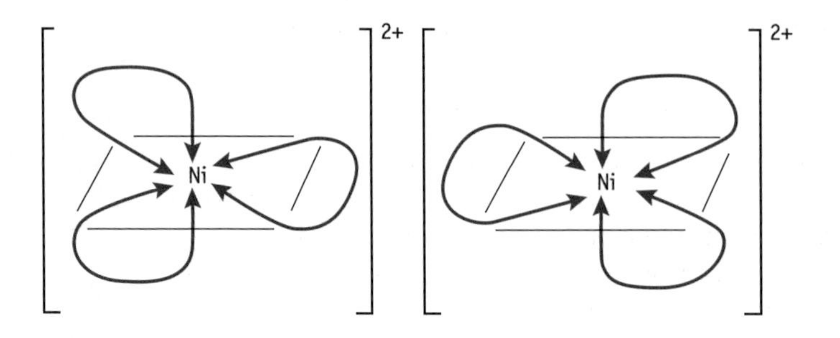

As is the case with organic molecules, it is the asymmetry of the structure that leads to optical isomerism as the two molecules cannot be superimposed.

Complex ions have various uses. A particularly interesting example is the *cis* (*Z*) form of the molecule $PtCl_2(NH_3)_2$. (The platinum is present as platinum(II), so the complex therefore has no overall charge). The structure of $PtCl_2(NH_3)_2$ is:

Cl, NH$_3$, Pt, Cl, NH$_3$

This compound is known as *cis*-platin. It is used during chemotherapy as an anti-cancer drug. It is a colourless liquid that is usually administered as a drip into a vein. It works by binding onto the DNA of cancerous cells and preventing their division.

The importance of the exact shape and structure of the molecule is emphasised by the fact that the *trans* isomer is ineffective.

Ligand substitution

The ligands that are attached to an ion can sometimes be exchanged for another ligand in a **ligand substitution** reaction. This occurs when the substituting ligand has a stronger attachment to the metal than does the original ligand.

Two colourful illustrations of this occur with the blue aqueous copper ion, $[Cu(H_2O)_6]^{2+}$. On adding concentrated ammonia solution, a ligand substitution occurs. Four of the water ligands are replaced by ammonia ligands, which creates an intensely dark blue solution of the ion $[Cu(NH_3)_4(H_2O)_2]^{2+}$. Both ions are octahedral.

$$NH_3 + H_2O \rightleftharpoons NH_4^+ + OH^-$$

$[Cu(H_2O)_6]^{2+}$ can also be converted to the complex ion, $[CuCl_4]^{2-}$, by the addition of concentrated hydrochloric acid. The $[CuCl_4]^{2-}$ is yellow, but the solution that is obtained appears green because the presence of water ensures that $[Cu(H_2O)_6]^{2+}$ is still present — a mixture of blue and yellow is therefore seen.

This experiment can also be carried out starting from dry solid copper(II) chloride. Its formula is usually written as $CuCl_2$, but it is actually $Cu^{2+}[CuCl_4]^{2-}$ in the anhydrous state and is yellow. As water is added, it appears initially as a dark green solution from the mixture of $[CuCl_4]^{2-}$ with a little $[Cu(H_2O)_6]^{2+}$. The mixture becomes progressively more blue as the chloride ions of the $[CuCl_4]^{2-}$ are substituted by H_2O until blue $[Cu(H_2O)_6]^{2+}$ ions predominate.

A similar experiment can be carried out using cobalt chloride which exists as $Co^{2+}[CoCl_4]^{2-}$ in the anhydrous state. $[CoCl_4]^{2-}$ is blue, but the addition of water results in the formation of a pink solution containing $[Co(H_2O)_6]^{2+}$ ions. This is the basis of a test for water using cobalt chloride paper which is blue but turns pink when it comes into contact with water.

e When this experiment is carried out using dilute ammonia, a precipitate of copper hydroxide is first obtained. This then dissolves to form the $[Cu(NH_3)_4(H_2O)_2]^{2+}$ ion. Dilute ammonia contains hydroxide ions from the ionisation.

NIGEL EVANS

Ligand substitution of $Cu^{2+}[CuCl_4]^{2-}$ by water to form $[Cu(H_2O)_6]^{2+}$

Many other ligand substitutions occur and complex ions are found widely in nature. Some ions are transported through plants as complexes, but a particularly interesting example is the transport of oxygen in blood by haemoglobin.

Haemoglobin and oxygen transportation

Haemoglobin is a complex molecule that contains an Fe^{2+} ion complexed to five permanent ligands but with a vacant sixth position that can bind with oxygen:

Histidine residue

When air is breathed in, oxygen is transferred into the bloodstream and binds to the sixth position. It is then transported through the body. The binding of the oxygen molecule is weak, which allows it to be removed to fuel oxidation reactions within the body. This is an efficient system, but it can be disrupted if another ligand were to bind so strongly at the sixth position that oxygen could not replace it. One such ligand is carbon monoxide. When a person inhales carbon monoxide, it occupies this sixth site and the molecule of haemoglobin can no longer transport oxygen. This results initially in the person feeling drowsy and then losing consciousness. Ultimately the poisoning can be fatal. When carbon monoxide occupies the sixth position the haemoglobin has a distinctive cherry-red colour, which is a tell-tale sign of its presence.

Stability constants

The strength of binding of a ligand to a cation can be represented quantitatively.

When a complex ion such as $[Cu(NH_3)_4(H_2O)_2]^{2+}$ is formed it exists in equilibrium with the hydrated copper ion, $[Cu(H_2O)_6]^{2+}$ from which it was made. The equation is:

$$Cu(H_2O)_6^{2+} + 4NH_3 \rightleftharpoons Cu(NH_3)_4(H_2O)_2^{2+} + 4H_2O$$

As with other equilibria, this has an equilibrium constant. In this case, it is called the **stability constant**:

$$K_{stab} = \frac{[Cu(NH_3)_4(H_2O)_2{}^{2+}]}{[Cu(H_2O)_6{}^{2+}][NH_3]^4}$$

It has the value $1.2 \times 10^{13}\,mol^{-4}\,dm^{12}$.

ⓔ This could be a good moment to make sure that you understand how to obtain the correct units for equilibrium constants such as this (see Chapter 10).

There are two important points to note about this equilibrium constant.

- The '$4H_2O$' on the right-hand side of the equation are not included in the equilibrium constant expression. This is because the reaction is carried out in aqueous solution, so the small amount of water produced in the reaction has very little effect.
- The square brackets mean concentration in $mol\,dm^{-3}$. They must not be confused with the brackets used in the formulae of complex ions to show that they are a complete unit. (To emphasise this, square brackets have been deliberately omitted from the equation.)

The value of this equilibrium constant, $1.2 \times 10^{13}\,mol^{-4}\,dm^{12}$, indicates that the reaction lies well to the right and gives a measure of the greater stability of $[Cu(NH_3)_4(H_2O)_2]^{2+}$ compared with $[Cu(H_2O)_6]^{2+}$.

There is no need to be able do calculations based on stability constant; it is their significance that must be understood. A few examples should make this clear.

Example 1 K_{stab} for $[CuCl_4]^{2-}$ is $4.2 \times 10^5\,mol^{-4}\,dm^{12}$. This is much smaller than the stability constant for the ammonia complex and reflects the fact that the $[CuCl_4]^{2-}$ complex is less stable.

Example 2 Silver ions form a soluble complex ion with two ammonia ligands, $[Ag(NH_3)_2]^+$.

This complex ion is formed when excess aqueous ammonia is added to a precipitate of silver chloride. It does not form if excess aqueous ammonia is added to a precipitate of silver iodide. It is a useful means of distinguishing between the two precipitates.

The complex ion has the stability constant $1.7 \times 10^7\,mol^{-2}\,dm^{+6}$. It is quite stable but it does require enough silver ions to be present to push the equilibrium:

$$Ag^+(aq) + 2NH_3(aq) \rightleftharpoons [Ag(NH_3)_2]^+$$

to the right.

Although silver chloride is not particularly soluble, it releases enough silver ions to force the complex to form. However, the less soluble silver iodide cannot do this and therefore remains insoluble.

Example 3: One of the reasons why cyanide ions are so poisonous is that they can form extremely stable complexes with a number of metal ions. For example, the ion, $[Fe(CN)_6]^{4-}$ which contains an Fe^{2+} ion has the stability constant $4.2 \times 10^{45}\,mol^{-6}\,dm^{18}$. Such complexes are understandably not usually able to undergo further ligand substitution reactions.

Redox reactions and titrations

Redox reactions

It has already been mentioned that a feature of transition metals is their ability to exist in more than one oxidation state. It is not surprising therefore that their redox chemistry is extensive. The importance of an understanding of electrode potentials in deciding on the feasibility of these processes was considered in Chapter 14. The principles established there can be used to predict the feasibility of the redox reactions of transition metals. Several examples were given in Chapter 14, so only a reminder is given here.

Worked example 1

Vanadium can exist in its compounds in four oxidation states:

- $V^{2+}(aq)$ is lilac
- $V^{3+}(aq)$ is green
- $VO^{2+}(aq)$ (oxidation state 4) is blue
- $VO_2^+(aq)$ (oxidation state 5) is yellow

Some electrode potentials are given in the table.

Reaction	$E^\ominus/V$
$VO_2^+(aq) + 2H^+(aq) + e^- \rightleftharpoons VO^{2+}(aq) + H_2O(l)$	+1.00
$VO^{2+}(aq) + 2H^+(aq) + e^- \rightleftharpoons V^{3+}(aq) + H_2O(l)$	+0.34
$MnO_4^-(aq) + 8H^+(aq) + 5e^- \rightleftharpoons Mn^{2+}(aq) + 4H_2O(l)$	+1.51

When $MnO_4^-(aq)$ is added to an acidified solution of $VO^{2+}(aq)$ will there be a reaction? If so, what would be observed?

Answer

The reaction $VO^{2+}(aq) + 2H^+(aq) + e^- \rightarrow V^{3+}(aq) + H_2O(l)$ cannot be involved because both $VO^{2+}(aq)$ and $MnO_4^-(aq)$, would require electrons to react.

The reaction $VO^{2+}(aq) + H_2O(l) \rightarrow VO_2^+(aq) + 2H^+(aq) + e^-$ (electrode potential – 1.00 V) could supply the electrons to $MnO_4^-(aq)$.

The variety of colours of its oxidation states led vanadium to be named after Vanadis, the Scandinavian goddess of beauty.

The reduction of $MnO_4^-(aq)$ to $Mn^{2+}(aq)$ has a potential +1.51 V.
This means that there is an overall electrode potential of +0.51 V, so the reaction is feasible.
Purple $MnO_4^-(aq)$ ions are reduced to pale pink $Mn^{2+}(aq)$ ions; blue $VO^{2+}(aq)$ ions are oxidised to yellow $VO_2^+(aq)$ ions. The final solution appears yellow as this is the more intense colour.
The overall equation is obtained by combining the two half-equations:
$MnO_4^-(aq) + 5\,VO^{2+}(aq) + H_2O(l) \rightarrow Mn^{2+}(aq) + 5VO_2^+(aq) + 2H^+(aq)$

In the overall equation some H^+ and H_2O species cancel out.

Other redox reactions of transition metal ions can be deduced in a similar way to that in the worked example..

Redox titrations

There is often a colour change when a transition element changes oxidation state, so it is possible to make use of this in redox titrations.

The reagent used most widely is a solution of potassium manganate(VII). This solution has an intense purple colour, but when it is reduced it is converted to pale pink (almost colourless) manganese(II) ions:

$$5e^- + MnO_4^-(aq) + 8H^+(aq) \rightarrow Mn^{2+}(aq) + 4H_2O(l)$$

Excess $H^+(aq)$ ions are required and these are usually supplied by adding dilute sulphuric acid.

A redox titration is possible if the reagent used to reduce $MnO_4^-(aq)$ is either colourless or only lightly coloured. An example is a solution containing pale green iron(II) ions. On oxidation, these are converted to yellow iron(III) ions. The equation for this reaction is:

$$5Fe^{2+}(aq) + MnO_4^-(aq) + 8H^+(aq) \rightarrow 5Fe^{3+}(aq) + Mn^{2+}(aq) + 4H_2O(l)$$

The concentration of a solution containing $Fe^{2+}(aq)$ ions is determined as follows:

- A solution of $MnO_4^-(aq)$ of known concentration is made up. The potassium salt is always used. (This solution has to be quite dilute because potassium manganate(VII) is not particularly soluble in water. It is wise not to exceed about 0.02 mol dm^{-3}.)
- This solution is placed in a burette.
- A fixed volume of the solution containing $Fe^{2+}(aq)$ is pipetted into a conical flask.
- Excess dilute sulphuric acid is added. The exact volume does not matter as long as sufficient $H^+(aq)$ ions are present to ensure that all the $MnO_4^-(aq)$ is reduced.
- The solution of $MnO_4^-(aq)$ is added from the burette. It is decolorised immediately.

Titration of iron(II) with manganate(VII)

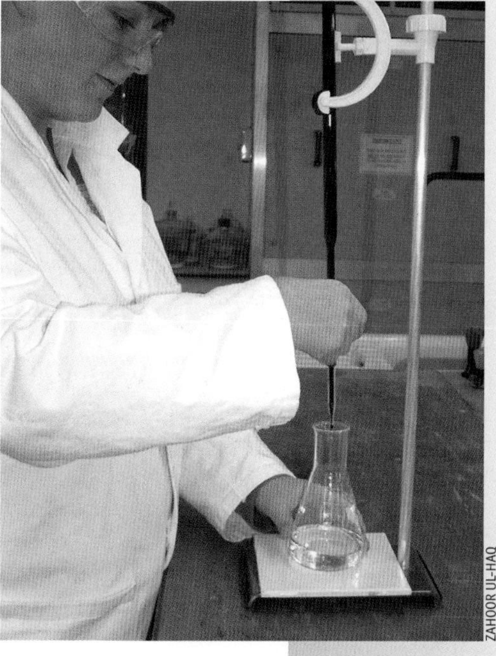
ZAHOOR UL-HAQ

- The end point of the titration is when the first sign of a permanent pink colour appears in the conical flask. This occurs when all the $Fe^{2+}(aq)$ has reacted and the first drop of $MnO_4^-(aq)$ in excess has been added. $MnO_4^-(aq)$ is coloured so intensely that the eye perceives this change almost instantly. Strictly, the end point has been passed, but the result will lie comfortably within the accuracy of the volumetric equipment used.
- The titration is repeated to obtain concordant results.
- The subsequent calculation is the same in principle as that used for an acid–base titration.

Worked example

25.00 cm^3 of a solution of iron(II) sulphate reacts exactly with 22.70 cm^3 of a 0.0200 mol dm^{-3} solution of potassium manganate(VII), according to the equation:

$$5Fe^{2+}(aq) + MnO_4^-(aq) + 8H^+(aq) \rightarrow 5Fe^{3+}(aq) + Mn^{2+}(aq) + 4H_2O(l)$$

Calculate the concentration in mol dm^{-3} of the iron(II) sulphate solution.

Answer

22.70 cm^3 of potassium manganate(VII) contains $(22.70/1000) \times 0.0200$
$= 4.54 \times 10^{-4}$ mol

From the equation, 1 mol of $MnO_4^-(aq)$ reacts with 5 mol of $Fe^{2+}(aq)$

amount (in moles) of $Fe^{2+}(aq)$ present in 25.00 cm^3 of iron(II) sulphate
$= 5 \times 4.54 \times 10^{-4} = 2.27 \times 10^{-3}$ mol

concentration (in mol dm^{-3}) $= (1000/25) \times 2.27 \times 10^{-3} = 0.0980$ mol dm^{-3}

Several other titrations involving manganate(VII) ions are possible and they all follow a similar procedure.

Although not exclusively related to transition metals, there is another titration that is employed widely. It is based on the redox reaction between iodine and thiosulphate ions. The equation for this reaction was considered on page 202.

$$2S_2O_3^{2-}(aq) + I_2(s) \rightarrow S_4O_6^{2-}(aq) + 2I^-(aq) \quad \text{equation (1)}$$

This titration is not usually used directly to determine the concentration of an iodine solution. Rather, it allows the determination of the concentration of a reagent that generates iodine as a result of a reaction.

An example is the determination of the concentration of a copper(II) sulphate solution. A known volume of copper (II) sulphate is reacted with excess potassium iodide:

$$2Cu^{2+}(aq) + 4I^-(aq) \rightarrow Cu_2I_2(s) + I_2(s) \quad \text{equation (2)}$$

Cu_2I_2 is copper(I) iodide, which forms as a grey–white precipitate. The reaction is, therefore, a redox process in which Cu^{2+} is reduced and I^- is oxidised.

The iodine produced is then titrated against a solution of sodium thiosulphate of known concentration.

At the start of the titration, the solution appears brown–purple because of the presence of the iodine. As the titration proceeds, this colour fades to yellow and the end point is reached when the solution is colourless. In practice, this colour change is quite difficult to see, mainly because of the presence of the copper(I) iodide precipitate. To help, some starch solution is added as the end point is approached. This gives rise to a dark-blue coloration that disappears sharply at the end point.

Calculations based on this titration may look rather daunting, but the equations provide information:

- equation (2) — 2 mol of Cu^{2+} react to produce 1 mol of I_2
- equation (1) — 1 mol of I_2 reacts with 2 mol of $S_2O_3^{2-}$

It follows therefore that for every 1 mole of Cu^{2+}, 1 mole of $S_2O_3^{2-}$ is required. So the amount (in moles) of thiosulphate used is equivalent to the amount (in moles) of copper(II) ions present originally.

Worked example 2

25.00 cm^3 of a solution of copper(II) sulphate reacts with excess potassium iodide solution to produce an amount of iodine that reacts exactly with 27.15 cm^3 of a 0.100 mol dm^{-3} solution of sodium thiosulphate. Calculate the concentration, in mol dm^{-3}, of the copper(II) sulphate solution.

Answer

Amount (in moles) of $S_2O_3^{2-}$ used in the titration $= (27.15/1000) \times 0.100$
$= 2.715 \times 10^{-3}$ mol

Amount (in moles) of copper sulphate in 25.00 cm^3 $= 2.715 \times 10^{-3}$ mol

Concentration (in mol dm^{-3}) of copper(II) sulphate $= (1000/25) \times 2.715 \times 10^{-3}$
$= 0.109$ mol dm^{-3}

e The reasons why the number of moles of copper(II) sulphate is the same as the number of moles of thiosulphate are given in the text above this worked example.

Iodine is produced readily in a number of reactions, so this type of titration has wide application.

Questions

1 Copy the formula for each of the following complex ions and deduce their net charge, if any:

a $[Zn(NH_3)_4(H_2O)_2]$

b $[Fe(CN)_6]$ containing Fe(II)

c $[Co(NH_3)_5Cl]$ containing Co(III)

d $[Co(C_2O_4)_3]$ formed from Co^{2+} ions and $K_2C_2O_4$

e $[Cr(CH_3COO)_2(H_2O)_2]$ containing Cr(III)

2 Explain the following:

a When 1,2-diaminoethane is added to light-blue aqueous copper(II) sulphate solution, the colour of solution intensifies to dark blue.

b When hydrochloric acid is added to the dark blue solution formed as a result of the experiment in part **(a)**, the colour returns to light blue.

3 Suggest explanations for each of the following:

a When concentrated hydrochloric acid is added to aqueous cobalt(II) chloride solution, the colour of the solution changes from pink to blue.

b If water is added to some of the blue solution formed in part **(a)**, the colour changes back to pink.

c If aqueous silver nitrate is added to some of the blue solution formed in part **(a)**, the solution changes to pink again and a precipitate is formed.

d If propanone is added carefully to the pink aqueous solution of cobalt(II) chloride, the propanone floats on the surface of the water. At the junction of the two liquids a blue layer is observed.

4 Salts that contain a 'complexed' cation with an 'uncomplexed' anion can be crystallised. For example, cobalt forms a salt, **A**, of formula $[Co(NH_3)_6]Cl_3$, which is orange in colour.

It is possible to prepare isomers of this cobalt complex that retain the same *number* of chloride ions but which have a chloride ion substituting one of the ammonia molecules in the ligand. One example is salt, **B**, which has the formula $[Co(NH_3)_5(Cl)]Cl_2$.

a If equal volumes and concentrations of salts **A** and **B** are reacted with excess aqueous silver nitrate, salt **A** produces a precipitate that has a mass 1.5 times greater than the precipitate produced by salt **B**.

Explain why.

b Two different salts, **C** and **D**, can be prepared that contain Co(III) coordinated with four ammonia ligands. Each salt also contains three Cl^- ions. When the experiment in part **(a)** is repeated separately with **C** and **D**, each produces a mass of precipitate that is half the mass obtained from compound **B**.

Suggest structures for **C** and **D** and explain your answer.

5 A solution is made that contains the $VO^{2+}(aq)$ ion. When 25.0 cm^3 of this solution is titrated against 0.0150 $mol\,dm^{-3}$ $MnO_4^-(aq)$ ions in the presence of excess sulphuric acid, it is found that 23.30 cm^3 is required to reach the end point.

The half-equation for the oxidation of the $VO^{2+}(aq)$ ions is:

$$VO^{2+}(aq) + 2H_2O(l) \rightarrow VO_3^-(aq) + 4H^+(aq) + e^-$$

Calculate the concentration (in $mol\,dm^{-3}$) of the solution containing $VO^{2+}(aq)$.

6 A general purpose solder contains antimony, lead and tin. When reacted with an acid, the solder dissolves forming a solution containing $Sb^{3+}(aq)$, $Pb^{2+}(aq)$ and $Sn^{2+}(aq)$. Neither $Sb^{3+}(aq)$ nor $Pb^{2+}(aq)$ reacts with dichromate ($Cr_2O_7^{2-}$) ions. However, $Sn^{2+}(aq)$ can be oxidised to $Sn^{4+}(aq)$ by an acidified solution of potassium dichromate(VI).

In an experiment, 10.00 g of solder is dissolved in acid to make 1.00 dm^3 of solution. When 25.0 cm^3 of this solution is titrated against an acidified potassium dichromate solution of concentration 0.0175 $mol\,dm^{-3}$, 20.00 cm^3 of the dichromate is required to reach the end point.

The half-equation for the reduction of $Cr_2O_7^{2-}(aq)$ is:

$$Cr_2O_7^{2-}(aq) + 14H^+(aq) + 6e^- \rightarrow 2Cr^{3+}(aq) + 7H_2O(l)$$

a Write an overall equation for the reaction of $Sn^{2+}(aq)$ with $Cr_2O_7^{2-}(aq)$

b Calculate the concentration of $Sn^{2+}(aq)$ in the solution in $mol\,dm^{-3}$.

c Calculate the percentage by mass of tin in the solder.

7 Rhubarb leaves contain poisonous ethanedioic acid, $(COOH)_2$. A dose of about 24 g of ethanedioic acid could be fatal if consumed by an adult. On heating, ethanedioate ions react with potassium manganate(VII) solution and so this forms the basis of a titration. The ethanedioate ions are oxidised to carbon dioxide.

$$(COO)_2^{2-}(aq) \rightarrow 2CO_2(g) + 2e^-$$

Four large rhubarb leaves are heated with water to extract the ethanedioic acid. The solution is filtered and then diluted to 250 cm^3 in a volumetric flask.

25.0 cm^3 samples of the solution are acidified with dilute sulphuric acid. 23.90 cm^3 of 0.0200 $mol\,dm^{-3}$ potassium manganate(VII) are required to reach the end point in a titration.

a Write an overall equation for the reaction between ethanedioate ions and acidified manganate(VII) ions

b Use the titration result to calculate the amount (in moles) of ethanedioic acid in the 250 cm^3 solution.

c Calculate the mass of ethanedioic acid in the rhubarb leaves.

d Suggest the number of rhubarb leaves that, when eaten, would be sufficient to kill an adult.

8 When excess potassium iodide solution is added to 25.0 cm^3 of a solution of bromine in water, the iodine formed reacts exactly with 24.75 cm^3 of a 0.800 $mol\,dm^{-3}$ solution of sodium thiosulphate.

Calculate the concentration of the bromine solution in $g\,dm^{-3}$.

9 To be safe, swimming-pool water should contain between 1 g and 2 g of chlorine in every 1000 dm^3 of water.

In an investigation, a 500.0 cm^3 sample of swimming-pool water is analysed by treating it first with excess potassium iodide solution and then titrating the iodine released against a 1.500×10^{-3} $mol\,dm^{-3}$ solution of sodium thiosulphate. 10.25 cm^3 of sodium thiosulphate is required to reach the end point.

Would it be safe to swim in this swimming pool?

10 An element, **X**, is able to form a compound of formula $NaXO_3$. An aqueous solution of $NaXO_3$ of concentration 0.0500 $mol\,dm^{-3}$ is prepared. When excess potassium iodide solution is added to 25.0 cm^3 of this solution, iodine is produced. When this is titrated against a solution of sodium thiosulphate of concentration 13.63 $g\,dm^{-3}$, 29.00 cm^3 of the solution is required to react completely with the iodine.

Deduce the change in oxidation state for element **X** in the reaction. (You may assume that element **X** is not iodine.)

Summary

You should now be able to:

- give the electron configuration of *d*-block elements and their ions
- explain what is meant by the term 'transition element'
- explain the typical transition metal properties as:
  - multiple oxidation states
  - the formation of coloured ions
  - their use as both heterogeneous and homogenous catalysts
- describe the precipitation reactions of $Cu^{2+}(aq)$, $Co^{2+}(aq)$, $Fe^{2+}(aq)$ and $Fe^{3+}(aq)$ with $OH^-(aq)$
- explain what is meant by:
  - a ligand
  - a complex ion
  - coordination number
- give examples of octahedral, tetrahedral and square planar coordinated transition metal ion complexes
- describe how *cis–trans* and optical isomerism occur in transition metal ion complexes
- describe the use of *cis*-platin in cancer treatment

- describe ligand substitution reactions and give some examples
- explain how oxygen is transported by the blood and the adverse effect of inhaling carbon monoxide
- define *stability constant* and explain how it gives a measure of the stability of a complex ion
- give examples and understand how electrode potentials can be used to explain or predict the redox reactions of transition metals
- describe the use of aqueous manganate(VII) ions and aqueous thiosulphate ions in redox titrations

Additional reading

15

A feature of transition metals is the range of colours shown by their complex ions. Invariably a ligand exchange is accompanied by a noticeable change in colour. As will be explained, the reason for this stems from the behaviour of the *d*-orbitals when surrounded by ligands.

First, it is necessary to understand how we perceive the colour of a solution. White light consists of a combination of colours — red, orange, yellow, green, blue, indigo and violet — each associated with particular wavelengths and energies. The spectrum of white light is continuous, so these wavelengths merge together and ours eyes fail to identify the individual components without splitting the light. Splitting occurs in a rainbow or, in the laboratory, by using a prism. However, when white light is passed through a substance, some components of the spectrum of the light are removed and certain wavelengths remain. So copper(II) sulphate appears blue because the wavelengths of the red end of white light are absorbed, leaving the blue end of the spectrum to predominate.

The wavelengths are no longer present because absorption of energy occurs as electrons are promoted from one energy state to another of higher energy. The energy required corresponds to the wavelength of one of the colours within the spectrum. Many jumps take place within an atom or ion but these do not always correspond to an energy that matches a colour. These jumps may be in the ultraviolet region of the spectrum and, although we cannot perceive them visually, apparatus is available that can provide a UV spectrum of a substance.

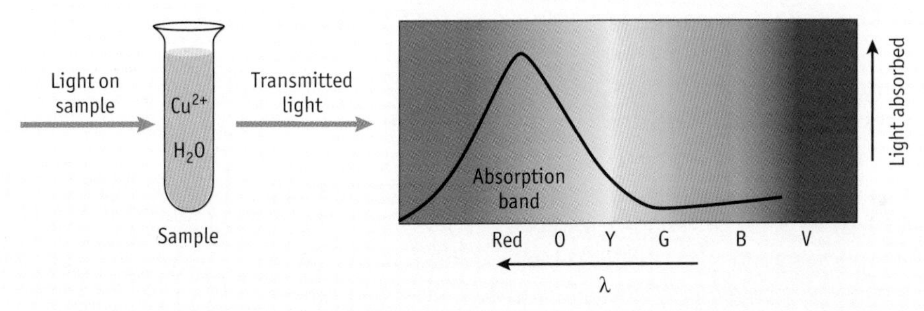

Returning to transition metals, it must be remembered that in the free ion the *d*-orbitals have equivalent energies even though their shapes are sometimes different. You may recall (see page 224) that three of the orbitals (d_{xy}, d_{xz} and d_{yz}) are identical and only differ in the axes that they lie between. The remaining two orbitals are different from these and from each other. The d_{z^2} orbital is obviously different; the $d_{x^2-y^2}$ has a different charge distribution from the three identical orbitals. When a complex ion is formed, the pairs of electrons binding the ligand to the transition metal ion do not attach directly to any particular orbital but instead adopt their own geometry around the metal ion. The examples discussed in the chapter include octahedral, square planar and tetrahedral shapes, although others are possible.

An octahedral complex is symmetrical with respect to the central ion. When the complex is formed the six ligands attach to the metal ion by approaching along the *x*, *y* and *z* axes. Therefore, the effect of the ligand electrons on the three equivalent orbitals is identical. The energy of all three orbitals is increased to the same extent. This new energy level is called the t_{2g} energy state. However, the energy of the two other orbitals is increased more because the electrostatic repulsion between the orbitals and the ligands is greater. In fact (although this may seem surprising) the enrgy of the d_{z^2}-orbital and the $d_{x^2-y^2}$ orbital is increased by the same amount to a new energy state called e_g. The situation can be summarised on an energy diagram as:

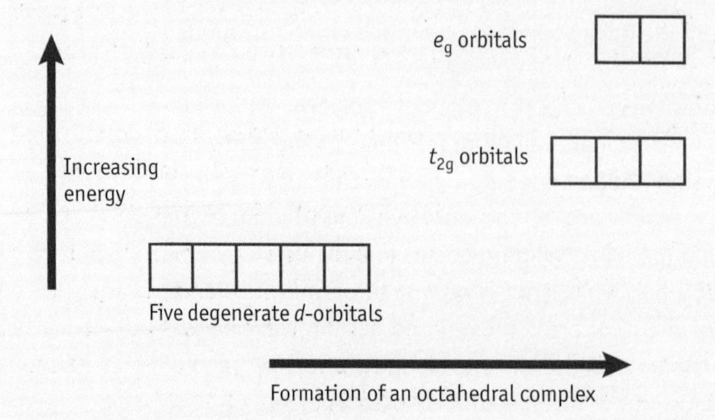

This splitting of the *d*-orbital energies is important because the energy difference between the t_{2g} and the e_g states corresponds to wavelengths within the visible spectrum. This means that electrons are able to jump from the t_{2g} state to the e_g state and absorb one of the components of white light. The colour of the complex is therefore due to the wavelengths of light that have not been removed.

The range of colours of different octahedral complexes shows that different ligands have different effects on the various transition metal ions. What changes is the energy difference between the t_{2g} state and the e_g state. If the energy difference is large, the more energetic end of the visible spectrum is removed. These are the blue–violet components, so such a complex would have a colour dominated by the red end of the spectrum.

A blue complex indicates that the energy separation is not so great. Although there is no exact way of predicting what will happen there is a rule of thumb which predicts that the energy separation is greatest when the ligand involves a group 4 element such as in CN^- and becomes progressively less through groups 5, 6 and 7.

The concept has been explained using an octahedral complex as the example, but the same principles apply to other geometries. In tetrahedral complexes, the d_{xy}, d_{xz} and d_{yz} orbitals are promoted to a higher energy level than the d_{z^2} and the $d_{x^2-y^2}$ orbitals; in square planar complexes a more involved splitting creates four different energy states, thus allowing a greater number of possible electron promotions.

Stretch and challenge

Ligand exchange can sometimes be used as a means of determining the end point of a complexiometric titration. An example of this is a titration to determine the concentration of calcium ions in water. The presence of magnesium ions is essential for the method — if they are not present naturally, some must be added.

This titration could be used to estimate the calcium ion concentration in a soil extract made by shaking a sample of soil with water and filtering. It is unnecessary to add magnesium ions in this case as enough will be present naturally. The method is as follows.

- A solution of the complexing agent EGTA is made up to a specified concentration.

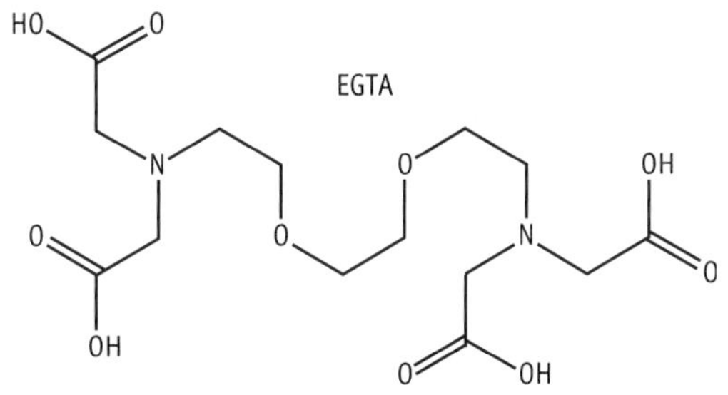

EGTA

- A known volume of EGTA solution is then added to a measured volume of the soil extract in a conical flask. The volume of EGTA must be large enough to ensure that all the calcium and magnesium ions in the soil extract are complexed with EGTA and that some free EGTA remains.
- A small volume of the complexing agent thymolphthalexone is added and the blue colour of a magnesium thymolphthalexone complex is observed.
- A titration is then carried out by running a solution of lead nitrate of known concentration from a burette into the mixture in the conical flask.
- During the course of this titration, the blue colour fades. The end point is when the solution becomes completely colourless.
- The amount (in moles) of calcium ions in the volume of soil extract is then equal to the amount (in moles) of EGTA added originally minus the amount (in moles) of lead ions added during the titration.

Use the description above to answer the following questions.

a Place each set of complex ions below in order of decreasing stability:

(i) calcium–water, calcium–EGTA

(ii) lead–water, lead–EGTA

(iii) magnesium–thymolphthalexone, lead–thymolphthalexone

(iv) magnesium–water, magnesium–EGTA, magnesium–thymolphthalexone

(v) magnesium–EGTA, calcium–EGTA, lead–EGTA

b What is the colour of a solution of thymolphthalexone?

c At the end of the titration, in what type of complex will magnesium ions, calcium ions and lead ions be? If they could be in more than one type of complex, give all the forms.

d Give a brief explanation of why the amount (in moles) of calcium ions is as explained in the final bullet point. State the assumptions that must be made.

Unit 6
Practical skills in chemistry 2

Practical skills

The assessed practicals that you carried out during the AS course covered qualitative, quantitative and evaluative tasks. At A2, the same skills are required but the range of experiments is wider and the depth of understanding expected is greater. In this chapter, each of the assessed skills is considered and some general advice is given that should allow you to approach the assessments with confidence. Many of the points covered here are reminders. However, you should remember the importance of the mark you obtain on this unit to your overall A-level result.

e This unit is worth 10% of the total A-level mark.

Safety in the laboratory

In the assessment of all three tasks, you could be asked for comment on safe working practice. Awareness of standard safety procedures and an understanding of essential safety precautions is expected and should be part of your normal working procedure.

Although it is important to wear a laboratory coat and use safety glasses when doing practical work, the emphasis in the assessments is on understanding the hazards associated with chemicals. These hazards do not have to be remembered. However, you should understand, and be able to select from information supplied, any hazards relevant to a particular experiment. It is a common mistake to provide irrelevant detail, rather than an essential precaution for the procedure. For example, handling concentrated hydrochloric acid requires great care because its fumes are dangerous and it is corrosive whereas dilute hydrochloric acid is an irritant which, although it should be handled carefully and not be spilt, is not especially hazardous. By contrast, strong alkalis at the same concentration require considerably more care. It is wise to handle all organic chemicals cautiously but it is not necessarily the case that they are extremely flammable. With experience, you will acquire the knowledge to decide which particular precautions should be taken.

Qualitative tasks

The level of skill expected for these tasks is not especially demanding. However, there is a danger that the importance of careful manipulation of apparatus and the recording of reliable accurate observations may be underestimated. Two types of experiment are likely.

The preparation of an organic or inorganic compound

There are several procedures that should have been mastered during laboratory sessions. In the main, great dexterity is not required, but it is easy, through thoughtlessness, to underestimate the skill required. Organisation is the key to success and the following should be considered carefully:

- If there are several steps in an experiment, it is essential to plan carefully so that the apparatus and chemicals are to hand when needed. Errors occur because students leave their workplaces to search for something, assuming that the reactions they are carrying out or the flasks they are heating will simply wait for their return. This can result in solutions boiling over or reactions continuing beyond the required stage.
- Is all the apparatus well supported and safe? Other students could accidentally upset apparatus that is poorly supported. Always check that no piece of apparatus is close to the edge of the bench or too close to a sink.
- Make sure that you read the instructions carefully. There is usually no going back if you add the wrong chemical or fail to follow the correct sequence.

In addition to the general advice above, there are specific skills that should be well-practised. Particular qualitative techniques that may go wrong are when:

- *filtering a precipitate from a solution*. This requires patience — filtering takes time. Do not fill a funnel so full that the solution spills over the top of the filter paper and down the inside of the funnel. Do not stir the contents of the filter paper in the hope of speeding up the filtration; this leads usually to the filter paper tearing and its contents (including the precipitate) pouring into the container below.

  You should also know how to filter under reduced pressure. There are different ways of doing this and you will be shown the method that is used in your laboratory.
- *evaporating a solution to form crystals*. This is sometimes necessary when inorganic crystals are being prepared. Care is needed not to boil off too much water because this results in poor quality crystals being obtained.
- *recrystallising from a solution*. This is usually necessary when preparing a solid product. The procedure separates the desired product (which must be soluble) from insoluble impurities and it is not always easy. An instruction to recrystallise a product by dissolving it in the minimum volume of solvent *must* be taken literally. A mistake often made is to use far too much solvent, which makes it difficult to form the crystals. When dissolving, the impure product should be stirred vigorously and more solvent should be added only if it is absolutely necessary.
- *taking a melting point*. The purpose of recrystallising a product is to obtain a pure sample that has a sharp melting point. Impurities cause melting to occur over a temperature range, rather than at a precise point. An impurity that is often ignored is water — before taking a melting point, the substance must be dried carefully. This requires a length of time in an oven set to an appropriate temperature.

Although it may not be part of the assessment of qualitative skills, the A2 specification expects knowledge and understanding of refluxing and distillation (see pages 24–25).

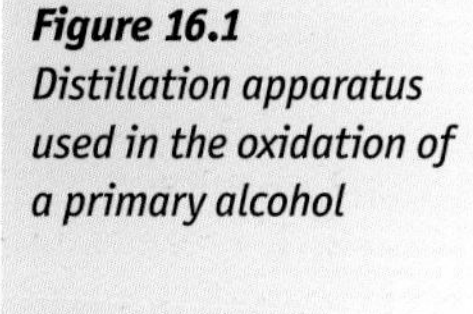

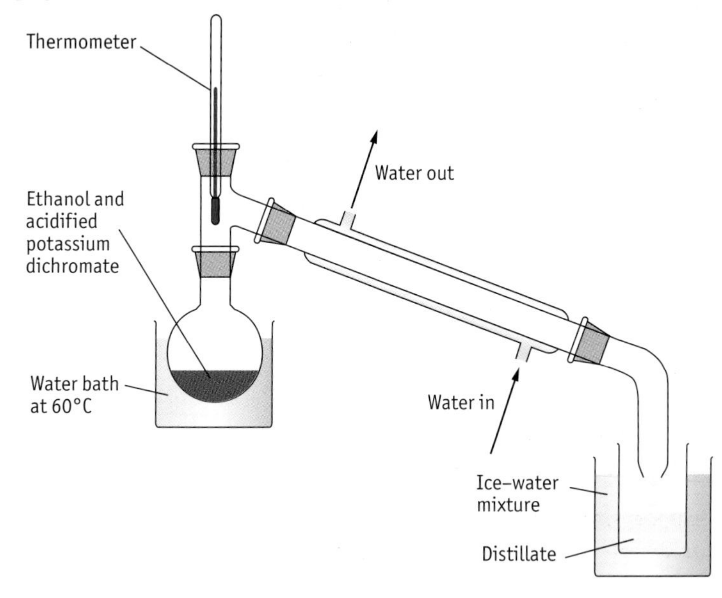

Figure 16.1 *Distillation apparatus used in the oxidation of a primary alcohol*

You should be familiar with the reasons for the use of these techniques. There are also several practical issues relevant to the assessments. An understanding of the following may be tested:

- For refluxing and distillation, the water in the outside jacket of the condenser should enter at the lower end of the apparatus. If this is not done, the inner tube will not be surrounded completely by the water and the condensing distillate will not be cooled efficiently.
- All joints in the apparatus must be checked to ensure that they are airtight. Escape of any volatile components is wasteful but, more importantly, the chemicals being used or formed may be flammable and/or toxic.
- Control of heating is often required. Precautions should always be taken to avoid overheating. This could be dangerous and can lead to organic compounds breaking down into unwanted products, particularly carbon. This is evident when colourless liquids become yellow or brown.
- Redistillation is employed to separate the desired product from contaminating chemicals. A thermometer is essential to ensure that the correct liquid is collected.

Observing and recording

This type of exercise ought to be simple. Nevertheless, it is easy to miss (or fail to record) simple changes. When noting down observations remember to include:

- the appearance of the substances before the test:
  - solid, liquid or solution?
  - colour?
  - characteristic smell?

- Observations during the test — for example:
  - colour change?
  - bubbling?
  - did anything change state (e.g. did a solid dissolve)?
  - obvious smell?
  - temperature increase?
- the appearance of the substances on completion of the test

Common reasons for the loss of marks include:

- suggesting that a reaction has occurred only when bubbling is observed
- giving only the final colour obtained after a reaction, rather than the colour change that has taken place
- failing to record a white solid dissolving to give a colourless solution — it is easy to regard this as 'no reaction',
- writing 'clear' when the word required is 'colourless',
- recording observations that occur during a reaction, but failing to record the appearance of products

Appropriate technical words should be used in descriptions. Although some marks may be awarded for inappropriate terminology, the word 'distillate' should be used rather than 'the liquid in the condenser' and 'precipitate' rather than 'a solid was formed'.

Reaching simple conclusions

Within the qualitative tasks, you may be asked to provide simple conclusions based on the observations you have made. You should be able to recognise the following:

- the gases hydrogen, oxygen and carbon dioxide
- the reactions of metals, bases, alkalis and carbonates with dilute acids
- the precipitation of silver halides — chlorides, bromides and iodides
- the presence of an unsaturated organic compound using aqueous bromine
- the behaviour of primary, secondary and tertiary alcohols when warmed with acidified potassium dichromate
- the different behaviour of phenols and alcohols on the addition of aqueous bromine
- the precipitation of carbonyl compounds with excess 2,4-dinitrophenylhydrazine solution
- the different behaviour of aldehydes and ketones with Tollens' reagent.

In addition, you should be able to categorise *types* of change even where the exact details of the reaction are unknown. These include:

- **Precipitation** It should be appreciated that a precipitate occurs when the aqueous cation and aqueous anion of an insoluble compound react together.
- **Redox reaction** This occurs when there is an exchange of electrons between the reagents in a reaction.
- **Acid–base reaction** There is an exchange of protons (hydrogen ions) between the reagents in a reaction
- **Ligand exchange** A ligand attached to a metal ion is replaced by another ion.

These reactions are usually identified by the occurrence of a colour change.

Quantitative tasks

As with the qualitative tasks, there is a limited number of likely assessments. However, the skills required are more demanding. It is essential that you have a clear understanding of:

- recording results to take account of the maximum error of the equipment
- calculating percentage errors
- using significant figures

This was covered in some detail in the AS textbook, but the essential points are also included here.

Recording results to an appropriate degree of accuracy

Any reading taken has a level of accuracy determined by the quality of the equipment used. When recording results this must be borne in mind. The reading must not be quoted in such a way as to suggest a greater level of accuracy.

For most pieces of equipment it is simply a matter of knowing what is possible. Although you should check the quality of the equipment provided, some typical examples are as follows:

- A volumetric flask when filled carefully so that the meniscus is on the line has a maximum error of $0.2\,cm^3$ (sometimes $0.3\,cm^3$). Therefore, the volume should be should be stated to one decimal place, e.g. $250.0\,cm^3$.
- The volume in a pipette (with a maximum error of $0.06\,cm^3$) should be given to one decimal place, e.g. $25.0\,cm^3$.
- A burette normally has a maximum error of $0.05\,cm^3$. Therefore, this is the maximum accuracy that should be specified. This accuracy also applies to the zero, which should be recorded as $0.00\,cm^3$. Values such as $26.25\,cm^3$, $26.00\,cm^3$ and $26.05\,cm^3$ are all valid readings. Values such as $24.0\,cm^3$ (only one decimal place) and $24.13\,cm^3$ (a greater level of accuracy than is justified) are not. When using a burette, the mean titre is required. It is worth emphasising that this should be obtained using the concordant readings only (i.e. those readings that agree most closely).
- Measuring cylinders vary considerably in their maximum error. A useful rule of thumb is that the maximum error corresponds to half the distance between the markings. For example, a $250\,cm^3$ cylinder marked in divisions of $2\,cm^3$ is accurate to the nearest $1\,cm^3$. It would therefore be inappropriate to include any decimal points in recording the volume of a liquid or solution. This applies to all measuring cylinders.
- Thermometers also have varying maximum error, so only the general rule 'accurate to half a division' can be given.
- When recording the results obtained when using digital equipment, it is usual to include all the decimal places displayed.

It is worth remembering that the maximum error of apparatus is usually inscribed on it by the manufacturer. You should get into the habit of checking this each time you use any piece of equipment.

Since the accuracy of equipment can vary, in questions asked in an assessment the maximum error will be stated clearly — for example, 'the volume was measured using a volumetric flask with a maximum error of 0.2 cm^3'. The stated value should be used in estimating percentage error.

Percentage errors

In both the quantitative and evaluative tasks, you may be asked to calculate a percentage error. If the measurement involves a single reading all that is required is to take the error quoted, divide it by the quantity measured and multiply by a hundred. For example:

> 25.0 cm^3 is measured using a pipette with a maximum error of 0.06 cm^3. The percentage error is:
>
> $(0.06/25.0) \times 100 = 0.24\%$

A difficulty occurs when a measurement is obtained from two readings as in, for example, using a burette. A volume measured from a burette is obtained by subtracting two readings both of which are subject to error. In this case, the maximum error is twice the error for a single reading.

Example A volume is delivered from a burette with a maximum error of 0.05 cm^3 by running a solution from 2.55 cm^3 to 27.20 cm^3. The volume delivered ($27.20 - 2.55 = 24.65$ cm^3) is subject to the error in both the original and the final reading.

The same issue arises when measuring masses and temperatures by difference.

Each of these errors is 0.05 cm^3. Therefore, the percentage error is:

$$\frac{2 \times 0.05}{24.65} \times 100 = 0.41\%$$

> **e** It is unlikely that the error would be as much as 2×0.05 cm^3 as this would be the result of the first reading being 0.05 cm^3 less than expected and the second reading being greater by 0.05 cm^3. There is a more sophisticated method of calculating the likely error, which is then found to be less than 0.1 cm^3. This is explained on page 253 of the AS textbook. However, this is neither expected nor required in the assessments.

The errors inherent in the measurements made using a burette make the appropriate quotation of the mean volume from a titration open to argument. It is common practice to quote the mean to an accuracy of 0.05 cm^3. However, a case could be made for a single decimal place being more correct. Your teacher will advise you what to do.

Significant figures

When writing a number, the intended accuracy is indicated by using an appropriate number of significant figures. For example, 14.23 indicates that the number should be taken to be accurate to the final digit. If there is doubt about the final digit (i.e. the 3) then it should be written as 14.2. Likewise, if the answer is only accurate to a whole number it should be written as 14.

- 14.23 is written to four significant figures;
- 14.2 is written to three significant figures
- 14 is written to two significant figures.

When quoting an answer to a particular number of significant figures, numbers may need to be rounded up or down, as in the following examples:

Example 1: 17.87 has four significant figures.

To three significant figures, it is 17.9 (as 17.87 is nearer to 17.9 than 17.8).

To two significant figures, it is 18 (17.87 is nearer to 18 than 17).

Example 2: a number ending in a '5' is raised to the higher number. So 32.5 is written as 33 to two significant figures. In other words, when a number is equidistant between two values, the higher value is conventionally quoted.

Example 3: if a number has zeros after the decimal point, these are not 'significant'.

- 0.004 has only one significant figure because the zeros do not count.
- 0.0526 is 0.0053 to two significant figures.

Example 4: if a number has zeros before the decimal point these are 'significant'.

For example, '1800' is ambiguous because it could represent a number with four, three or two significant figures. The way round this is to write the number in standard index form (sometimes just called standard form). Written in this way, 1.8×10^3 has two significant figures. If four significant figures were intended, it would be written as 1.800×10^3.

Quoting an answer to an appropriate number of significant figures

When a quantity has been calculated its accuracy is subject to the measurement errors that have occurred. This may make it difficult to decide on the appropriate number of significant figures to include in the final answer. This can be quite complex, but the rule that should be adopted in the assessments is to quote the calculated quantity to the same number of significant figures as the least accurate measurement made.

For example, in an enthalpy experiment, if a mass of 1.06 g of sodium carbonate reacts with 50.0 g of excess acid and a temperature rise of 3.9°C is measured then the heat produced could be calculated thus:

$$q = mc\Delta T$$

$$q = 50.0 \times 4.18 \times 3.9 = 815.1\,\text{J or } 0.8151\,\text{kJ}$$

where $4.18\,\text{J}\,\text{g}^{-1}\,\text{K}^{-1}$ is the specific heat capacity of the solution.

However, since the temperature rise was only measured to an accuracy of two significant figures, the answer should be quoted to two significant figures i.e. 0.82 kJ.

e It could be argued that the heat was distributed to the 51.06 g of solution (if the calcium carbonate completely dissolved), rather than just the 50.0 g of acid. Both approaches would be allowed in an assessment mark scheme.

Rounding numbers in a calculation

A calculation may involve several steps. The question arises, should the answers to the steps be rounded during the course of the calculation or only when the final answer is quoted? It is important to understand that it is only the final answer that *must* be quoted to a suitable number of significant figures. If steps on the way to producing the answer are also quoted to a suitable number of significant figures that is fine, provided that the full number obtained by the calculator is used in each subsequent step. Numbers should not be rounded before the next step in the calculation.

To illustrate the importance of this, consider an experiment in which 0.28 g of calcium is reacted with water and the gas evolved at room temperature has a measured volume of 174 cm^3. Use the experimental data to calculate the relative atomic mass of calcium.

The equation for the reaction is:

$$Ca(s) + 2H_2O(l) \rightarrow Ca(OH)_2(aq) + H_2(g)$$

The mass is measured to two significant figures and so this should be the accuracy of the answer.

1 mol of gas has a volume of 24 000 cm^3 at room temperature, so the amount (in moles) of hydrogen collected is 174/24 000 = 0.00725 mol

If this is adjusted to two significant figures it becomes 0.0073 mol.

From the equation, 1 mol of calcium produces 1 mol of hydrogen.

So, the relative atomic mass of calcium = mass of calcium/amount (in moles) of Ca.

Using the rounded figure 0.0073 in the calculation gives:

0.28/0.0073 = 38.3562

which, rounded to two significant figures, is 38.

Using the unrounded figure 0.00725 in the calculation gives:

0.28/0.00725 = 38.6207

which, rounded to two significant figures, is 39.

The answer 38 is obtained by rounding numbers twice to two significant figures; 39 is obtained when the number is rounded only at the final step. This is the correct thing to do and the answer should be given as 39. The answer 38 would not earn the mark.

In an assessment, 1 mark would have been lost. At the intermediate stage, either 0.00725 or 0.0073 would be accepted.

Quantitative experiments

There are a number of possible tasks that you could be asked to complete. You should be familiar with:

- the use of titrations to determine the concentration of an unknown solution and to determine a quantity such as relative molecular mass from the results. The titrations include the use of potassium manganate(VII) and sodium thiosulphate. As with the AS course, familiarity with the correct use of pipettes, burettes and volumetric flasks is required.
- the direct and indirect (using Hess's law) calculation of ΔH from an enthalpy experiment. Experiments similar to those met at AS could be asked.
- measuring the volume of a gas with a measuring cylinder or a syringe
- an experiment to interpret thin-layer or paper chromatography
- the interpretation of rate experiments (both 'initial rate' and 'continuous' methods) to establish orders of reaction. The necessary understanding required for this is covered in Chapter 9.
- the experimental determination of an equilibrium constant
- the use of electrode potentials to establish the feasibility of a reaction (Chapter 14)
- experiments involving acids and bases
- the interpretation of mass spectra, infrared spectra and NMR spectra

Either a calculation or plotting a graph (particularly for rate experiments) will be required. These skills must be practised before the assessment.

e Graphs are often plotted poorly by students. Make sure that the axes are such that the sheet of graph paper is used fully and that axes have labels that include the correct units.

Evaluative tasks

The third type of assessment requires evaluation of the procedure used and the measurements made to decide whether the results (and any quantity calculated from them) are reliable and accurate. This is a difficult skill to master and good results in this assessment will only be achieved by regular practice. Questions may either:

- require analysis of an experiment to determine significant weaknesses in experimental procedures and measurements

or

- require you to suggest changes to the procedure and state whether they would lead to a more accurate result

You are also expected to be able to suggest simple improvements to the procedure and measurements.

It is important that you appreciate the reasons behind procedures adopted in the experiments that you have carried out. The emphasis is on understanding, rather than on simply memorising facts or applying formulae.

Some examples are:

- Are you clear that there is a distinction between a compound being thermodynamically stable and kinetically stable? The former refers to a product having a lower free energy than the reactants; the latter means that the activation energy is high enough to slow down the reaction to such an extent that it may appear not occur.
- Do you fully understand why the enthalpy of neutralisation of a weak acid with a strong base is less than that of a strong acid with a strong base? As has

been explained on pages 166–167, it is not simply a matter of saying that this occurs because the weak acid produces fewer hydrogen ions.

- Are you clear about the importance of temperature control when measuring an equilibrium constant? Are you aware that many reactions take a long time to reach equilibrium?
- Do you understand how to interpret the results of rate experiments obtained by the initial rate method and those from a continuous run? Why is 1/time a suitable approximation for the measurement of an initial rate? Why is the approximation unsatisfactory when the time taken is longer? These questions are all answered in Chapter 9 of this book.
- Can you confidently construct overall redox equations from the relevant half-equations? Given redox (electrode) potentials, can you predict the feasibility of a reaction?
- Can you interpret mass, infrared and NMR spectra?

As you will almost certainly do the evaluative assessment after you have been taught the relevant topic, there will be no need to remember the answers to all these questions for any one occasion. (However, all are relevant to the A2 theory paper, Unit 5.)

Measurement and procedural errors

Measurement errors in the use of a particular piece of apparatus are identified by calculating the percentage error. A given experiment could involve several measurements, but it is not necessary to know rules to calculate the total error. However, you should appreciate that there may be uncertainty in any conclusions drawn from the data. In evaluating an experiment, you will only be asked to focus on significant errors. A potential trap for those who simply calculate percentage errors and use them thoughtlessly is that sometimes not all measurements need to be accurate. Two examples will illustrate the point.

Example 1 When a potassium manganate(VII) titration is carried out, it is necessary to add some hydrogen ions to ensure that the reduction to manganese(II) ions takes place. All that is required is that excess acid is present. Using an approximate volume measured with measuring cylinder is adequate. Using a pipette is unnecessary and would not make the titration any more accurate.

Example 2 Many iodine–thiosulphate titrations involve the generation of iodine from another compound. This is usually accomplished by adding excess aqueous potassium iodide to the compound and, as long as there is sufficient iodide present, it is unnecessary to measure its volume accurately.

When reviewing a collection of results (such as those from a rate experiment), care should be taken to identify anomalous values. This is particularly relevant to drawing graphs, where it may be difficult to decide on a line of best fit. When evaluating data you should ignore values that are obviously the result of unexplained error and are not a systematic problem with the readings.

Procedural errors are varied and often come down to commonsense. There is a tendency for students to blame poor results wholly on poor quality apparatus.

Although this might be true, the procedures adopted could be just as influential. No general rules can be given, but you should go through each part of an experiment and consider whether anything was particularly awkward to set up or carry out, and whether the conditions for the experiment were hard to control.

Some examples of situations where procedural errors might be significant are:

- A rate experiment that requires the addition of two chemicals at the same time as a stopwatch must be started. This is difficult to do reliably. Similarly, establishing exactly when to stop the stopwatch may be uncertain.
- To obtain reliable results from an experiment may require a constant temperature. This applies particularly to rate experiments and the measurement of equilibrium constants.
- Control of temperature during the course of an experiment might be unnecessary, but subsequent measurements might require returning to the same initial conditions. Measuring the volume of a gas is an example where a change in temperature would affect the result.
- If an experiment involves decomposition of a substance, the control of the temperature may be important. An insufficiently high temperature could mean that the decomposition is incomplete; an excessively high temperature might cause the product to decompose. For example, heating iron(II) sulphate crystals produces anhydrous iron(II) sulphate, which breaks down readily to iron(III) oxide and sulphur oxides.
- The reactants or products may react with oxygen in the air and affect the measurements:
  - With iron(II) sulphate, oxidation to an iron(III) compound could be a problem.
  - Reactive metals are difficult to handle because of the ease with which they react.
  - Absorption of water from the surrounding air often occurs.
  - Alkalis can absorb carbon dioxide.
- The physical state of the chemicals may affect a reaction. Both enthalpy and rate experiments are affected by the surface area of any solid used.

These, and other, points should be considered and a qualitative judgement made as to their likely significance. Remember that it is always wrong to regard mistakes made as a result of carelessness in performing an experiment as procedural errors. For example, during an experiment some chemical may have been spilt but this is not an inherent weakness in the procedure.

Improving experiments

The important point to appreciate when suggesting improvements is that they should be simple. In general, all that is expected is a straightforward adjustment to the method or, if appropriate, the use of more accurate measuring equipment.

In the former case, consider the points mentioned above (and others you consider relevant) and modify the procedure as required:

- If the temperature needs controlling, a water bath might help.

- If a substance absorbs water, a dry environment should be used (perhaps a desiccator).
- If the surface area is important, grinding the solids using a pestle and mortar may be required.

Do not suggest impractical solutions such as working in an oxygen-free environment or using remote-controlled robots.

For measurement errors, a more accurate piece of equipment usually suggests itself. However, it is important to be specific. Do not write 'use a better thermometer' when 'a thermometer that can be read to two decimal places' is what you mean. What is possible must also be borne in mind. For example, a balance weighing to an accuracy of six decimal places might, in theory, produce an improved result, but the reality is that a balance with a maximum of four decimal places is all that will be available and even these balances are very expensive.

Questions

1 There are two serious errors in the following diagram of a reflux apparatus.

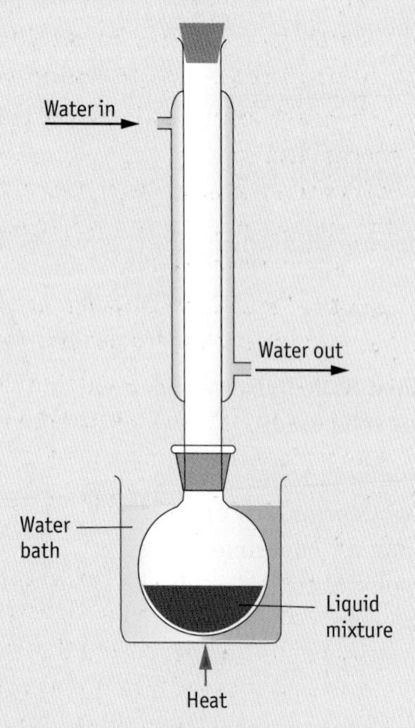

Explain what they are.

2 Correct each of the following sentences.

a The colour of aqueous sodium chloride is clear but aqueous copper sulphate is a blue liquid.

b Nothing is observed when sodium chloride is dissolved in water but when silver nitrate is added to it, it goes white.

3 a Study the following table showing the results of a titration. Redraw it, correcting any errors.

Titration	Rough	1	2	3
Final burette reading	28.00	27.8	27.35	27.25
Initial burette reading	0.00	0.14	0.2	0.00
Volume used	28.00	27.66	27.15	27.25

b What is the mean value of the volume used in the titration?

4 Calculate the percentage error when $25.0\,cm^3$ is measured using:

a A $25.0\,cm^3$ pipette with a maximum error of $0.06\,cm^3$.

b A burette with a maximum error of $0.05\,cm^3$.

5 a Calculate the percentage error in measuring a temperature rise from 18.5°C to 22.0°C using a thermometer with a maximum error of 0.5°C.

b What would the percentage error be if the thermometer used had a maximum error of 0.1°C?

6 **a** Quote 34.756 to:
  (i) four significant figures
  (ii) three significant figures
  (iii) two significant figures

b Quote 0.00438 to:
  (i) three significant figures
  (ii) two significant figures

c Quote 0.499 to:
  (i) two significant figures
  (ii) one significant figure

d Quote 1567 to three significant figures.

7 The mass of a solid is determined as follows:

- The solid is weighed in a container using a balance with a maximum error of 0.005 g. Its mass is found to be 68.67 g.
- The solid is removed and the empty container is re-weighed using the same balance. It is found to have a mass of 66.78 g.

What is the maximum percentage error in the mass of solid?

8 The temperature rise in a reaction between calcium carbonate and hydrochloric acid is measured using the following procedure:

- 2.26 g of calcium carbonate measured with a maximum error of 0.01 g
- 50 cm^3 of hydrochloric acid measured with a maximum error of 1 cm^3
- a thermometer with a maximum error of 0.5°C

The initial temperature of the acid is measured as 18.0°C and the final temperature as 23.5°C.

Assume the hydrochloric acid has a density of 1 $g\,cm^{-3}$ and that its specific heat capacity is 4.18 $J\,g^{-1}\,K^{-1}$.

a Ignoring the maximum errors in the measurements, use the results obtained to calculate the amount of heat produced in the experiment.

b Using the limits of the possible errors, calculate:
  (i) the minimum value that might be obtained for the amount of heat produced in the experiment
  (ii) the maximum value that might be obtained for the amount of heat produced in the experiment

c Comment on the appropriate number of significant figures to use in your answer to part **a**.

9 In an experiment, the mass of zinc in a sample of 0.50 g of brass is determined.

- The sample is warmed with 25 cm^3 of 2.00 $mol\,dm^{-3}$ hydrochloric acid until all the zinc has dissolved. None of the other components of brass dissolves, so a residue remains.
- The solution of zinc chloride is filtered to remove the residue.
- 30 cm^3 (an excess) of aqueous sodium carbonate is added to the zinc chloride solution. A precipitate of zinc carbonate forms.
- The precipitate is filtered off, dried and weighed. 0.32 g of zinc carbonate is obtained.

a Write the equations, giving state symbols, for the reactions:
  (i) zinc with aqueous hydrochloric acid
  (ii) aqueous zinc chloride with aqueous sodium carbonate

b Calculate the amount (in moles) of hydrochloric acid used in the experiment.

c Calculate the amount (in moles) of zinc carbonate precipitated. Hence determine the mass of zinc contained in the brass.

d What is the percentage of zinc in the sample of brass?

e The measurements made in the experiment are not absolutely accurate. The masses are measured with a balance such that they have a maximum error of 0.01 g. The volumes of solution are measured in a measuring cylinder such they have a maximum error of 1 cm^3.

Use this information to explain what improvements to the measurements would most improve the accuracy of the determination of the mass of zinc.

10 A student plans to determine the number of moles of water of crystallisation in 1 mol of hydrated copper(II) sulphate. The method proposed is as follows.

- Weigh a sample of copper(II) sulphate crystals in a crucible.
- Heat the crucible strongly for a few minutes to remove the water of crystallisation.
- Allow the crucible to cool.
- Reweigh the crucible to determine the mass of anhydrous copper sulphate formed.

Suggest precautions that should be taken to ensure that the *procedure* gives the most accurate result possible.

Stretch and challenge

1 A student plans to determine the order of reaction with respect to hydrochloric acid when it reacts with marble chips.

The student proposed the following method:

- Weigh out six samples of marble chips, each weighing 2.00 g.
- Prepare solutions of hydrochloric acid of concentration 4.00 mol dm^{-3}, 2.00 mol dm^{-3}, 1.00 mol dm^{-3}, 0.500 mol dm^{-3}, 0.250 mol dm^{-3} and 0.125 mol dm^{-3}.
- Perform six experiments reacting a 2.00 g sample of marble chips with 50.0 cm^3 of each of the different concentrations of aqueous hydrochloric acid. In each case, record the time, t, taken for bubbling to cease as a measure of the time taken for the reaction to be completed.

The student hopes to plot a graph of $1/t$ against concentration of the acid and from its shape to determine the order with respect to the acid.

List the main reasons why this experiment will not provide a conclusive result.

Index

C

F

G

H

I

K

O

P

Q

R

S

The periodic table

Key:

Relative atomic mass
Atomic symbol
name
Atomic (proton) number

Period	Group 1	Group 2											Group 3	Group 4	Group 5	Group 6	Group 7	Group 0
1	1.0 H hydrogen 1																	4.0 He helium 2
2	6.9 Li lithium 3	9.0 Be beryllium 4											10.8 B boron 5	12.0 C carbon 6	14.0 N nitrogen 7	16.0 O oxygen 8	19.0 F fluorine 9	20.2 Ne neon 10
3	23.0 Na sodium 11	24.3 Mg magnesium 12											27.0 Al aluminium 13	28.1 Si silicon 14	31.0 P phosphorus 15	32.1 S sulphur 16	35.5 Cl chlorine 17	39.9 Ar argon 18
4	39.1 K potassium 19	40.1 Ca calcium 20	45.0 Sc scandium 21	47.9 Ti titanium 22	50.9 V vanadium 23	52.0 Cr chromium 24	54.9 Mn manganese 25	55.8 Fe iron 26	58.9 Co cobalt 27	58.7 Ni nickel 28	63.5 Cu copper 29	65.4 Zn zinc 30	69.7 Ga gallium 31	72.6 Ge germanium 32	74.9 As arsenic 33	79.0 Se selenium 34	79.9 Br bromine 35	83.8 Kr krypton 36
5	85.5 Rb rubidium 37	87.6 Sr strontium 38	88.9 Y yttrium 39	91.2 Zr zirconium 40	92.9 Nb niobium 41	95.9 Mo molybdenum 42	[98] Tc technetium 43	101.1 Ru ruthenium 44	102.9 Rh rhodium 45	106.4 Pd palladium 46	107.9 Ag silver 47	112.4 Cd cadmium 48	114.8 In indium 49	118.7 Sn tin 50	121.8 Sb antimony 51	127.6 Te tellurium 52	126.9 I iodine 53	131.3 Xe xenon 54
6	132.9 Cs caesium 55	137.3 Ba barium 56	138.9 La lathanum 57	178.5 Hf hafnium 72	180.9 Ta tantalum 73	183.8 W tungsten 74	186.2 Re rhenium 75	190.2 Os osmium 76	192.2 Ir iridium 77	195.1 Pt platinum 78	197.0 Au gold 79	200.6 Hg mercury 80	204.4 Tl thallium 81	207.2 Pb lead 82	209.0 Bi bismuth 83	[209] Po polonium 84	[210] At astatine 85	[222] Rn radon 86
7	[223] Fr francium 87	[226] Ra radium 88	[227] Ac actinium 89	[261] Rf rutherfordium 104	[262] Db dubnium 105	[266] Sg seaborgium 106	[264] Bh bohrium 107	[277] Hs hassium 108	[268] Mt meitnerium 109	[271] Ds darmstadtium 110	[272] Rg roentgenium 111	Elements with atomic numbers 112–116 have been reported but not fully authenticated						
				140.1 Ce cerium 58	140.9 Pr praseodymium 59	144.2 Nd neodymium 60	144.9 Pm promethium 61	150.4 Sm samarium 62	152.0 Eu europium 63	157.2 Gd gadolinium 64	158.9 Tb terbium 65	162.5 Dy dysprosium 66	164.9 Ho holmium 67	167.3 Er erbium 68	168.9 Tm thulium 69	173.0 Yb ytterbium 70	175.0 Lu lutetium 71	
				232 Th thorium 90	[231] Pa protactinium 91	238.1 U uranium 92	[237] Np neptunium 93	[242] Pu plutonium 94	[243] Am americium 95	[247] Cm curium 96	[245] Bk berkelium 97	[251] Cf californium 98	[254] Es einsteinium 99	[253] Fm fermium 100	[256] Md mendelevium 101	[254] No nodelium 102	[257] Lr lawrencium 103	